The Phonological

Humans instinctively form words by weaving patterns of meaningless speech elements. Moreover, we do so in specific, regular ways. We contrast *dogs* and *gods*, favor *blogs* over *lbogs*. We begin forming sound-patterns at birth and, like songbirds, we do so spontaneously, even in the absence of an adult model. We even impose these phonological patterns on invented cultural technologies such as reading and writing. But why are humans compelled to generate phonological patterns? And why do different phonological systems – signed and spoken – share aspects of their design? Drawing on findings from a broad range of disciplines including linguistics, experimental psychology, neuroscience, and comparative animal studies, Iris Berent explores these questions and proposes a new hypothesis about the architecture of the phonological mind.

IRIS BERENT is a Professor of Psychology at Northeastern University, Boston. Her research concerns phonology, morphology, and reading. She has published extensively in psychological and linguistic journals.

The Phonological Mind

Iris Berent

CAMBRIDGE UNIVERSITY PRESS
Cambridge, New York, Melbourne, Madrid, Cape Town,
Singapore, São Paulo, Delhi, Mexico City

Cambridge University Press
The Edinburgh Building, Cambridge CB2 8RU, UK

Published in the United States of America by Cambridge University Press, New York

www.cambridge.org
Information on this title: www.cambridge.org/9780521149709

© Iris Berent 2013

This publication is in copyright. Subject to statutory exception
and to the provisions of relevant collective licensing agreements,
no reproduction of any part may take place without the written
permission of Cambridge University Press.

First published 2013

Printed and Bound in the United Kingdom by the MPG Books Group

A catalogue record for this publication is available from the British Library

Library of Congress Cataloguing in Publication data
Berent, Iris, 1960–
The phonological mind / Iris Berent.
pages cm
Includes bibliographical references and index.
ISBN 978-0-521-76940-2 (hardback) – ISBN 978-0-521-14970-9 (paperback)
1. Grammar, Comparative and general – Phonology. 2. Phonetics. 3. Cognitive grammar. I. Title.
P217.3.B47   2012
414–dc23
                                                         2012017898

ISBN 978-0-521-76940-2 Hardback
ISBN 978-0-521-14970-9 Paperback

Cambridge University Press has no responsibility for the persistence or
accuracy of URLs for external or third-party internet websites referred to
in this publication, and does not guarantee that any content on such
websites is, or will remain, accurate or appropriate.

Contents

List of figures	*page* viii
List of tables	x
Copyright acknowledgements	xi
Preface	xiii

I Introduction

1 Genesis   3

2 Instinctive phonology   9
- 2.1 People possess knowledge of sound patterns   9
- 2.2 Productivity   11
- 2.3 Regenesis   12
- 2.4 Shared design   18
- 2.5 Unique design   28
- 2.6 Phonological knowledge lays the foundation for the cultural invention of writing and reading   32

3 The anatomy of the phonological mind   35
- 3.1 The phonological grammar is a core algebraic system   35
- 3.2 Phonology is a core system   44
- 3.3 Domain-general and non-algebraic alternatives   49
- 3.4 Rebuttals and open questions   55
- 3.5 A roadmap   58

II Algebraic phonology

4 How phonological categories are represented: the role of equivalence classes   63
- 4.1 What are phonological patterns made of?   63
- 4.2 The role of syllables   65
- 4.3 The dissociations between consonants and vowels   73
- 4.4 Conclusions and caveats   82

5 How phonological patterns are assembled: the role of algebraic variables in phonology — 84

- 5.1 How do phonological categories combine to form patterns? — 84
- 5.2 A case study: the restriction on identical root consonants in Hebrew — 87
- 5.3 The restriction on identical consonants generalizes to native Hebrew consonants — 91
- 5.4 The restriction on identical consonants generalizes across the board — 97
- 5.5 Coda: on the role of lexical analogies — 111
- 5.6 Conclusion — 113

III Universal design: phonological universals and their role in individual grammars

6 Phonological universals: typological evidence and grammatical explanations — 117

- 6.1 Phonological universals in typology: primitives and combinatorial principles — 119
- 6.2 Grammatical accounts for typological universals — 123
- 6.3 Non-grammatical explanations for language universals — 131
- 6.4 Why are phonological universals non-absolute? — 132
- 6.5 Algebraic, phonological universals are autonomous from phonetic pressures — 139
- 6.6 Conclusion — 147

7 Phonological universals are mirrored in behavior: evidence from artificial language learning — 149

- 7.1 Phonological interactions target segments that share features — 151
- 7.2 Learners favor directional phonological changes — 155
- 7.3 Learners favor phonetically grounded interactions — 158
- 7.4 Discussion — 160

8 Phonological universals are core knowledge: evidence from sonority restrictions — 165

- 8.1 Grammatical universals and experimental results: correlation or causation? — 165
- 8.2 Sonority restrictions are active in spoken languages: linguistic and typological evidence — 166
- 8.3 Broad sonority restrictions are active in the grammars of individual speakers: experimental evidence — 176
- 8.4 Summary and conclusions — 196

IV Ontogeny, phylogeny, phonological hardware, and technology

9 Out of the mouths of babes — 201

- 9.1 Computational machinery — 202
- 9.2 Gauging core phonology: some ground rules — 204
- 9.3 Phonological primitives — 205
- 9.4 Universal combinatorial principles: some markedness reflexes — 213
- 9.5 Conclusions — 223

10 The phonological mind evolves — 226
- 10.1 The human phonological instinct from a comparative perspective — 226
- 10.2 Is phonological patterning special? — 228
- 10.3 The evolution of the phonological mind — 247

11 The phonological brain — 251
- 11.1 Individuating cognitive functions: functional specialization vs. hardware segregation — 251
- 11.2 The phonological network of spoken language — 254
- 11.3 Is the phonological network dedicated to phonological computation? — 265
- 11.4 Minds, and brains, and core phonology — 275

12 Phonological technologies: reading and writing — 280
- 12.1 Core knowledge as a scaffold for mature knowledge systems — 280
- 12.2 Writing systems recapitulate core phonology — 283
- 12.3 Reading recovers phonological form from print — 287
- 12.4 Reading recruits the phonological brain network — 295
- 12.5 Grammatical phonological reflexes in reading — 296
- 12.6 Conclusion — 305

13 Conclusions, caveats, questions — 307
- 13.1 Phonological instincts: what needs to be explained — 307
- 13.2 Some explanations — 309
- 13.3 The core phonology hypothesis: some open questions — 311

References — 316
Index — 352

Figures

2.1	The emergence of movement in ABSL (from Sandler, 2011)	*page* 15
2.2	Two classifiers for object vs. handling of an object	16
2.3	Twinkle, Twinkle, Little Star	31
3.1	The use of atomic shapes as symbols for singleton phonemes, either specific phoneme instances (a) or phoneme categories (b)	40
3.2	The use of atomic shapes to encode geminates	41
3.3	The use of complex shapes to encode geminates	41
3.4	The representation of semantic complexity using forms that are either syntactically complex (on the left) or simple (on the right)	50
4.1	The prosodic structure of multisyllabic words	67
4.2	An illustration of the cohorts activated by the initial syllable of two Spanish words	71
4.3	Color naming as a function of the CV-skeletal structure (from Marom & Berent, 2010, Experiments 1 & 3)	78
5.1	The formation of the root *smm* from *sm*	88
5.2	Rating result for novel roots generated from nonnative consonants (from Berent et al., 2002, Experiment 2)	103
5.3	Rating result of novel roots generated from roots with the nonnative phoneme /θ/ (Data from Berent et al., 2002, Experiment 2)	105
6.1	The distinction between syllable and morphological structure in American Sign Language	144
8.1	Response accuracy in the syllable count task (from Berent et al., 2007a)	182
8.2	Response accuracy and response time to non-identity trials in the identity-judgment task (from Berent et al., 2007a)	183
8.3	The phonetic vs. phonological accounts of misidentification	188
8.4	The effect of task demands on the misidentification of ill-formed onsets (from Berent et al., 2012a)	191

8.5	The effect of phonological ill-formedness on the identification of printed materials (from Berent & Lennertz, 2010, Experiment 1)	193
8.6	The sensitivity of Korean speakers to the sonority hierarchy in an identity-judgment task (from Berent et al., 2008)	195
9.1	The effect of markedness on response accuracy to unattested onsets in the "unsuccessful imitation" condition (Berent et al., 2011a)	223
10.1	The hierarchical structure of the Zebra Finch song (from Berwick et al., 2011)	228
10.2	Learned variations in song patterns of Swamp Sparrows (from Balaban, 1988a)	236
11.1	Two cartoon accounts of the relationship between two cognitive functions – phonology and audition – and their hardware implementation	252
11.2	Functional anatomy of left hemisphere areas engaged in the phonological processing in spoken language and their interconnectivity (from Hickok & Poeppel, 2007)	255
11.3	The design of Phillips et al.'s experiments (2000)	258
11.4	Brain responses to the phonological and acoustic control conditions in Phillips et al.'s (2000) experiments.	258
12.1	Lexical access from print	288
12.2	Reading without phonology	289
12.3	Two routes to phonology from print: assembled and addressed phonology	291

Tables

3.1	The contingency between geminate consonants and their singleton counterparts in the *UCLA Phonological Segment Inventory Database*	page 43
4.1	An illustration of the materials in illusory conjunctions	67
4.2	An illustration of the materials in Marom & Berent (2009)	77
5.1	The structure of Hebrew words	87
5.2	An illustration of various word classes, generated by inserting a root in various word patterns	93
6.1	Tone-bearing capacity of syllables in Standard Thai and Navajo as a function of the duration of the nucleus, coda, and rhyme (in ms) (from Zhang, 2004)	138
6.2	The distinction between syllable structure and morphological structure in spoken language	143
7.1	English phonemes and diphthongs (following Hayes, 2009)	152
7.2	The design of Finley and Badecker's experiments (from Finley & Badecker, 2008; 2010)	157
7.3	The design of Wilson's (2006) palatalization experiment	159

Copyright acknowledgements

I would like to thank the publishers for permission to reproduce the following material:

Figures 2.1 and 2.2 source: Figures 24.10 and 24.11 in W. Sandler (2011). The phonology of movement in sign language. In M. Oostendorp, C. Ewen, B. Hume & K. Rice (eds.), *The Blackwell Companion to Phonology* (pp. 577–603). Oxford: John Wiley and Sons.

Figure 4.3 source: Figure 3 in M. Marom & I. Berent (2010). Phonological constraints on the assembly of skeletal structure in reading. *Journal of Psycholinguistic Research*, 39, 67–88.

Figures 5.2 and 5.3 source: Results in I. Berent, G. F. Marcus, J. Shimron & A. I. Gafos (2002). The scope of linguistic generalizations: evidence from Hebrew word formation. *Cognition*, 83, 113–139

Ch. 8 (7) & (8) source: Illustrations 152 and 154 in P. Smolensky (2006). Optimality in phonology II: harmonic completeness, local constraint conjunction, and feature domain markedness. In P. Smolensky & G. Legendre (eds.), *The Harmonic Mind: From Neural Computation to Optimality-Theoretic Grammar (Vol. II: Linguistic and Philosophical Implications*, pp. 27–160). Cambridge, MA: MIT Press.

Figures 8.1 and 8.2 source: data from I. Berent, D. Steriade, T. Lennertz & V. Vaknin (2007). What we know about what we have never heard: evidence from perceptual illusions. *Cognition*, 104, 591–630.

Figure 8.4 source: Figure 5 in I. Berent, T. Lennertz & E. Balaban (2012). Language universals and misidentification: a two way street. *Language and Speech*, 55(3), 1–20.

Figure 8.5 source: Figure 1 in I. Berent & T. Lennertz (2010). Universal constraints on the sound structure of language: phonological or acoustic? *Journal of Experimental Psychology: Human Perception & Performance*, 36, 212–223.

Figure 8.6 source: Figure 2 from I. Berent, T. Lennertz, J. Jun, M. A. Moreno & P. Smolensky (2008). Language universals in human brains. *Proceedings of the National Academy of Sciences*, 105, 5321–5325.

Figure 9.1 source: Table 2 in I. Berent, K. Harder & T. Lennertz (2011). Phonological universals in early childhood: evidence from sonority restrictions. *Language Acquisition*, 18, 281–293.

Figure 10.1 source: Figure 1 in R. C. Berwick, K. Okanoya, G. J. L. Beckers & J. J. Bolhuis (2011). Songs to syntax: the linguistics of birdsong. *Trends in Cognitive Sciences*, 15, 113–121.

Figure 10.2 source: Figure 2 in E. Balaban (1988). Bird song syntax: learned intraspecific variation is meaningful. *Proceedings of the National Academy of Sciences of the United States of America*, 85, 3657–3660.

Figure 11.2 source: Figure 1 in G. Hickok & D. Poeppel (2007). The cortical organization of speech processing. *Nature Reviews Neuroscience*, 8, 393–402.

Figures 11.3 & 11.4 source: Figures 1 and 7 in C. Phillips, T. Pellathy, A. Marantz, E. Yellin, K. Wexler, D. Poeppel et al. (2000). Auditory cortex accesses phonological categories: an MEG mismatch study. *Journal of Cognitive Neuroscience*, 12, 1038–1055.

Ch. 6 (11) source: Table 36, 'Chamicuro I', in P. de Lacy (2006). *Markedness: Reduction and Preservation in Phonology* (p. 107). Cambridge; New York: Cambridge University Press.

Table 7.1 source: Table 2.2 in B. Hayes (2009). *Introductory Phonology*. Malden, MA; Oxford: Wiley Blackwell.

Ch. 9 (8) source: Table 1 in L. Davidson, P. Jusczyk & P. Smolensky (2006). Optimality in language acquisition I: the initial and final state of the phonological grammar. In P. Smolensky & G. Legendre (eds.), *The Harmonic Mind: From Neural Computation to Optimality-Theoretic Grammar (Vol. II: Linguistic and Philosophical Implications*, pp. 231–278). Cambridge, MA: MIT Press.

Preface

This book concerns a linguistic human compulsion – our tendency to assemble words that comprise internal patterns. All natural languages manifest such patterns – no known human tongue uses only single atomic sounds as words (e.g., "*a o u*" for 'I love you'). Rather, words are intricately woven from smaller meaningless elements that form systematic patterns – we contrast *god* with *dog* and *blog* with *globe*. We begin spinning these webs in the womb, and we do so prodigiously, not only for familiar words but also for ones that we have never heard before. Our instinct to form those meaningless patterns is so robust that children have been shown to generate them spontaneously, even if they have witnessed no such patterns in their own linguistic community. In fact, people impose these patterns not only on their natural linguistic communication but also on their invented cultural technologies – reading and writing. This book seeks to unveil the basis of this human compulsion.

The human capacity to weave linguistic messages from patterns of meaningless elements (typically, speech sound) is *phonology*. Phonology has been the subject of much previous research, mostly in linguistics and psychology. For the most part, however, these efforts have proceeded in parallel lines across different disciplines, and as a result our understanding of the phonological mind remains fragmentary. Linguists (specifically, those in the field of formal phonology) have mostly concerned themselves with the structure of the phonological grammar, but the cognitive mechanisms underlying phonological patterns are rarely considered. Psychologists, for their part, have assumed without question that phonological patterns can be adequately handled by rather simple, non-specialized computational systems, but these investigations remain largely divorced from the progress made in formal phonological theory in recent decades. This book seeks to bridge the interdisciplinary divide and reconsider phonology in a new light.

At the center of this book is a novel hypothesis regarding the architecture of the phonological mind. The discussion evaluates this hypothesis against recent advances in formal linguistics, cognitive science, neuroscience, and genetics and reviews these literatures in a manner that is accessible to readers across various disciplines. In so doing, I hope to spark renewed interest in the design of phonological patterns and to demonstrate the benefits of an interdisciplinary approach to the study of this intricate human capacity. To facilitate dialog across disciplines, I have tried to present the material in a manner that is accessible to professionals and advanced students in either field – psychology or linguistics – who lack expertise in the neighboring discipline. This approach necessarily requires some measure of simplification. I have thus attempted to minimize the use of technical jargon; in as much as possible, I have deliberately attempted to avoid the use of phonetic transcription, and, when background information is absolutely necessary, I provide it in "Box" inserts.

Readers can choose to selectively focus on distinct portions of this book, depending on their interests. The Introduction (Part I, Chapters 1–3) provides an accessible overview of the main thesis of the book. The subsequent three parts provide more technical discussion of the different aspects of the thesis, and these sections can be read independently. Part II (Algebraic phonology, Chapters 4–5) examines the basis of the human capacity to generalize phonological knowledge by investigating the computational properties of the phonological mind. Part III (Chapters 6–8, Phonological universals) considers the design of phonological systems and the extent that they are constrained to putatively universal principles. Chapter 6 reviews linguistic evidence for phonological universals. Although the discussion targets readers with minimal linguistic expertise, this chapter is probably the heaviest on linguistic theory. Readers can therefore pick and choose, as subsequent chapters do not require detailed understanding of this one. Chapters 7–8 consider the role of grammatical phonological universals in light of experimental evidence; Chapter 7 evaluates numerous case studies, whereas Chapter 8 focuses in depth on a single case. The final part of the book, Chapters 9–12, examines phonological ontogeny (the development of phonological competence with special emphasis on the first year of life), phylogeny (a comparative analysis of "phonological" abilities across species and their evolution), hardware (brain areas mediating phonological computation and their genetic regulation) and technology (i.e., reading and writing – both typical and impaired, in dyslexia). Conclusions and caveats are presented in Chapter 13.

This book is the product of many years of research. The ideas have grown out of my interactions with several close collaborators. Steven Pinker and Gary Marcus have shaped my understanding of how the mind works, Paul Smolensky has sparked my interest in the problem of language universals, and Donca Steriade has challenged my thinking about phonology and its interactions with phonetics. These ideas, however, probably would not have materialized in a book if it weren't for Andrew Winnard, my editor at Cambridge, who saw this volume coming well before I did. Evan Balaban, Lisa Barrett, Bronwyn Bjorkman, Judit Gervain, Bruce Hayes, Ray Jackendoff, Paul de Lacy, Joanne Miller, Steven Pinker, Wendy Sandler, and Paul Smolensky offered valuable

Preface xv

comments on significant portions of this manuscript – I am immensely grateful for their suggestions and encouragement. Saul Bitran and Monica Bennett have patiently proofread earlier drafts; Kristina McCarthy assisted on various technical matters; Vered Vaknin-Nusbaum has offered constant support; my students and lab assistants Athulya Aravind, Amanda Dupuis, Kimi LaSalle, Katalin Tamasi, Marriah Warren, and Xu Zhao, and two anonymous Cambridge readers have added many useful comments. I am indebted to Jacqueline French, who copyedited the entire manuscript with uncanny intelligence, sharp eye, and warm heart. Finally, Saul, Amir, and Alma Bitran have shared this journey with me. The book is dedicated to them.

Part I

Introduction

1 Genesis

> What does an embryo resemble when it is in the bowels of its mother? Folded writing tablets. Its hands rest on its two temples respectively, its two elbows on its two legs and its two heels against its buttocks ... A light burns above its head and it looks and sees from one end of the world to the other, as it is said, then his lamp shined above my head, and by His light I walked through darkness (Job XXIX, 3) ... It is also taught all the Torah from beginning to end, for it is said, And he taught me, and said unto me: "Let thy heart hold fast my words, keep my commandments and live" (Prov. IV, 4) ... As soon as it sees the light, an angel approaches, slaps it on its mouth and causes it to forget all the Torah completely ...
> (*Babylonian Talmud: Tractate Niddah, folio 30b* "Niddah," 1947)

Of the various aspects of human nature, the biology of our knowledge systems is an area we struggle to grasp. The possibility that our knowledge might be predetermined by our organic makeup is something we find difficult to accept. This is not because we resist our condition as biological organisms – living breathing bodies whose design is shaped by natural laws and evolution. We rarely give a second thought to our lack of fur or our inability to fly and swim underwater. We are not even disturbed by many obvious shortcomings of our mental faculties – our inability to perceive infrared light, the fallibility of our memory, and the haphazard fleeting character of our attention. Those fickle quirks of our neural machinery are surely inconvenient, but they rarely leave us pondering the confinements of our fate.

Inborn knowledge systems, however, are a whole different matter. Inborn knowledge systems are biologically determined frameworks of knowledge. The animal literature presents countless examples of partly inborn knowledge systems, ranging from birdsong and ape calls to the amazing ability of bees to recruit an inborn code in communicating the location of the nectar to their sisters, and the astonishing capacity of the Indigo Bunting to find its navigational path guided by the stars and the earth's magnetic field (Gallistel, 2007; Hauser, 1996). But when it comes to our own species, such inborn frameworks of knowledge raise many difficulties (Pinker, 2002). Inborn knowledge systems constrain our capacity to recognize ourselves and grasp the world around us.

Their existence implies that there are truths we are bound to hold and others we are destined to expunge. Some of us might find these confinements too disturbing to accept. Others suggest that innate truths are privileges of which we, humans, are not worthy. Subsequent discussions of the cited Talmudic passage indeed explain that it is the stubborn refusal of the embryo to leave the womb that forced the angel to slap her face, thereby causing her to forget her inborn knowledge of the Torah (Tanhuma, Exodus, Pekudei, III). But regardless of whether innate knowledge is a burden we are bound to carry or a precious gift that we are morally unfit to embrace, the prospective of such knowledge systems is unsettling.

And yet, modern cognitive science suggests that, like their animal counterparts, human infants come to the world equipped with several systems of rudimentary knowledge. While no mortal is born knowing the Bible or the Koran, infants seem to have basic knowledge of physics, math, biology, psychology, and even morality. They know, for example, that objects are cohesive entities that can only move by contact (Carey, 2009; Spelke, 1994), and they have a rudimentary concept of number that allows them to distinguish two from three objects (for example Feigenson et al., 2002). Young children also understand that, unlike artifacts, living things have an essence that is immutable even when their appearance is changed (Keil, 1986), that humans are agents that have thoughts and beliefs of their own (Onishi & Baillargeon, 2005), and they distinguish between agents with benevolent intentions and those with sinister goals (Hamlin et al., 2007). While the content of such knowledge systems is quite coarse, these systems nonetheless fix our early grasp of the world and pave the road for all subsequent learning.

Of the various candidates for inborn knowledge systems, language has a central role (Chomsky, 1957; 1972; Pinker, 1994). Much research suggests that the capacity for language is not only universal to humans but also unique to us. But the nature of our language mechanisms remains controversial. Moreover, the debate concerning the origins of language has focused almost exclusively on a single aspect of our linguistic competence – our capacity to structure words into sentences (Jackendoff, 2002). The narrow focus on syntax does not do full justice to our linguistic ability. One of the most striking features of human languages is that they all include two distinct levels of organization (Hockett, 1960). One level is the patterning of words to form sentences. A second, less familiar, level, however, generates words (meaningful elements) from patterns of meaningless elements, typically sounds. It is this second level that is the topic of this book.

When we consider our own language, it is usually meaning, rather than sound patterns, that first catches our attention. But think of yourself hearing spoken announcements in a foreign airport, or stumbling upon a foreign-language clip on YouTube, and the pattern of sounds will immediately become apparent. Even if you speak neither French, Russian, nor Arabic, you can still tell these languages are different from each other. Perhaps you can even guess what they are. And since you cannot do so by tracking the syntactic structure of the language or the contents of the conversations around you, the only clues available to you are linguistic sound patterns – the inventory of sounds that make up each language and the unique ways in which those sounds combine.

Every human language patterns words from meaningless elements. In spoken languages like English, those meaningless elements are sounds. The words *dog* and *god* for instance, comprise three sound elements – the vowel *o* and the two consonants, *d g*. Taken on its own, none of these elements carries a meaning, but together, these meaningless elements form words. And the difference between these two words stems only from the ordering of their sounds – their sound pattern. If you are an English speaker, you recognize that the sounds *d,o,g* are English sounds, whereas the *ch* of *chutzpa* isn't. English speakers also notice that patterns such as *dog* are typical of English words, unlike permutations such as *dgo*, which sound foreign. Being an English speaker entails knowledge about the sound structure of this language: its inventory of meaningless elements (sounds), and how these sounds pattern together. This knowledge is called phonology.

We humans are extremely good at tracking the phonological structure of our language. When an infant arrives into the world, language, in his or her mind, is a sound pattern. There are no sentences or words. Just sounds spoken by people – sounds and sound combinations. But linguistic sounds are special for infants. Newborn infants are preferentially tuned to the patterns of human speech (Vouloumanos & Werker, 2007; Vouloumanos et al., 2010). Moreover, newborns recognize the characteristic rhythm of their native language (e.g., French, which they have heard in the womb for several months) and distinguish it from foreign languages (e.g., Russian) even when spoken by the same bilingual talker (Mehler et al., 1988). They can pick up the abstract pattern in a speech-stream (e.g., the ABB pattern in *mokiki, ponini, solili*) after only a few minutes of exposure (Gervain et al., 2008) and automatically generalize it to novel items (e.g., *wafefe*). And by the time an infant reaches her first birthday, she becomes familiar with the particular sounds and sound combinations characteristic of her language (e.g., Jusczyk et al., 1994; Kuhl et al., 1992; Mattys et al., 1999; Saffran et al., 1996; Werker & Tees, 1984).

Why does every human language exhibit phonological patterns? Why are people so adept at weaving and tracking the sound structure of their language? And why do languages have the specific phonological patterns that they do?

For many people, laymen and experts alike, the answer is patent. The patterns we produce mimic the ones we hear. An English-learning infant produces words like *dog* rather than *perro* because this is what his or her English-speaking community says. And when further pressed to consider why language communities employ these particular sound patterns – *dog*, for English, rather than *dgo*, for instance – most people would shrug the "obvious" answer: *dog* is just easier to articulate. Together, such statements capture a widely held sentiment. Phonological patterns, in this view, are determined by the properties of our memory, ears, and mouths. Memory leads us to follow the patterns we have heard in the speech of people around us, and our ears and mouths favor certain patterns over others. Our skill at weaving phonological patterns stems from those generic abilities. Indeed, memory, audition, and motor control are not specific to language or humans. These same abilities allow us to track and memorize linguistic sound patterns much in the way we track any other configurations – visual motifs on wallpaper, patterns of sounds in our favorite musical piece, or the statistical trends in the stock market frenzy. Similarly, the aural and oral restrictions on linguistic sequences are indistinguishable from the ones shaping the perception of noises and music, or the aural command we exercise in kissing or chewing. In short, phonological patterns require no special linguistic talents. And to the extent our phonological patterns differ from those of other species, the difference can only reflect the anatomy of those shared mechanisms or their control.

While nonlinguistic pressures (e.g., memory, attention, auditory and motor limitations) undoubtedly influence the design of phonological patterns, these forces are not necessarily their makers. Memory, for instance, does not explain why it is that all human languages exhibit phonological patterns. A phonological system is indeed not logically mandatory for communication. Speakers could certainly convey concepts by holistic sounds: one sound (e.g., "a") for lion, and another "o" for eating, would suffice to generate sentences ("a o" for lions eat; "o a" for eating a lion, etc).

Memorization not only fails to explain why phonological patterning exists but also cannot account for the characteristics of attested patterns. Our phonological capacity is prolific and robust. We do not merely parrot the sound patterns we hear in our linguistic environment. Rather, we instinctively extend those patterns to new words that we have never heard before. And in rare cases where people have been raised deprived of any phonological system, they have been shown to spontaneously generate one on their own.

The most striking feature of phonological systems, however, is their unique, nearly universal design. Linguistic research has shown that the phonological systems attested across languages exhibit some common characteristics. These similarities in design are important because they imply a common pattern maker that imposes broad, perhaps universal restrictions on all languages (e.g., Jakobson, 1968; Prince & Smolensky, 1993/2004). These common restrictions, moreover, are reflected not only in statistical regularities across languages, but also in the behavior of individual speakers. Given two structural variants, such that one variant A is more "popular" across languages than the other,

B, people will reliably prefer A to B even when neither occurs in their language (e.g., Jusczyk et al., 2002; Moreton, 2002). And when a new language is born, it eventually recapitulates the design of existing phonological systems (Sandler et al., 2011).

The shared design of phonological systems – existing and recently nascent – cannot be trivially explained by general principles of oral or aural patterning. First, like all correlations, the link between ease of articulation/perception and phonological structure is ambiguous. While certain patterns might be preferred because they are easier to produce and comprehend, the causal link could also go in the opposite direction: patterns might be easier to perceive and produce because they abide by the demands of the language system itself. And indeed, people's sense of articulatory ease greatly varies depending on their language. While English speakers find a sequence like *dgo* impossible to utter, Hebrew and Russian speakers produce it without blinking an eye, whereas Japanese speakers would stumble not only on the "exotic" *dgo* but even on the plain English *dog*. Phonological patterns, moreover, are not restricted to articulatory sequences. People extend their phonological sequences to the perception of language in either oral or printed rendition. In fact, phonological patterns are not even confined to aural language. Since phonology is the patterning of meaningless elements, phonological patterns can extend to the visual modality as well. Indeed, every known sign language manifests a phonological system that includes meaningless units of manual linguistic gestures, and, despite the different modalities, signed phonological systems share some important similarities with spoken language phonologies (Sandler, 2008). Phonological design, moreover, is not only quite general but arguably unique – it differs in significant ways from both other systems that use the auditory modality (i.e., music) and the auditory communication systems used by nonhuman animals (Jackendoff & Lerdahl, 2006; Pinker & Jackendoff, 2005).

My claim, to reiterate, is not that the properties of the communication channel – ears and mouths – are irrelevant to the design of phonological patterns. In fact, subsequent chapters show that the tailoring of the phonological mind to its preferred channel of communication – the aural/oral medium – is a critical feature of its adaptive design. But the fit between phonological patterns and their channel does not necessarily mean that the channel is itself the pattern-maker. Rather than weaving phonological patterns directly, the aural/oral channel could have shaped our phonological abilities in a nuanced oblique fashion.

Phonological design is indeed evident not only in our instinctive natural language but also in its encoding via writing, and its decoding, in reading. Unlike language, reading and writing are cultural inventions that are not invariably shared by every human society, just as the sciences of math and physics are not universal. But just as math and physics are founded on our rudimentary inborn systems of number and physics, so are our inventions reading and writing erected upon the phonological principles of our spoken language (DeFrancis, 1989; Perfetti, 1985).

(1) Some interesting properties of phonological patterns
   a. *Generality*: All established languages exhibit phonological patterns.
   b. *Generalization*: Phonological patterns are not confined to the memorization of familiar patterns.
      (i) People generalize the patterns of their language to novel words.
      (ii) Phonological systems reemerge spontaneously.
   c. *Design*: Phonological patterns manifest a shared design.
      (i) The phonological patterns of different languages share a common design.
      (ii) The design of phonological systems is partly shared across modalities – for signed and spoken language.
   d. *Uniqueness*: The design of phonological systems is potentially unique.
      (i) It differs from the design of nonlinguistic auditory forms of communication.
      (ii) It differs from the structure of auditory communication systems used by nonhuman species.
   e. *Scaffolding*: The design of the linguistic phonological system lays the foundation for the invention of reading and writing.

The generality of phonological patterns, their regenesis, their potentially universal, unique design and centrality to cultural inventions (summarized in 1) all suggest an instinctive capacity for phonology, supported by a specialized, partly inborn knowledge system. This book explores this possibility. Doing so will require that we take a closer look at what we mean, precisely, by "knowledge systems," "specialization," and "inborn." But before we consider the mental and brain mechanisms that support phonological patterning, it might be useful to first review some of the instinctive phonological talents of humans. Chapter 2 uses a rather broad brush to paint some of the most intriguing aspects of the phonological mind. Inasmuch as it is possible, this introduction separates the explanandum – the properties of phonological patterns – from the explanation, the mental system that generates them. Some accounts of this system are discussed in Chapter 3 and evaluated in subsequent chapters.

2 Instinctive phonology

> Humans have some special phonological talents. We instinctively intuit that certain phonological patterns are preferred to others even if we have never heard them before, and we will weave phonological patterns regardless of whether our language uses oral speech or manual gestures. Phonological instincts are so robust that people spontaneously generate a whole phonological system anew, and when human cultures invent systems of reading and writing, they impose those patterns on their design. Phonological patterns, however, are not arbitrary: they conform to some recurrent principles of design. These principles are broadly shared across many languages, but they are quite distinct from those found in animal communication or music. This chapter documents those instinctive talents of our species, and in so doing, it lays down the foundation for discussing the architecture of the phonological system in subsequent chapters.

2.1 People possess knowledge of sound patterns

All human communities have natural languages that impose detailed, systematic restrictions on phonological patterns. Unlike traffic laws or the US Constitution, the restrictions on language structure, in general, and phonological patterning, specifically, are not known explicitly. Most people are not aware of those restrictions, and even when professional linguists desperately try to unveil them, these regularities are not readily patent to them. Yet, all healthy human beings know these restrictions tacitly – we encode them in our brain and mind and we religiously follow them in our everyday speech despite our inability to state them consciously. And indeed, we all have strong intuitions that certain sound structures are systematically preferable to others (see 1–3). For example, English speakers generally agree that *blog* is better-sounding than *lbog*; they prefer *apt* to *tpa*; they consider *came* as rhyming with *same* or even *rain* (indicated by ~), but not *ripe*; and they have precise intuitions on the parsing of words into smaller constituents. A frustrated motorist might refer to their noisy car exhaust as an *eg-freaking-zaust*, but not an *e-freaking-gzhaust* (a fact marked by the * sign, which conventionally indicates ill-formed linguistic structures).

(1) Syllable-structure intuitions
 a. blog *lbog
 b. apt *tpa, *pta
 c. apt *apd
 d. box, *bocz
(2) Rhyme
 a. came~same
 b. came~rain
 c. came≁ripe
(3) Parsing *exhaust*
 a. eg- freaking -zaust
 b. *e- freaking -gzaust

Not only do people have strong intuitions regarding the sound patterns of their language, but they also take steps to repair pattern-violators. Phonological repairs are usually too rapid and automatic to be noticed when applied to words in one's own language, but careful analyses demonstrate that repairs take place routinely. English speakers frequently talk about *keys*, *bees*, and *dogs* (pronouncing all *s*'s as *z*), but when it comes to *ducks*, they render the *s* sounding like *us*, not *buzz*. And when presented with novel singular nouns like *bokz* and *mukz* (with the *k* of *buck* and *z* of *buzz*, see 1d and 4c), these, too, are strange sounding (Mester & Ito, 1989). It thus appears that the plural suffix of *duck* should have been *z* (as in *dogs*), but the "badness" of the -*kz* sequence leads people to automatically adjust it to yield *ducks*.

And indeed, speakers tend to confuse illicit sound sequences with licit ones (e.g., Hallé et al., 1998; Massaro & Cohen, 1983). For example, when presented with the illicit *tla*, English speakers incorrectly report that they have heard a disyllabic form, the licit *tela* (Pitt, 1998). Similarly, Japanese speakers misidentify *ebzo* (with the syllable -*eb*, illicit in Japanese) as *ebuzo*, whereas speakers of French (which tolerates *eb*-type syllables) recognize it accurately (Dehaene-Lambertz et al., 2000; Dupoux et al., 1999; Dupoux et al., 2011; Jacquemot et al., 2003). The fact that such confusions are detected very early in life – at the age of 14 months (Mazuka et al., 2012) – and persist despite people's best efforts to distinguish between those forms all suggest that the errors are not the product of some prescriptive conventions. Rather, these linguistic illusions occur because we instinctively extend the phonological pattern of our language to all inputs, and when violators are detected, we automatically recode them as licit forms.

Phonological repairs are indeed readily noticeable when we hear nonnative speakers of our language. English speakers, for example, immediately notice that some speakers of Spanish turn the English *stress* into *estrés* (repairing the sequence *st*, which is illicit in Spanish (Hallé et al., 2008; Theodore & Schmidt, 2003) and giggle at an online classified ad for "*tubo de estriptease para pole dance*," but they are far less likely to notice their own adaptation of words borrowed into English: the foreign consonant combination *bn* in *bnei Israel* (from the Hebrew words *bnei*; sons of) is automatically separated by a schwa (a short vowel), yielding *benei Israel*. Likewise, because many English speakers automatically add a glide to repair the final syllable in José, they take it to rhyme with *way*, whereas to Spanish ears, this rhyme sounds about as good as *way* and *wait*.

(4) English repairs
    a. Dog+s→dogz
    b. Duck+s→ducks (*duckz)
    c. Muk+s→muks (*mukz)
    d. No way José

(5) Spanish repair
    a. Stress→estrés
    b. Striptease→estriptease

2.2 Productivity

Recognizing *dogz* and *ducks* is useful, but not terribly impressive. After all, people have heard such patterns countless times, so it's little wonder that they sound better than the unfamiliar *Xenops* (a South American bird). But our instinct for phonological patterning is not confined to familiar words. Although people possess a rich memory for words, ranging from abstract meaning to their renditions by distinct talkers (e.g., Goldinger, 1998), phonological systems are not passive repositories for fossilized memory traces. The hallmark of phonological patterning is its productivity – the capacity of speakers to systematically and instinctively generalize their phonological knowledge to novel examples.

People have clear intuitions that certain novel words "sound better" than others despite never having heard either (see 6–8): *bnif* "sounds" better than *nbif*, and *mux* better than *mukz*; *hame* rhymes with *rame* and *bain*, but not *duck*; and an annoying *enbot* is probably an *en-freaking-bot*, but not an *e-freaking-nbot*.

(6) Syllable-structure intuitions
    a. bnif *nbif
    b. mux *mukz

(7) Rhyme
    a. hame~rame
    b. hame~bain
    c. hame≁dake

(8) Parsing *enbot*
   a. en- freaking -bot
   b. *e- freaking -nbot

Such generalizations demonstrate that phonological knowledge entails broad principles that extend to novel items. Like driving a car, or navigating the maze of streets leading to one's home, phonological knowledge is largely tacit, and it is acquired without explicit instruction. Indeed, most people who agree with the above-mentioned intuitions cannot offer a systematic explanation for their preferences. But the fact that people, infants or adults, manifest reliable preferences concerning linguistic patterns that they have never heard before suggests that they possess systematic knowledge of phonological patterning.

2.3 Regenesis

Not only do people automatically extract phonological principles that extend to novel words, but they can even spontaneously invent an entire phonological system of their own. The following discussion presents two cases of newly born phonological systems. Both instances concern children who are deprived of access to any phonological input. Remarkably, such children have been shown to develop phonological patterns of their own, thereby demonstrating the astonishing capacity of the phonological mind for spontaneous regenesis.

In addition to their main feature – the regenesis of a phonological system – these two cases also share another common denominator that is salient, but secondary to my argument. In both cases, the phonological system created by the child involves signed, rather than spoken, language. Most people are unaware that sign languages include a phonological system, so before discussing the invention of signed phonological systems, a brief comment on why those patterns are considered phonological should be made.

Phonology concerns our knowledge regarding the patterning of meaningless linguistic elements. While most hearing communities prefer to weave phonological patterns from speech sounds, nothing in this definition requires that those meaningless elements constitute speech. And indeed, absent the capacity to process aural speech, deaf people rely on languages transmitted along the visual modality. Beginning with the pioneering work of William Stokoe (1960), linguistic research has repeatedly shown that sign languages are equipped with fully developed phonological systems, comparable in complexity to spoken language phonologies (e.g., Brentari, 1998; Liddell & Johnson, 1989; Padden & Perlmutter, 1987; Perlmutter, 1992; Sandler, 1989; 1993; Sandler & Lillo-Martin, 2006). Signed and spoken phonologies, moreover, also share many structural characteristics. Just as English phonology patterns syllables, so do syllables form part of the phonology of American Sign Language. In both modalities, syllables are distinct, meaningless units whose structure is constrained by the language. The English syllable *ven*, for example, is a meaningless unit that forms part of *venture* and *ventilate*, and while English allows syllables such as *ven*, it disallows syllables such as *nve*. Similarly, words in American Sign Language comprise meaningless units, and each such unit is subject to multiple restrictions – chief of which is the demand that a syllable must minimally include a hand movement (see Sandler & Lillo-Martin, 2006 for an overview).

As sign languages manifest full-fledged phonological systems, they can potentially gauge the capacity of the phonological mind for regenesis. The fact that its typical native users – deaf individuals – cannot access spoken language renders this case particularly strong. In order for a phonological system to be born spontaneously, children must be deprived of linguistic experience. In the case of hearing individuals, the lack of a linguistic model is an extremely rare event that is invariably accompanied by serious social and emotional deprivation that makes the role of linguistic factors difficult to evaluate separately. By contrast, deaf children who are raised in a purely oral environment lack access to a linguistic model even if their hearing families are loving and nurturing, so the unique contribution of linguistic input can be evaluated with greater precision. Remarkably, rather than remaining robbed of language, such children have been repeatedly shown to spontaneously develop sign systems that exhibit many of the morpho-syntactic hallmarks of well-developed languages, both spoken and signed (Goldin-Meadow & Mylander, 1983; 1998; Senghas & Coppola, 2001; Senghas et al., 2004). The following discussion demonstrates that the regenesis of the linguistic structure extends to the phonological system. We discuss two such cases. The first concerns the birth of an entire phonological system in a new language that lacks phonological patterning; the second illustrates the spontaneous emergence of one specific aspect of phonological patterning. While this particular aspect is novel to the phonological system in question, its design recapitulates features that are found in many other sign systems, but absent in nonlinguistic gestures. These cases demonstrate that phonological patterning has the capacity for spontaneous regenesis, and that the design of newly emerging patterns recapitulates the structure of attested phonological systems.

2.3.1 Case 1: the birth of phonology in the Al-Sayyid Bedouin Sign Language

As discussed above, all languages manifest two levels of patterning – one level concerns meaningful units (e.g., words), whereas a second level – phonology – concerns meaningless units (e.g., phonemes, features). Accordingly, duality of patterning is considered an inviolable design feature of human languages (Hockett, 1960). But the recent discovery of the Al-Sayyid Bedouin Sign Language (ABSL) by Wendy Sandler and colleagues (Sandler, 2011; Sandler et al., 2011) would seem to challenge this assertion. At its onset, ABSL was a language without phonology – the only known case of its kind, and a blatant exception to the duality of patterning principle. Very rapidly, however, this young language has spontaneously morphed to give rise to a phonological system. Sandler and colleagues were able to capture ABSL in this stage of flux and document the birth of a phonological system in the signs of its youngest members.

ABSL is a sign language that emerged very recently (seventy-five years ago) in a Bedouin village in the Israeli Negev. Consanguineous marriage resulted in a rapid spread of congenital deafness, but the close ties among members of the community encouraged the emergence of manual linguistic communication shared by all members, deaf and hearing alike. Far from being a gesturing system, however, ABSL is a language in its own right. It includes various productive grammatical devices, such as syntactic constraints on word order and morphological compounding (Sandler et al., 2005; Sandler et al., 2011). But one component is still missing from this young nascent language: ABSL lacks phonological structure. Most adult ABSL signers produce signs that are iconic and holistic, with no internal systematic patterning, frank violations of phonological restrictions attested in other sign languages, and large variability in the production of signs compared to other sign languages (Israel & Sandler, 2009; Sandler et al., 2011).

Given that ABSL mostly lacks phonological patterning, one would expect children exposed to ABSL to exhibit a similarly impoverished system. Remarkably, however, these children surpass their elders. Rather than using unphonologized gestures, ABSL children manifest several aspects of phonological patterning.

Consider, for example, the signs used by a family of ABSL signers for "kettle" (Sandler, 2011). The adult version denotes "kettle" with a compound, including two signs – one sign for a CUP, followed by another for ROUND-OBJECT (see the left panel in Figure 2.1). But the sign for ROUND-OBJECT has an unusual property – it is produced without a movement of the hand. The lack of movement is unusual because in every documented well-established sign language, hand movement is obligatory. This is not because movement-less signs are impossible to produce – the generation of such signs by adult ABSL speakers clearly shows that static gestures are amply possible. Rather, movement is an obligatory aspect of every sign language phonology: all syllables require hand movement, and those that lack movement are ill formed. While adult speakers of ABSL violate this phonological constraint that is universally present in all documented sign languages, children spontaneously obey it. As shown in the right part of Figure 2.1, the child's signing of ROUND-OBJECT includes a movement of the two hands toward each other.

Sandler and colleagues demonstrate that the signs of ABSL children spontaneously observe many other phonological properties that are absent from the signs of their elders, but present in other sign languages. In particular, the

Figure 2.1 The emergence of movement in ABSL (from Sandler, 2011)

children's signs manifest symmetry, reduplication (the copying of a sign, either fully or partially); they also typically include a single movement, and they manifest assimilation – a process that spreads phonological features among adjacent phonological elements (akin to the English process that transforms *in+possible*→*impossible*). The possibility that children,[1] rather than adults, are the engines of phonological patterning is significant for two reasons. First, it indicates that the capacity for grammatical regenesis might be age-sensitive – children acquire and generate linguistic structure more readily than adults (Senghas et al., 2004). Second, the fact that adults' signs typically lack phonological patterns demonstrates that patterns are not necessary for the manual production of signs, nor are they required for communication. Accordingly, the spontaneous invention of phonological patterning by children suggests that phonological patterning is a human reflex.

2.3.2 Case 2: phonological patterning in home signs

Another demonstration of spontaneous phonological regenesis is presented in home signs. Home signs are sign systems used by deaf people (mostly children)

---

[1] The available evidence does not make it clear how, precisely, the phonological system has emerged in ABSL. While children clearly exhibit phonological characteristics that are absent in the signs of adult members of their community, these phonological kernels are often shared with the child's immediate family. This observation, in turn, raises the question of whether it is effectively the child or older family members (parents, or older siblings) who are the engines of phonological patterns. While other cases of emerging home signs have specifically identified children as the originators of regenesis (Goldin-Meadow & Mylander, 1983, 1998; Senghas et al., 2004), this possibility awaits further research in the case of ABSL.

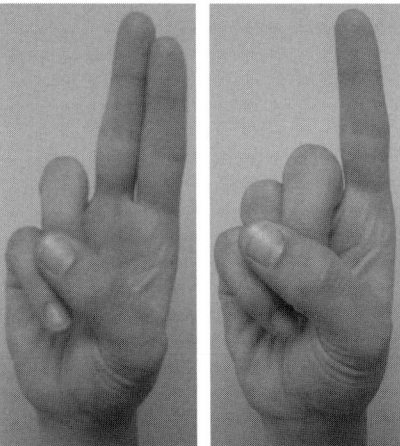

Figure 2.2 Two classifiers for object vs. handling of an object

to communicate with hearing members of their family and community. But since these community members are not signers, the deaf children cannot model their linguistic communication after the adult. Rather, it is the child who is the spontaneous generator of linguistic structure. Careful analyses have indeed documented numerous morpho-syntactic elements devised by children. Remarkably, while these linguistic devices are absent in caregivers' gestures, they are present in many other languages – spoken and signed – to which the child is not privy (e.g., Goldin-Meadow & Mylander, 1983; 1998). A recent study by Diane Brentari and colleagues (2012) suggests that this spontaneous regenesis extends to the phonological system.

The specific case study concerns the phonological elements that mark morphological classifiers. Morphological classifiers are grammatical elements that distinguish between lexical categories. For example, in many sign languages, object (e.g., book) and the handling of an object (e.g., handling a book) form distinct categories, marked by different classifiers. Our interest is in the phonological form of those classifiers. Across sign languages, object classifiers typically have phonological forms that are more complex than handling classifiers. For example, the object classifier on the left of Figure 2.2 is marked by two selected fingers, so its handshape is more complex than the handling classifier, depicted on the right, with one selected finger. Handshape, in general, and the number of selected fingers, specifically, indicate phonological complexity because it is a meaningless attribute that can contrast among meaningful signs (words), just as the selection of an oral articulator can contrast among spoken words (e.g., the lips, for *pea* vs. the tongue tip or blade, for *tea*).

Accordingly, the contrast between the object and handling classifiers reflects a systematic phonological pattern shared among numerous sign languages.

Why, then, do different sign languages converge on their phonological patterning of object and handling? One possibility is that the convergence reflects an instinctive phonological universal. However, distinct languages might also converge for various nonlinguistic reasons. For example, the greater finger complexity of object signs could be dictated by universal properties of the human conceptual systems, not ones that are specific to language. If this interpretation is correct, then object gestures should invariably exhibit greater complexity (compared with handling), even when gestures do not form the primary means of linguistic communication. But a careful comparison of the gestures of hearing people and signers refutes this possibility. In the study, people were asked to manually describe a video display depicting either objects (e.g., a number of airplanes arranged in a row) or the handling of these objects (e.g., putting those airplanes in a row). As anticipated, signers from two different languages (American and Italian sign languages) exhibited greater complexity in their signs for objects compared to handling. Remarkably, however, non-signers showed the opposite pattern – their handling gestures were more complex than objects.

This striking contrast between signs and gestures narrows the range of explanations for the object vs. handling conspiracy evident across sign languages, but it does not entirely nail down the case. Indeed, different languages might converge for historical reasons that have little to do with instinctive linguistic principles that are active in the brains and minds of living breathing speakers.

A final piece of the puzzle, however, rules out this possibility as well. The critical evidence comes from home signs produced by deaf individuals. These individuals had no contact with existing sign languages nor did they interact with each other, so their linguistic production could not possibly mimic any model in their linguistic environment. But despite the absence of a model, home-signers spontaneously regenerated phonological patterns that converged with existing sign languages. Like their counterparts, conventional signers, home-signers exhibited greater phonological complexity in the marking of objects than of handling. The contrast between the structure of home signs and the nonlinguistic gestures of non-signers, on the one hand, and the convergence of home signs with the structure of conventional sign languages, on the other, suggests that this emergent pattern specifically reflects a phonological reflex, rather than a historical accident or nonlinguistic conceptual structure.

Together, the findings from home signs and the signs of ABSL children show that phonological patterning can emerge spontaneously, in the absence of a model. Not only does such regenesis result in the emergence of a phonological pattern, but it further recapitulates the design of existing phonological systems. These observations suggest that the design of all phonological systems – actual and potential – might be shaped by a common set of linguistic constraints. The next section further considers this possibility.

2.4 Shared design

The pervasive instinct to form phonological patterns and their spontaneous regenesis are certainly remarkable, but is phonological patterning truly special? After all, patterning is not unique to phonology. People are adept at recognizing patterns of various kinds – we instantly identify regularities in music, patterns of light and in social interactions, and we share our patterning talents with many other species. It is not only our ability to extend sound patterns to novel instances or even generate entire phonological systems anew that is special about phonological patterning. Rather, it is the fact that once phonological patterns are generated, they tend to all follow a common *design* – a design shared by many human languages, but potentially absent from other systems of communication, both human and nonhuman. The presence of this unique, shared design is extremely significant because it suggests a specialized pattern-maker as its origin. Whether or not the phonological system is in fact specialized is a question that is addressed in the next chapter. Right now, our goal is to describe some of its hallmarks.

The shared design of phonological patterns is evident at various levels of analysis. At the immediate level of formal analysis, phonological systems exhibit primitives and combinatorial principles that are putatively universal. Some of these principles have already been illustrated in the design of ABSL and home signs; additional examples from spoken language phonology are offered next. Viewed from yet a wider functional perspective, however, phonological systems also share a broader master principle of design: Phonology is a combinatorial system, designed to adapt to its channel – the human production and perceptual systems. In what follows, this master principle is first reviewed; some specific candidates for formal universal phonological primitives and constraints are illustrated in the next section.

2.4.1 *Broad principles of phonological patterning: phonology is a combinatorial system designed to optimize phonetic pressures*

Phonological systems must abide by two conflicting sets of demands. They must be sufficiently general and abstract to support the vast productivity of human language. These abstract restrictions, however, must be executed in a human body. So if the phonological system is to survive cultural evolution across numerous generations of speakers, then it had better conform to the limitations imposed by the production and perceptual systems that mediate language transmission. Our question here is how the phonological system negotiates these conflicting demands. To address this question, we must take a closer look at the anatomy of the phonological system – its raw elements and method of patterning. Of interest is how the phonological system conveys information, and whether its method of information-transmission is similar in kind to the acoustic and articulatory channels. But before addressing this question, the notion of information must first be clarified and an explanation given for how it is expressed in various types of information processing systems.

2.4.1.1 Two types of information processing systems: combinatorial vs. blending systems

Our brain is a system of information processing – we use signals, external and internal, to extract information. Information allows us to predict the occurrence of events. If you are about to read two English words (XXX XXXX), your uncertainty regarding what you are about to read is rather high, as the number of possibilities is enormous. But once you know the first word is *the* (e.g., *the XXXX*), your uncertainty has decreased, as some words (verbs, e.g., **the went*, prepositions, **the to*) are unlikely to follow. The decrease in your uncertainty indicates that the signal, the word *the*, carries information (Shannon, 1948; see also Gallistel & King, 2009). This signal is informative because its occurrence manifests some lawful correlation with other signals occurring within a given system (Pinker, 1997). The English orthography, for example, manifests a lawful correlation between the shape of printed letters and words' meanings. Similarly, the amount of sugar correlates with a cake's taste, and the ratio of yellow to blue ingredients correlates with the intensity of the resulting green. All these signals convey information because they form part of a system. Furthermore, in these systems, the ingredients interact to form new combinations. Systems differ, however, on how their ingredients convey information and how they interact (Abler, 1989; Pinker, 1994; Pylyshyn, 1984).

One type of system takes non-discrete elements as raw materials and generates new ones by blending old ingredients together – the formation of green from blue and yellow, the baking of cakes from sugar and flour, and the building of a house from adobe (the mixture of clay, sand, and water) are all examples of blending systems. In these systems, the ingredients (e.g., sugar, the color yellow) are substances, rather than discrete entities, and each such ingredient signals information in an analog manner – the more sugar we pour in, the sweeter the taste. Moreover, when these substances are put together, each individual ingredient is no longer recognizable by itself, so its separate contribution is difficult to evaluate. We cannot identify the yellow bit in the green color, nor can we discern the sand in an adobe house. These cases are examples of *blending systems*.

At the other extreme are *combinatorial systems* – systems such as our number system, DNA, or a Lego tower. Unlike sugar and sand, the ingredients of those systems are discrete. A digit, a DNA base, and a Lego block are elements with clear boundaries, and they signal information digitally – the information associated with each individual signal is either present or absent, rather than varying continuously with the physical properties of a signal (e.g., the information conveyed by a digit is independent of its font size – large (1) or small (1)). Moreover, when discrete ingredients are combined, the signaling element and the information it conveys are both recognizable. One can identify the blue Lego in the tower, the 1 digit in 11, etc. Accordingly, each such ingredient can make a discernible and systematic contribution to the whole. For example, the effect of adding a 1 to any number is fully predictable because 1 makes precisely the same contribution when added to 1 as it does when added to 100,000,000.

This systematicity of combinatorial systems is absolutely crucial for their ability to convey information. Systematicity allows combinatorial systems to generate many new informative expressions. Because we can precisely predict the consequences of adding 1 to 100,000,000, we can use the system to generate an infinite number of new numbers, and the number of such novel combination can be very large. Since productivity is the hallmark of language (Chomsky, 1957), many researchers argue that the language system is discrete and combinatorial (e.g., Chomsky, 1957; Pinker, 1994). But the systematicity of discrete combinatorial systems can also exert a cost for their transmission. Our next question is how the phonological mind negotiates these conflicting pressures.

2.4.1.2 The systematicity-transmissibility dilemma

The conflict between systematicity and efficient transmission presents a dilemma for the evolution of language. While a combinatorial system mandates that each of its building blocks (e.g., *d*) is maintained intact, irrespective of context, the perception and production of speech requires flexibility. To transmit the acoustic signal rapidly and reliably, a well-engineered speech production system modulates the transmission of any segment (e.g., *d*) depending on its context (e.g., *di vs. du*) and speech rate, and indeed, human speech is a blending system (Abler, 1989; Liberman et al., 1967). So the phonological system faces a conflict between the demands for productivity and efficient transmission.

One response to this dilemma makes a unilateral choice between one of two extremes. Some researchers have indeed portrayed phonology as a blending system that mostly follows the dictates of the speech channel, and a few have even gone as far as questioning that phonology is an independent system. Proponents of this view eschew discrete digital phonological units. Syllables, phonemes, and features are merely convenient labels invented by linguists to capture chunks of acoustic stuff or units of motor control (e.g., MacNeilage, 2008; Ohala, 1990). At another extreme, others view phonology not only as a discrete combinatorial system, but as one that is opaque to the limitations imposed by the speech channel – the so-called "substance." Viewed from that perspective, any attempt to link phonological principles to transmissibility pressures is considered "substance abuse" (e.g., Hale & Reiss, 2008).

In between these two extremes lies the possibility that phonology is both discrete and combinatorial and functionally adaptive (e.g., Hayes, 1999; Hyman, 2001; Pierrehumbert, 1975; Zsiga, 2000). In this intermediate view, the phonological mind fits the limitations imposed by the production/perception channels. But an adaptation to the channel does not imply that the channel subsumes the phonological system. While the input to the phonological system is continuous and analog, phonological building blocks are digital and discrete, and the principles that put them together are combinatorial and autonomous from the perception/production channels. *Phonological optimization* – the computational ability to use discrete combinatorial means to optimize phonetic pressures – is a significant hallmark of all phonological systems.

The following sections begin to explore this possibility, first by showing that some phonological representations are discrete and combinatorial (two attributes of an "algebraic" computational framework, detailed in Chapter 3). Next, it is demonstrated that these properties reflect the design of the phonological system itself, rather than the perceptual and auditory channels. The final section demonstrates how phonological principles, while autonomous from the perception/production channel, are nonetheless designed to fit the channel's properties.

2.4.1.3 *Phonological patterns combine discrete building blocks, distinct from their phonetic raw materials*

Many linguists would agree that the phonological system is capable of forming representations that are discrete (e.g., Chomsky & Halle, 1968; Keating, 1988; Pierrehumbert, 1990). Here, this fact will be illustrated by examining the minimal segments of phonological patterns, but similar arguments can be made with respect to other phonological units (e.g., syllables).

Consider the English words *bill* and *pill*. English speakers identify these two words as different, and the difference is attributed to a single sound. The minimal sound unit that contrasts two words is called a *phoneme*. The English /b/ and /p/ are different phonemes because they are the minimal sound unit that distinguishes *bill* from *pill*. What counts as "minimal" and "contrastive," however, intricately depends on one's linguistic knowledge (Steriade, 2007). While English speakers might consider the *ts* sequence in *cats* as two phonemes (e.g., *tip* vs. *sip*), in Hebrew, it is one, as the *ts* sound can appear in word contexts that require a single phoneme (e.g. *tsad* 'side' vs. *bad* 'garment'). And even when two languages employ a sound unit of the same size, they may not necessarily agree on its function. English speakers produce different /p/ variants in *pie* ([p^{h}ai], with a detectable puff of air after the /p/) and *spy* ([spai], with no equivalent puff), but they mentally represent them as a single phoneme as no English words contrast on this dimension (these variant manifestations of a single phoneme are called allophones, and they are traditionally notated with square brackets). Thai speakers, however, use the same sounds to distinguish between words (e.g., *pâ* 'aunt' vs. *pʰâ* 'cloth'; Ladefoged, 1975), so for them, these are distinct phonemes (transcribed using slanted brackets, /p/ vs. /pʰ/). The fact that the contrast between sounds depends on one's linguistic knowledge, not acoustics, demonstrates that a phoneme is a *mental* linguistic representation.

Crucially, the representations of phonemes are discrete. Just as women are never half-pregnant, an English phoneme can be either /p/ or /b/, but never in between. The discreteness of our mental representation of phonemes is particularly remarkable given that they are extracted from physical signals that are continuous and analog. The English syllables /ba/ and /pa/ are both produced by the release of air pressure by the lips, followed by vibration of the vocal folds – an event known as voicing. The distinction between /b/ and /p/ depends on the lag between the release of the air pressure in the lips and the onset of voicing, namely, voice onset time (VOT; Lisker & Abramson, 1964). VOT, however, is an acoustic continuum. For an English /b/, voicing typically occurs up to 10 ms after the release of the consonant (and sometimes with the release, or even just before it), whereas in /p/, voicing typically occurs 50–60 ms after the release (Ladefoged, 1975). But each segment can also take intermediate values that vary according to the specific context, the speech rate, and individual differences among talkers (e.g., Liberman et al., 1967; Miller & Grosjean, 1981). Unlike the continuous acoustic input, however, the phonological system is discrete: English speakers identify sounds along the continuum as either /b/ or /p/, and ignore the differences between intermediate values (e.g., between two /b/ variants) even when people are explicitly instructed to discriminate between them (Liberman et al., 1961). Similarly, phonological processes that target voicing disregard such gradations. Recall, for example, that English words ending with a voiceless stop take a voiceless suffix – they allow *tips* and *bidz* (/tɪps/, /bɪdz/) not *tipz* and *bids* (e.g., /tɪpz/, /bɪds/). This phonological constraint will apply to any instance of a voiceless consonant alike, regardless of its specific VOT. Finer distinctions pertaining to the acoustic and articulatory realizations of a phoneme (e.g., the VOT value produced by any given talker at a particular speaking rate) form part of a separate, phonetic system.

The indifference of phonological processes to fine-grained phonetic variations is not simply due to the limitations of the human ear. As noted above, languages differ on the precise units that they discretely contrast – one language's phonetic distinction (i.e., one that cannot differentiate words) can be phonemic (i.e., one that can contrast among words) in another. Thus, speakers' indifference to phonetic contrasts is not due to the inability of the human auditory system to register these distinctions. Moreover, speakers routinely encode phonetic distinctions and use them online in the process of speech perception. English speakers, for instance, favor typical exemplars of a phonetic category (e.g., a typical

English *p*) over atypical ones (Miller, 2001; Miller & Volaitis, 1989;), and they even register the particular VOT value that is characteristic of an individual talker (Theodore & Miller, 2010). But despite the ability of the human ear to register distinctions along these gradient phonetic continua, phonological systems ignore them. For the purpose of a given phonological system, "a *b* is a *b* is a *b*" – no matter whether their VOT values are all the same or different. Discreteness, then, is neither a property of the acoustic input nor a limitation of the human ear. Rather, it reflects a design property of the phonological system itself.

2.4.1.4 Phonological principles are combinatorial and autonomous from phonetic pressures

Not only does the phonological system rely on building blocks that are discrete, but it also puts them together by relying on principles that are combinatorial. When phonemes are combined, each phoneme makes a systematic, predictable contribution. Adding *b* to *ill* (b+ill→bill), for example, makes precisely the same contribution as adding *b* to *ell* (b+ell→bell). In this way, phonological principles differ in kind from the phonetic system, which largely operates as a blending system. For example, the phonetic distinction between *b* and *p* is informed by multiple acoustic cues, and these cues interact in complex tradeoff relations. While the *b-p* distinction is reliably signaled by VOT, the precise value of the VOT contrast varies as a function of speaking rate: a slower speaking rate is typically associated with an increase in VOT (Miller & Volaitis, 1989). If the analog phonetic component is likened to playdough, amassing sound structures by molding together analog components, then phonology is a Lego system – it assembles discrete parts according to combinatorial principles.

The combinatorial nature of phonological principles and their complex relation to the phonetic system is clearly illustrated in an example suggested by Bruce Hayes (1999). In this example, Hayes shows how phonological processes are not arbitrary, but are rather shaped by phonetic pressures. Nonetheless, these phonetic pressures do not pattern phonological elements directly, but instead, they are "reincarnated" as independent phonological principles. And while these principles certainly make phonetic "sense," once they are represented in the phonological system, they acquire a life of their own, so much so that they can sometimes betray their original phonetic purpose.

The specific case here comes from Egyptian Arabic, and it concerns the voicing of stop consonants. Segments like *p*, *b*, *g*, *and k* are called stops because their production obstructs the flow of air through the vocal cavity. Those four stops can be classified according to two dimensions (see 11): whether or not they are voiced – that is, whether their production is accompanied by vibrations of the vocal folds (*b* and *g* are voiced, *p* and *k* aren't); and the articulator that is involved in the constriction (*b* and *p* are produced by the lips, so they are called *labials*; *k* and *g* are produced by the velum, so these are velar sounds). The place of articulation (e.g., at the lips vs. velum), its manner (e.g., whether or not the airflow is completely stopped), and the voicing are articulatory dimensions that define phonological features (a broader overview including additional phonological features mentioned in subsequent chapters is presented in Box 2.1).

Box 2.1 A brief overview of some major phonological features

Phonemes can be organized into classes according to their constituent features. Like the periodic table of chemical elements, the feature-classification of phonemes is important because it explains their behavior – their susceptibility to interact in phonological processes. While features and phonemes are both discrete phonological entities, the definition of features is intimately linked to the articulatory events associated with the production of the phoneme. Consonants, specifically, are defined by three classes of features: place of articulation, manner of articulation, and voicing.

All consonants are produced by constricting the airflow along the vocal tract. Place-of-articulation features indicate the approximate area of constriction. *Labials* are produced by the lips; *coronals* are articulated by the tongue blade; and *dorsals* are produced by the tongue body, the dorsum (the above-mentioned *velars*, like *k* and *g*, are the subclass of dorsal components involving the tongue dorsum and the velum, the soft palate; other dorsal consonants, like /χ/ of Chanukkah, involve the uvula, and a third subclass of dorsals, the palatals, like /j/ in *yes*, involve the hard palate). Within each such class, one can further distinguish between segments whose production is accompanied by vibrations of the vocal folds – the so-called *voiced* segments (e.g., *b,d*) – and ones that are *voiceless* (e.g., *p,t*). Voicing is a second major feature that classifies consonants.

The third class of features indicates the manner of constriction. **Obstruents** form the subclass of phonemes that are produced while temporarily obstructing airflow – either fully, creating a *stop* consonant (e.g., *p,b,t,d*), or momentarily, resulting in a *fricative* (e.g., *f,v,s,z*). A third type of obstruents, *affricates*, comprise sequences of stop-fricative consonants that share the same place of articulation, such as tʃ and dʒ. The stricture associated with obstruents tends to inhibit spontaneous voicing (because vibration of the vocal folds requires continuous airflow), so their voicing requires a deliberate adjustment. In contrast, *sonorants* (which do not obstruct the airflow), including nasals and approximants, are naturally voiced. *Nasals* are produced by lowering the velum so that air escapes through the nose. *Approximants* are constrictions that do not create air turbulence or trilling. Central approximants allow airflow at the center of the tongue (e.g., *w,r*); lateral approximants allow for airflow at its side (e.g. *l*). Some of these features are listed in (9) using

International Phonetic Alphabet symbols to represent phonemes; these symbols are further illustrated in (10).

(9) Some of the phonological features of English consonants

		Labials		Coronals		Dorsals	
		Voiceless	Voiced	Voiceless	Voiced	Voiceless	Voiced
Obstruents	Stops	p	b	t	d	k	g
	Fricatives	f	v	s ,ʃ, θ	z, ʒ, ð		
Sonorants	Nasals		m		n		ŋ
	Central approximant		w		r, j		w
	Lateral approximant				l		

(10) An illustration of some International Phonetic Alphabet symbols that do not correspond to English spelling
θ **thin**
ð **the**
ʃ **she**
ʒ **pleasure**
j **yes**
ŋ **king**

(11) Some features of stop consonants

		Voicing	
		Voiced	Voiceless
Place of articulation	Labial	b	p
	Velar	g	k

With these facts at hand, consider now Egyptian Arabic. This language bans the voiceless bilabial stop *p*. It manifests words like *katab* (he wrote), but disallows *katap*. For velars, in contrast, the voiced and the voiceless counterparts are both allowed. So the labial voiceless stop is single-handedly banned (indicated by the shading in 11), whereas the other three cells are admitted. This ban on *p* indeed makes good phonetic sense, as it is easier to maintain the voicelessness feature for velars than labials, especially when surrounded by vowels (for detailed explanation, see Hayes & Steriade, 2004; Hayes, 1999; Ohala & Riordan, 1979). In fact, one can rank the phonetic difficulty of producing these various voiceless stops as follows (ignoring the double consonants for the moment):

(12) The phonetic difficulty of producing voiceless stops (Hayes & Steriade, 2004):

    p   k   pp   kk
        bb pp
hard   →   easy

But given that the exclusion of *p* obeys phonetic pressures, one might wonder who runs the show: Is the distribution of these segments governed by abstract phonological principles that are independently represented in the mind, or is it directly regulated by the phonetic system, depending on their ease of production?

A second aspect of Egyptian Arabic allows us to distinguish between these possibilities. This fact concerns geminate consonants. Geminates (e.g., *bb*) are longer versions of their singleton counterparts (e.g., *b*), but the difference in length is significant phonologically, as it can distinguish between words (e.g., *sabaha* 'to swim' vs. *sabbaha* 'to praise'). As it turns out, Egyptian Arabic allows the geminate *bb*, but it bans its voiceless counterpart *pp* (see 13). But unlike the original ban of voiced singletons (*apa), the ban on geminates (*appa) has no phonetic rhyme or reason. While, as noted above, voicelessness is hard to maintain for the labial singleton *p*, for geminates it is the reverse that is true: the voiceless geminates *pp* are **easy** to produce. So the illicit *pp* are easier than the licit *bb*, and as shown in (12), *pp* geminates are also easier to produce than the licit singleton *k*. Accordingly, the ban on *pp* must have a phonological source that is independent of the phonetic factors governing voicing.

(13) Phonological units are preserved under combinations: the case of Egyptian Arabic.
    Voiceless stops are disallowed (*ap)
    Voiceless geminate stops are disallowed (*app)

The systematic link between the presence of geminates and their singleton counterparts in Arabic is not an isolated case. An analysis of a diverse language sample (see Box 3.1) suggests that it is statistically significant and robust: Languages that include some geminate *xx* will also include its singleton counterpart *x*. Such examples show that phonological restrictions are distinct from phonetic pressures, and they each obey different principles. While the phonetic system is a blending system in which the various ingredients interact in complex manners that can obliterate their individual roles, phonological processes are discrete and combinatorial. The phoneme *p* is a discrete element, sampled from the VOT continuum, and the identity of the *p* unit will be preserved when combined to form a geminate, *pp*. While the original ban on *p* is phonetically motivated, it is independently represented at the phonological level. Since phonological principles are governed by abstract combinatorial principles, and given that *p* is recognizable as a part of *pp*, the phonological ban on *p* can extend from singletons (e.g., *ap) to geminates (e.g., *app) despite its greater phonetic cost.

Shared design

Such systematic reflexes, I argue, directly follow from the architecture of the phonological system. Phonological edifices follow good engineering practices inasmuch as they often (perhaps always) optimize phonetic pressures, but their building materials are all discrete, and they are assembled according to principles that are combinatorial. This principle is a broad master plan in the design of phonological systems.

2.4.2 Specific design principles: shared primitives and combinatorial constraints

Phonological systems not only share a broad master plan – the use of discrete combinatorial means to optimize phonetic pressures – but also converge on the specific designs that they ultimately manifest – the set of phonological primitives that recur across languages, and the principles that govern their combinations. This shared design (discussed in Chapters 6–8) offers some of the most decisive arguments for the specialization of the phonological mind. For now, the specialization of the pattern-maker will not be addressed, but some of these common patterns will, however, be illustrated using a few examples from spoken language (commonalities with sign languages are discussed in the next section).

Phonological systems share a set of phonological primitives that are putatively universal. All spoken languages include segments (e.g., *p*) patterned from features (e.g., labial, voiceless, stop); they contrast consonants and vowel segments, and combine them to form syllables (e.g., *pen.cil*), which, in turn, are grouped into hierarchically metrical feet (e.g., [[*red*] [[*pen.cil*]]]. While the specific instances of those categories vary across languages (e.g., not all languages include the segment *b*), the categories themselves are largely shared.

Spoken languages also exhibit common, perhaps universal, constraints on the patterning of those primitives. Consider, for example, the constraints governing the internal patterning of syllables (Prince & Smolensky, 1993/2004). Every syllable has a nucleus (usually, a vowel) at its core. In addition to the obligatory nucleus, syllables may also include one or more consonants at each margin – the initial consonant(s) is called an onset (e.g., *dog*), whereas the final one(s) is called the coda (e.g., do*g*). While syllables may come in different shapes, not all shapes are equally desirable (see 14). Across languages, syllables that include an onset are far preferred (e.g., more frequent) compared to those that lack one (e.g., *ba*≻*a*, where ≻ indicates preference), simple onsets (with only one consonant) are preferred to complex ones (e.g., *ba*≻*bla*), and finally, open syllables (syllables that lack a coda) are preferred to ones with a coda (*ba*≻*bag*).

(14) Cross-linguistic preferences concerning syllable structure
 a. Onsets are preferred: Syllables with onsets are preferred to those without them (*ba*≻*a*).
 b. Complex onsets are dispreferred to simple ones (e.g., *ba*≻*bla*).

c. Codas are dispreferred: syllables with codas are dispreferred to those without them (*ba*≻*bag*)

These cross-linguistic tendencies could suggest the existence of universal phonological principles that render certain structures better formed than others. Indeed, certain structures are not only systematically underrepresented (e.g., *ab* is less frequent than *ba*), but they are also less likely to result from productive phonological processes. A string like *aba*, for example, is far more likely to be syllabified as *a.ba* (including a well-formed *ba* syllable) than *ab.a* (with the comparatively ill-formed *ab*). Moreover, a growing body of experimental evidence demonstrates striking parallels between these cross-linguistic tendencies and the behavior of individual speakers. These findings show that people tend to favor structures that are preferred across languages to ones that are cross-linguistically dispreferred. Crucially, these preferences are documented even when both types of structures are absent in participants' language. Subsequent chapters review the evidence for phonological universals, their developmental onset and neural implementation.

2.5 Unique design

Phonological systems not only share a common design, but their design also differs from other forms of communication, including both the natural communication systems of nonhumans and the nonlinguistic forms of human communications. At first blush, these observations would seem to follow trivially from the properties of the auditory and articulatory channels. Since the human sensory and articulatory systems differ substantially even from those of our closest ape relatives (Lieberman, 2006), the obliviousness of chimps to onset structure (e.g., *ba*≻*bla*), for example, comes as no surprise.

But while the intimate link between the phonetic channel and phonological patterns is undeniable, the possibility that the channel alone is responsible for the structure of phonological patterns runs into two obstacles. First, channel properties, though certainly necessary, are insufficient to explain the design of phonological patterns, as channel and design can doubly dissociate from each other. Despite different modalities, spoken phonological systems share some important primitives and constraints with sign languages. Conversely, phonological and musical patterns share an aural channel, but differ on their structure. Second, the uniqueness of human phonological patterns is evident even beyond their choice of specific primitives and constraints. It is the broad master principle of phonological patterning that is unusual, if not uniquely human. The following section illustrates some of the broad features that distinguish phonological systems from animal communication. We next touch on some of the properties that distinguish phonological patterns from patterns of musical sounds.

2.5.1 Negotiating productivity and channel pressures in animal communication

At their broadest form, phonological patterns assemble discrete meaningless elements according to combinatorial principles that optimize phonetic pressures. None of these ingredients – meaningless patterning, phonetic constraints, or even the representation of a discrete combinatorial system – is unique to humans. But while the ingredients of the phonological mind might be shared with other species, their combination is quite rare. The detailed evidence is presented in Chapter 10. Here, we consider a few illustrations.

Consider the case of birdsong – the quintessential example of vocal patterning in animal communication. Many birdsongs exhibit systematic constraints on the patterning of elements that are apparently meaningless (i.e., lack precise reference). Swamp Sparrow songs, for example, are made of discrete "notes," combined in a particular manner that varies across communities: the New York population of Swamp Sparrows chain their notes in one order (I–VI sequences, where – stands for an intermediate note) whereas Minnesota birds favor the opposite ordering (e.g., VI–I), and individuals from the New York community prefer their local note ordering to the Minnesota syntax (Balaban, 1988a). While these geographic variations must be learned, other organizational principles (e.g., the inventory of notes and some of the restrictions on their combinations) appear to be universal and innate (Lachlan et al., 2010). Moreover, like the adaptation of human phonological systems to phonetic pressures, birdsongs are similarly shaped by motor articulatory constraints (Suthers & Zollinger, 2004).

Birdsong, then, manifests two important hallmarks of human phonology: It exhibits discrete combinatorial structure, and it adaptively fits its transmission channel. But remarkably, the possession of these two capacities does not necessarily give rise to phonological patterning. None of our great ape relatives manifests natural phonological patterns despite demonstrating the capacity for combinatorial structure in laboratory settings. And even when those two ingredients of phonological patterning – combinatorial structure and adaptive fit to the channel – are each deployed in natural birdsong, they do not appear to spontaneously combine together, giving rise to powerful combinatorial principles that are grounded in the properties of the communication channel.

The presence of the ingredients – combinatorial structure and adaptive design – in the absence of their product – algebraic optimization – is significant because it suggests that the product does not trivially fall out from its parts. Merely having the capacity to represent combinatorial structure and to fit the transmission channel does not guarantee the capacity to put complex combinatorial machinery to the service of those phonetic pressures. The rarity of this combination in animal communication and its absence in our phylogenetic relatives, specifically, suggest that it is the result of genetic/neural modification in the human lineage, possibly due to its role in human language. Whether this capacity turns out to be uniquely human remains to be seen (humpback whales might present one notable exception; Payne & McVay, 1971; Suzuki, Buck & Tyack, 2006). But at the very least, it appears that human phonological patterns are highly specialized, and that this specialization does not emerge spontaneously, nor does it spontaneously emerge from the properties of the phonetic channel.

2.5.2 *Phonology and music: similar channels, different designs*

Another perspective on the potential uniqueness of human phonological patterns is gained from the comparison of their design to that of other forms of human communication. If the design of phonological systems were solely determined by the properties of the human auditory and motor systems, coupled, perhaps, with a generic capacity for discrete combinatorial structure, then two predictions should follow. First, the phonological design of spoken languages should differ markedly from that of signed languages. Second, the design of phonological systems should converge with musical systems – systems that rely on the same auditory and articulatory interfaces. Neither of those predictions is borne out.

Although the phonologies of signed and spoken languages differ in important ways, they nonetheless share numerous primitives and combinatorial principles (Sandler & Lillo-Martin, 2006). As in spoken languages, the phonological patterns of sign languages hierarchically combine features to form two binary classes (Location and Movements), which, in turn, yield syllables – meaningless units that are demonstrably distinct from the meaningful morphemes. Moreover, signed and spoken phonologies also share some phonological constraints, including sonority and identity restrictions (Brentari, 1998; Sandler, 1993). Thus, the putative phonological universals are not invariably modality-specific.

The complementary aspect of the dissociation between channel and design is evident from the comparison of phonological and musical patterns – systems that share the aural channel, but differ in important aspects of their design (Patel, 2008). Since a full exposition of those differences falls beyond the scope of this discussion, we will resort to one telling illustration concerning the properties of hierarchical structures in the two domains.

Music and phonology both represent hierarchical structures. In both cases, events are related to each other by virtue of their role in an overall hierarchy. But those hierarchies encoded in the two domains are different in kind. Phonological hierarchies define *containment*: A syllable (e.g., *bag*) contains an onset (e.g., *b*) and a rhyme (e.g., *ag*), which, in turn, contains a nucleus (*a*) and a coda (*g*), each of which comprise feature hierarchies. Furthermore, phonological hierarchies outline domains used to restrict the co-occurrence of elements. Syllables are the

Unique design 31

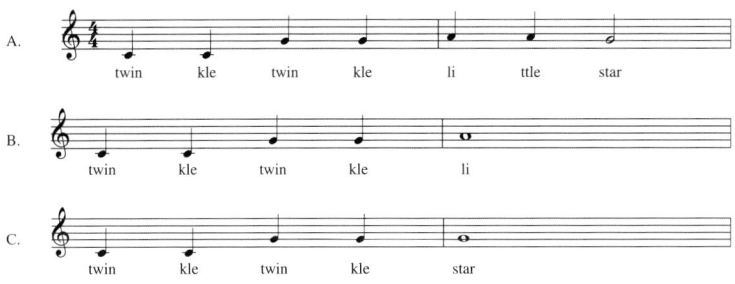

Figure 2.3 Twinkle, Twinkle, Little Star

domains of restrictions on consonant co-occurrence: sequences like *lb* are allowed across syllables (e.g., *el.bow*) but not within them (e.g., *lbow*).

Like phonological patterns, musical pitch systems constrain the sequencing of auditory events, and some of those representations encode containment hierarchically (i.e., grouping hierarchies, see Lerdahl & Jackendoff, 1983). But the organization of musical events also supports constraints of a markedly different type – the representation of relative *stability*.

Consider, for example, "Twinkle, twinkle, little star" (see Figure 2.3). Let us focus on the second bar. This bar includes two pitch events: an A (of *little*) and a G (of *star*). Now, suppose you were asked to perform a little "Solomon judgment": You may retain only one of the two pitches and you must give up the other. Which one would you choose?

When push comes to shove, most listeners would sacrifice the A (of *little*) over the G (of *star*) since the G sounds more stable and hence provides a better ending than an A. Moreover, the unstable A is perceived as an elaboration of the more stable G. Indeed, tonal events are encoded in terms of their prominence-stability and elaboration. Certain events are represented as more stable, and hence, more prominent, than others, and unstable events are perceived as the elaboration of more stable ones (Jackendoff & Lerdahl, 2006; Lerdahl & Jackendoff, 1983). By contrast, ill-formed syllables (e.g., *lba*) are not unstable and neither do they elaborate better-formed syllables (e.g., *ba*) – stability and elaboration uniquely pertain to the domain of musical tonality, they have no role in phonological patterning. Thus, despite their common reliance on hierarchical organization, musical and phonological hierarchies are different in kind.

Summarizing, then, phonological patterning is intimately linked to the phonetic channel, yet the phonetic channel alone is insufficient to explain the design of phonological systems. Music and phonological systems share an auditory channel but differ in design, whereas the phonological patterns of signed and spoken languages exhibit significant similarities in design despite relying on different channels. It is thus unlikely that the special status of phonology relative to the various manifestations of human aural patterns is solely due to the phonetic channel. The human phonetic channel, including its auditory and articulatory characteristics, is also unlikely to solely account for the profound differences between human phonology and the natural communication systems of nonhumans. While many nonhuman animals exhibit phonetically adaptive patterns of meaningless elements, and several species might even possess discrete combinatorial means, no other species is known to deploy the powerful combinatorial arsenal characteristic of phonological patterning in the service of its natural communication. These observations suggest that the design of phonological systems may be not only universal but possibly unique.

2.6 Phonological knowledge lays the foundation for the cultural invention of writing and reading

The human propensity for phonological patterning manifests an interesting quirk. Not only does it instinctively apply to our primary form of linguistic communication but it also extends to reading and writing. Unlike language, a natural biological reflex that is universally present in any human community, writing and reading are cultural inventions. They deploy elaborate invented technologies that use visual symbols to convey linguistic information – inventions that emerge only in certain select human cultures. But as detailed in Chapter 12, even these language technologies are phonologically based. The precise phonological element depicted by the writing system varies – some orthographies use graphemes to stand for syllables (e.g., Chinese); others (e.g., English) encode consonants and vowels; while some register only a bare skeleton of segmental structure using mostly consonants (e.g., Hebrew). Nonetheless, all fully developed writing systems encode phonological units of some level (DeFrancis, 1989).

Phonological patterning defines not only the encoding of linguistic messages in writing, but also their decoding in reading. Reading invariably entails the decoding of phonological forms from print. Phonological decoding is clearly evident in the laborious, intentional decoding of beginning readers, but it is not limited to the initial stages of reading acquisition. Although many skilled readers believe they extract words' meaning "directly" from letters, without any phonological mediation, appearances can be misleading. A large body of experimental research demonstrates that phonological recoding is quite robust. For example, adult skilled readers tend to incorrectly classify *rows* as a flower, reliably more than a spelling control *robs* (Van Orden, 1987; Van Orden et al., 1988). The difficulty with *rows* cannot be due to its spelling, as *rows* and *robs* are both matched for their letter-overlap with the intended homophone, *rose*. Accordingly, the errors with *rows* show that skilled readers extract the phonological form of printed words, and they do so automatically, even though the task clearly calls for spelling verification. Such phonological effects have been demonstrated in numerous orthographies, ranging from English to Chinese (Perfetti & Zhang, 1995). Readers' reliance on phonological computation in different orthographies suggests that, despite their different methods of decoding phonological structure, the computation of phonological representations from print might be universal (Perfetti et al., 1992).

Further evidence for the link between people's phonological competence and reading ability is offered by developmental dyslexia. Developmental dyslexia is a heritable reading disorder characterized by the failure to attain age-appropriate reading level despite normal intelligence, motivation, affect, and schooling opportunities (Shaywitz, 1998). This very broad definition does not specify the etiology of dyslexia or its precise manifestations, and indeed, dyslexia might acquire several distinct forms resulting from distinct underlying deficits. Although there is clearly not a single cause for dyslexia, research in this area has consistently shown that many dyslexic readers exhibit deficits in decoding the phonological structure of printed words, which, in turn, can be linked to subtle, heritable deficits in processing spoken language that are detectable even in early infancy (Leppänen et al., 2002; Molfese, 2000).

Taken at face value, these findings are puzzling. Why should writing and reading rely on phonology? The reliance on phonological principles is not logically necessary for the visual encoding of language – new words could conceivably be formed by combining semantic, rather than phonological, attributes (a *girl* could be expressed by the combination of *female* and *young*, a *book* by combining signs for *language* and *sight*, etc.) and semantic features do, in fact, play a role in some writing systems (e.g., in Chinese). Yet, no full writing system uses it to the exclusion of phonological patterning. Even more puzzling is the recoding of phonological structure in skilled reading given that doing so is even detrimental, as evident in the phenomenon of homophone confusion (e.g., a *rose is a rows is a roze*).

The compulsive patterning of print, however, is far better understood within the broader context of phonological patterning in natural language. All human languages, both spoken and signed, manifest phonological patterning, the design of phonological patterns exhibits some common universals that distinguish it from the structure of nonlinguistic systems, and humans are instinctively tuned to discover the phonological patterning of their native language: they begin doing so practically at birth, and when no phonological structure exists, they invent a phonological system that bears some similarity to the structure of existing phonological systems. Seen in that light, the phonological basis of reading and writing follows naturally from the design of oral language. Although reading and writing are invented systems, these systems encode linguistic information, and consequently, they must rely on some of the linguistic machinery that is already in place. If phonological patterning is instinctive and indispensable for the representation of language, then it stands to reason that the language system cannot handle printed linguistic symbols unless they are first recoded in phonological format. The recapitulation of phonological principles in the invention of writing and reading thus underscores the instinctive nature of phonological patterning.

3 The anatomy of the phonological mind

The special phonological talents of humans, reviewed in the previous chapter, demand an explanation. This chapter articulates two rival accounts for these facts. One view asserts that humans are biologically equipped with a specialized system for phonological patterning, the phonological grammar. The productivity of phonological patterns, their spontaneous emergence and universality all spring from two broad properties of the system: its algebraic computational machinery, and the presence of substantive universal constraints on the structure of potential phonological patterns. On an alternative explanation, the phonological talents of humans result from systems that are not specialized for phonological patterning. The following discussion outlines these two competing hypotheses as the basis for their evaluation, in subsequent chapters.

3.1 The phonological grammar is a core algebraic system

Humans are equipped with remarkable phonological talents. We instinctively recognize phonological patterns in the structure of our language, we spontaneously generate phonological systems anew, and the patterns we produce have some recurrent and potentially unique design properties.

What is the basis for the pervasive phonological talents of humans – the reflexive tendency of the very young and old to engage in phonological patterning, to systematically extend the patterns of their native language to novel forms that they have never heard before, and to generate phonological systems anew in the absence of a model? Why do different languages exhibit common phonological patterns that are intricate and complex, but distinct from other expressive communication systems used by humans and nonhumans? And why do the phonological reflexes of language extend to the invented cultural technologies of reading and writing?

In this chapter, I consider the possibility that these special human talents are the product of a special system, dedicated to the computation of phonological structure. At the heart of the phonological system is a set of constraints that favor certain phonological structures over others (e.g., ba≻a). These constraints further manifest two important properties. First, they are highly productive and systematic, and consequently, they allow speakers to extend their phonological knowledge to novel forms. Second, those constraints are shared across many languages, perhaps even universally. These two properties define a system of the brain and mind that is specialized for phonological patterning – the *phonological grammar*.

As in previous chapters, terms like "constraints" and "grammar" are used here to refer to a set of principles that are tacit and instinctive, not to be confused with the set of normative restrictions on how one *should* use one's language, as determined by some self-designated language mavens. Just as young infants instinctively know that a dropped object will fall down, rather than fly up in the air, so do people, young and old, have instinctive knowledge of the sound structure of their language. They acquire this knowledge spontaneously, without any explicit tutoring, and they encode it as a set of constraints on the shape of possible phonological patterns. Those tacit constraints are the phonological grammar. Our questions here, then, are what allows phonological grammars to extend generalizations across the board, and why the phonological grammars of different languages manifest shared properties.

In this book, I trace these features to two broad characteristics of the phonological system. The productivity of phonological patterns is derived from several computational properties of the phonological grammar, which are collectively dubbed "algebraic." The second characteristic of the phonological grammar, namely universality, suggests a core knowledge system whose design is partly innate. Core knowledge systems (e.g., our instinctive knowledge of number, physics, biology, and the mind of others) were briefly mentioned in Chapter 1. Each such system manifests a unique design that is relatively invariant across all individuals and forms the scaffold for the acquisition of all subsequent knowledge. Here, I suggest that the phonological grammar likewise forms a system of core knowledge. Combining these two properties (see 1), then, the phonological grammar is an algebraic system of core knowledge.

(1) The anatomy of the phonological grammar
    a. The phonological grammar is an algebraic computational system.
    b. The phonological grammar is a system of core knowledge.

The hypothesis of an algebraic core system of phonology, however, is certainly not the only possible explanation for phonological patterns. On an alternative account, phonological patterns are the outcome of several nonlinguistic systems (e.g., auditory perception, motor control, general intelligence) that are neither algebraic nor specialized for phonological computation. Attested phonological systems are shaped by the linguistic experience available to speakers. And to the extent that different phonological systems happen to share some of their properties, such convergence is ascribed to historical forces and generic aspects of human perception and cognition, not the reflexes of a specialized phonological system per se.

The following discussion outlines the hypothesis of a specialized algebraic phonological system. We next consider some of the challenges to this view. Although the empirical observations that motivate those challenges are undeniable, there are several compelling reasons to reconsider the hypothesis of a specialized phonological system. Subsequent chapters evaluate this hypothesis in detail.

3.1.1 The phonological grammar is an algebraic system

Phonological knowledge entails the ability to recognize and generate novel patterns of meaningless linguistic elements. As we saw in Chapter 2, speakers will extend many of the phonological patterns of their language to novel forms. When presented with new words that he or she has never heard before, an English speaker will recognize rhymes (e.g., *hane* vs. *rane*), parse words into syllables (e.g., *en-freaking-bot* vs. *e-freaking-nbot*), and enforce voicing agreement (e.g., *Bachs*, but not *Bachz*). In fact, people will systematically generalize phonological patterns even when doing so gives rise to forms that are harder to produce and perceive. Such cases are instructive because they demonstrate that this productive force pertains to the phonological system itself. To explore the mechanisms that support grammatical generalizations, we will focus on one such case, the Egyptian ban on geminate *pp*, mentioned in Chapter 2.

To briefly review the relevant facts (from Hayes, 1999; Hayes & Steriade, 2004), Egyptian Arabic bans the singleton voiceless *p* (e.g., *apa*), but allows its voiced counterpart *b* (e.g., *aba*). Taken at face value, the ban on *p* could potentially result from either phonetic or phonological constraints, as forms like *apa* are harder to produce than *aba*. But remarkably, the ban on the singleton *p* extends even to the geminate *pp*. Because the illicit *appa* is actually easier to produce than the attested *abba*, its exclusion must result from the phonological system proper, rather than the subsidiary phonetic component. And, as shown in Box 3.1 below, the contingency of geminates on singletons is a robust phenomenon, present in many languages. Our question here, then, is how the phonological grammar forms such generalizations. Specifically, why does a ban on *p* automatically extend to ban *pp?*

Generalization, of course, is not unique to phonological systems. Many aspects of our cognition exhibit systematic leaps from evidence to conclusions (see (2)). Given premises such as *Socrates is human* and *Humans are mortal*, people routinely conclude that *Socrates is mortal*. Similarly, if *Socrates is barred from heaven* and *Plato is barred from heaven*, we gather that *Socrates and Plato* will not be admitted either. The question of why the "no *p*" ban extends to "no *pp*" is a special case of this broader phenomenon. But what allows the human brains to exhibit systematic inferences?

(2) Some systematic algebraic inferences from inputs to outputs
   a. *{Socrates is human; Humans are mortal}*→*Socrates is mortal*
   b. *{Socrates is barred from heaven; Plato is barred from heaven}*→ *{Socrates and Plato are barred from heaven}*;
   c. **p* →**pp*

That our brain, a physical machine, is capable of such feats should not be taken lightly. Although modern life presents us with countless thinking machines, physical devices that perform systematic inferences, the capacity of physical systems – brains or mechanical devices – to solve such problems presents a fundamental challenge. Every time our brain lawfully connects a set of inputs (e.g., *Socrates is human; Humans are mortal*) and outputs (e.g., *Socrates is mortal*), it effectively traverses the abyss between mind and matter. In our mind, there is a natural link between premises and inferences, and this link is formulated at the semantic level: We sense that the semantic content of the inference follows from the semantic contents of the premises. What is remarkable is that our brain can establish this link. Like all physical devices, the operation of the brain can only follow natural laws, guided by the physical properties of matter, not some elusive notion of semantics. How, then, is a physical device such as our brain capable of establishing systematic semantic relations (see 2)? And why is it compelled to deduce certain inferences from premises – either logical (e.g., *Socrates is human*) or phonological (no *p*)?

The solution put forward by the philosopher Jerry Fodor (Fodor, 1975; Fodor & Pylyshyn, 1988) is known as the Computational Theory of Mind (CTM). Building on Alan Turing's (1936) analysis of thinking machines (i.e., a Turing machine – a theoretical device capable of performing any computer algorithm by sequentially reading and writing symbols on a tape, one symbol at a time; Haugeland, 1985), Fodor suggested that mental processes are computational – they are sensitive to the formal organization of data structures, and they operate by manipulating the structure of those representations (see 3). I review those distinct assumptions in turn.

(3) The Computational Theory of Mind (CTM): some core tenets.
   a. Structured representations
      (i) Mental representations are symbols, either simple (atomic) or complex (molecular).
      (ii) Complex mental representations have syntactic form and semantic content.

(iii) The semantic content of complex mental representations depends on the syntactic structure of their parts and their atomic meanings.
b. Structure-sensitive processes
   (i) Mental processes manipulate the syntactic form of representation in a manner that is blind to their semantic content.
   (ii) Mental processes operate on variables.

3.1.1.1 Mental representations are discrete symbols

The first set of assumptions in (3) concerns the structure of **mental representations**. Mental representations, according to the CTM, are symbols: arbitrary pairings of discrete forms and semantic contents. To illustrate this assertion, let us first consider words where the distinction between form and meaning is easy to grasp. English, Spanish, and Hebrew speakers all share a concept of DOG, but they each express it using different forms: English speakers say *dog*; Spanish speakers use *perro*, and for Hebrew speakers, it's a *kelev*. The phonological forms *dog*, *perro*, and *kelev* are all discrete, they are each linked to the concept DOG, and the link is arbitrarily set in each such language.

The pairing of form and semantic content plays a similar role in the representation of phonology. As discussed in Chapter 2, phonemes are mental representations that are discrete and digital. Within the phonological grammar, each phoneme is thus represented by a symbol that carries a distinct semantic content – the content /p/, for instance, is different from /k/. This semantic content is not any specific concept – phonological symbols carry no conceptual meaning. Rather, it is the interpretation given by the phonological system to the forms that represent each phoneme (e.g., the form that encodes /p/). The actual physical representation of phonemes in the brain remains unknown, but if the CTM is true, then we should expect the forms encoding /p/ and /k/ to be different. For now, we will illustrate this hypothesis using different shapes to stand for different phonemes – a circle for /p/, a square for /k/ (see Figure 3.1). Obviously, I do not claim that the brain represents phonemes by geometrical shapes. Rather, these shapes simply illustrate the hypothesis that different semantic contents (those of /p/ and /k/) are represented by different forms (here, circle and square). Also, note that forms can represent either an individual instance *i* (e.g., /p/ vs. /b/, members of the class of "labial consonants," see Figure 3.1a) or an entire class (e.g., any labial consonant, and velar consonant, etc., see Figure 3.1b). Our goal here is to explore the use of form to convey semantic distinctions. Doing so, as we will next see, requires, *inter alia*, the capacity to distinguish simple and complex representations, and to operate on variables.

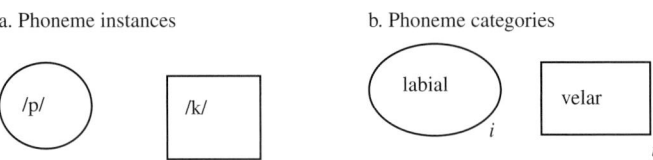

Figure 3.1 The use of atomic shapes as symbols for singleton phonemes, either specific phoneme instances (a) or phoneme categories (b)

3.1.1.2 Complex representations and structure-sensitive processes
A single phoneme (e.g., /p/) or a feature (e.g., labial) can each be considered an atom – their form is simple (unstructured), and so is their semantic content.[1] We now consider how those simple atomic representations give rise to semantically complex representations, such as the geminates /pp/. We consider two encoding schemes for such semantically complex representations: ones that are either formally simple or complex (see Figure 3.2 vs. 3.3)

The first approach represents the semantic complexity of /pp/ by coining a new atomic form – an octagon (see Figure 3.2). The resulting system is clearly adequate, inasmuch as it captures the distinction between /pp/, /p/, and /k/, but there is nonetheless a problem. To unveil it, consider how this system might represent the ban on specific phonemes (e.g., /p/ is disallowed). Let us encode such bans by a star. A star followed by a circle will ban /p/, a star followed by a square bans /k/, and a star followed by an octagon bans /pp/. While, at face value, this system captures all the relevant facts, its method of doing so is fundamentally flawed. Each fact must be encoded separately: given */p/, one has to independently stipulate the ban */pp/. The problem, here, is not that it is impossible to ban /pp/; evidently, this constraint can be represented quite easily. Rather, the concern is that doing so requires a separate stipulation, distinct and independent from the ban on /p/. If one didn't encode it separately, the */pp/ ban wouldn't automatically follow from an existing */p/ ban.

Human minds, however, do not work this way. Thinking of a semantically complex proposition such as *Socrates & Plato* does not merely *allow* for the possibility of thinking of *Plato* but effectively requires it. So if *Socrates* is barred from heaven, then, other things being equal, he will remain shunned even when accompanied by *Plato* ("*Plato & Socrates are barred from heaven*"). Like logical inferences, phonological generalizations exhibit systematicity. The Egyptian Arabic ban on /p/ systematically transfers to /pp/, and the transfer occurs on formal grounds that are internal to the phonological grammar – there

---

[1] The view of phonemes is adopted here for purely expository reasons. In reality, phonemes are bundles of features whose precise composition is critical for most phonological processes.

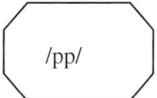

Figure 3.2 The use of atomic shapes to encode geminates

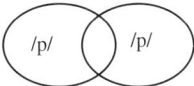

Figure 3.3 The use of complex shapes to encode geminates

is no phonetic motivation for such generalizations. The question, then, is what property of the grammar ensures that such systematic inferences *must* follow.

The solution proposed by the CTM makes semantic complexity follow from the syntactic complexity of forms. The semantic complexity of *Socrates & Plato* is represented by the combination of two distinct forms, for *Socrates* and *Plato*, respectively. Similarly, the semantic complexity of geminates /pp/ is expressed by the reduplication of the form standing for the /p/ atom (Figure 3.3). Since mental processes are sensitive to the formal structure of mental representations, syntactic complexity mandates semantic content in a systematic and obligatory fashion: The syntactically complex *Socrates & Plato* entails *Plato*, /pp/ entails /p/. In the same vein, a ban on /p/ automatically transfers to each constituent of the complex */pp/ symbol, so */p/ entails */pp/, as is indeed the case in Egyptian Arabic.

3.1.1.3 The role of variables

Geminates, such as /pp/, are special cases of a general relation between two phonological categories – the relation of identity. And while the machinery discussed so far would support the formation of systematic links between simple and complex representations, it is still limited in its ability to capture the identity function itself.

The problem is, once again, systematicity. Recall that forms can stand for either an instance (e.g., /p/) or an entire category (e.g., *any labial consonant*). While the specific restriction */pp/ applies to a specific instance /p/, identity restrictions typically apply to entire categories (e.g., *labials*). Because all members of the class are represented by the same symbol, they form an **equivalence class**. And by referring to equivalence classes, identity restrictions in turn can extend to any of their members, either familiar or novel. But while restrictions over entire categories (e.g., *any labial*) can support powerful generalizations, they also raise some new challenges. To see those challenges, consider, for example, a phonological system that requires any two labials to be identical. Given /p/, the system should produce /pp/, and given /b/, the system will produce /bb/. Forms like /pb/, with nonidentical labials, are banned. Our goal, then, is to allow the formation of identical consonants (e.g., /pp/, /bb/) while excluding non-identical members (e.g., /pb/).

The formation of identical elements (e.g., X➔XX) is the mirror image of the ban on identical items (e.g., *XX), and both cases are identity functions. Identity functions apply to any member of a class X, and this class is instantiated multiple times within a single expression (e.g., *XX). The challenge is to ensure that these instances are all identical: If a particular labial /p/ is selected for the initial category X, then this very same labial (and not /b/) should be selected in all others. Variables, akin to the algebraic variables in mathematical expressions (X➔XX), can address this challenge. Variables can stand for a category of instances as a whole. This, in turn, allows us to encode a given category multiple times and bind these multiple occurrences together. For example, if we take X to stand for "any labial," variables will ensure that we select the same labial instance (e.g., /p/) whenever the labial category is called. Moreover, once such variables are introduced, then the generalization of relations, such as identity, will follow automatically. A ban on any singleton will not merely allow for the ban on its geminate counterpart – it will effectively require it.

3.1.1.4 Summary

The explanation proposed by the Computational Theory of Mind, in general, and in its present application to the case of phonology, discussed here, makes some broad predictions with respect to the types of representations available to people and the types of inferences they support (see 4). First, as discussed in Chapter 2, it assumes that phonological representations are discrete and combinatorial. Second, such representations distinguish between instances of a category (e.g., *p*, *k*) and the category (e.g., *any consonant*). But because instances are all represented by forms, the system can treat all members of a category alike, and ignore their differences. Such equivalence classes are important because they allow the system to encode generalizations that apply to any class member, either existing or potential, and as such, they can extend generalizations across the board. Finally, because phonological systems have the capacity to operate over variables, they can encode abstract relations among categories, such as their identity. Following Marcus (2001), we will refer to systems that exhibit these properties as algebraic. Our question here is whether the phonological grammar is in fact an algebraic system. Subsequent chapters will address this question.

(4) Some features of algebraic systems
    a. Algebraic representations are discrete and combinatorial.

b. Algebraic relations define equivalence classes that support across-the-board generalizations to any class member.
c. Algebraic systems allow for the encoding of abstract formal relations, such as identity (e.g. *bb*).

Box 3.1 A typology of geminates

The algebraic account of phonology, discussed in section 3.1.1, predicts that languages with geminate consonants should also exhibit their singleton counterparts. The evidence reviewed so far, however, was based on a single case, Egyptian Arabic. Accordingly, one might wonder how general is this phenomenon: Is the phonological representation of geminates typically complex, or might geminates be represented in an atomic fashion, unrelated to singletons?

To examine this question, one can compare the occurrence of geminate consonants and their singleton counterparts in a representative language sample. This survey uses the *UCLA Phonological Segment Inventory Database*, a sample including 919 segments from 451 languages complied by Ian Maddieson (1984); (the html interface by Henning Reetz can be found on http://web.phonetik.uni-frankfurt.de/upsid_info.html). If geminates are represented algebraically, as complex segments, then a language that allows a geminate (e.g., /pp/) will necessarily have to also allow its singleton counterpart (e.g., /p/); an atomic representation predicts no such contingency.

Table 3.1 lists the number of occurrences of geminates and singleton segments. An inspection of these figures suggests that geminates are not very frequent in the sample (a total of seventy-five occurrences overall, summed across all languages that manifest geminates (a total of twelve languages) and all geminate segments within each language). Remarkably,

Table 3.1 *The contingency between geminate consonants and their singleton counterparts in the* UCLA Phonological Segment Inventory Database

	Singleton present	Singleton absent	*Total*
Geminate present	63	12	*75*
Geminate absent	3,129	13,943	*17,072*
Total	*3,192*	*13,955*	***17,147***

however, when a geminate does occur in a language, in most cases (63/75=0.84) so does its singleton counterpart. For example if /pp/ occurs, so does /p/. There are only twelve counterexamples (occurrences of geminates without their singleton counterparts, such as /pp/ without /p/), and their probability *(p=.165)* is far lower than the probability that these singletons are independently absent. To calculate the probability of those cases, we consider all singleton sounds for which a geminate counterpart does exist in the sample (e.g., /p/, whose geminate, /pp/, exists in some languages). For each such sound, we next count the number of languages in which only the singleton sound was present. For example, although the /mm/ geminate is attested in the sample (there are three such languages), there are 423 languages in which /m/ occurs in the absence of /mm/, and when all these singleton-only occurrences are summed across segments, the total reaches 3,129 cases. For each of these singletons, we next tally the number of languages that lack both the singleton and its geminate counterparts. For the /m/-/mm/ example, there were twenty-five languages in which neither /m/ nor /mm/ is attested, and when summed across all segments, the total cases in which both singleton and geminate are absent is 13,943. The probability of these independently unattested singletons relative to the total segment occurrences in the sample (17,147) is thus 13,955/17,147, *p=.81*. By contrast, the probability that singletons are absent when their geminate is present is significantly lower (12/75, *p=.165*). A statistical comparison of these two proportions (a binomial test, $\chi^2(1)=208.26$, p<.0001) indicates that the presence of a geminate implies its singleton. Thus, if a language includes a geminate, it is also likely to include its singleton counterpart. While this analysis does not address phonetic explanations for these facts, it does suggest that the singleton-geminate contingency in Egyptian Arabic may not be an isolated case.

3.2 Phonology is a core system

While algebraic machinery might be necessary to attain phonological generalizations, algebraic machinery alone cannot be the entire story. The evidence reviewed in Chapter 2 suggests that phonological patterns are special. They have some common design features that hold across languages. Moreover, this design is not found in the natural communication of nonhuman species, nor is it invariably present in all aural patterns generated by humans (e.g., music). The existence of unique, idiosyncratic phonological patterning would suggest an idiosyncratic pattern-maker – a system that is *specialized* for the computation of phonology.

Specialized mental systems, however, are often confused with various straw-man claims that have attracted lots of bad press (see 5). Some people view specialized systems as encapsulated processors whose operation is blind to any contextually relevant information. Others have presented specialized systems as "neural hermits" – brain substrates sharing none of their components with any other cognitive mechanism, regulated by genes that are expressed exclusively in those regions. Finally, domain-specificity is sometimes equated with a radical nativist perspective that requires the relevant knowledge to be fully formed at birth, immune to any relevant experience.

(5) Some presumed properties of domain-specificity: a cartoon version
   a. Processing autonomy: Domain-specific knowledge is fully encapsulated from contextually relevant information.
   b. Hardware independence: Domain-specific knowledge is encoded in neural substrates that share none of their components with other domains, and are regulated by genes that are expressed exclusively in those regions.
   c. Radical nativism: Domain-specific knowledge is fully formed at birth, and is immune to experience.

None of these claims fares very well against the available evidence. Decades of intense experimental tests have failed to unequivocally identify any cognitive system that is fully encapsulated. And since some degree of processing encapsulation is a general hallmark of automaticity – the reflex-like character of highly practiced skills, such as driving or typing – encapsulation, if shown, would hardly demonstrate specialization anyway. Likewise, no known cognitive system, including language, is realized by neural substrates that are fully or even partially segregated; no known brain substrate comes "preloaded," blind to its neighboring cells and immune to experience, and no known gene is expressed solely in "language" brain areas. Finally, while some hereditary traits are present at birth, others (e.g., secondary sexual characteristics) manifest themselves only in later development. Regardless of their development onset, however, inherited traits are exquisitely sensitive to variations in experience, subtle changes in environmental conditions, and plain chance (Balaban, 2006). Accordingly, the requirement that domain-specific knowledge be fully immune to experience is biologically untenable.

Although there is plenty of evidence to refute the cartoon version in (5), such observations do not in fact address cognitive specialization per se. To appreciate this fact, consider a case of putative specialization in a nonhuman species. Zebra Finches are famous for their characteristic song, which manifests several hallmarks of a specialized biological system: Its acquisition is typically limited to a specific "window of opportunity" (Immelmann, 1969), it is regulated by a well-defined brain network that is genetically controlled (Warren et al., 2010), and the song is highly invariant across members of the species. In fact, Zebra

Finches have been shown to converge spontaneously on this song pattern even when they have never heard it before (Fehér et al., 2009). Nonetheless, a full acquisition of the song requires learning, the brain substrates engaged in birdsong might include components shared with other behaviors, and they are regulated by genes that are expressed in multiple sites (e.g., Haesler et al., 2007; Scharff & Haesler, 2005; Warren et al., 2010). But despite these blatant violations of "hardware independence" and "radical nativism," most people still consider birdsong as a good candidate for a specialized cognitive system. In fact, Marcus (2006) notes that such violations are all but expected. Because evolution proceeds by tinkering with existing systems, one would expect the biological products of natural selection to be erected upon existing systems and share many of their neural substrates and regulating genes. Just as hardware independence and radical nativism are not necessary conditions for specialization in birds, so there is no reason to expect them to define domain-specificity in humans.

But if none of the conditions in (5) are necessary for specialization, then what features might define domain-specific mechanisms? In what follows, I will not offer an all-encompassing theory of mental architecture. But when it comes to early knowledge systems, such as language, one feature I believe is necessary, and a few others are likely. Following the proposals of Susan Carey and Elizabeth Spelke (Carey, 2009; Carey & Spelke, 1996; Hauser & Spelke, 2004; Spelke, 1994; Spelke & Kinzler, 2007), I will refer to these knowledge systems as systems of *core knowledge*. Many features of core knowledge, however, are shared with systems dubbed modules (Fodor, 1983), mental organs (Chomsky, 1980), learning organs (Gallistel, 2007), instincts (Pinker, 1994), and domain-specific mechanisms (Cosmides & Tooby, 1994).

(6) Some properties of core knowledge systems
  a. Core knowledge manifests a *unique universal design* that is largely invariant across individuals despite wide variations in experience.
  b. Core knowledge systems are likely to manifest an *adaptive design*: They solve a task that benefits its carrier, their design serves their specific function, and it fits the architecture of its carrier and its environmental niche.
  c. Core knowledge systems are preferentially implemented by an *invariant neural substrate* whose assembly is genetically regulated.
  d. Core knowledge systems are likely to be active in *early development*.
  e. Core knowledge lays down the foundation for *cultural inventions* and discoveries.

The one single feature that truly defines core knowledge is design (see 6a). A core knowledge system manifests a unique universal design, characterized by a set of shared representational primitives and combinatorial principles. While some of these principles are grounded in external pressures, most notably perceptual ones, the design as a whole is unique. For example, recall (from Chapter 2) that many of the constraints on the voicing of stop consonants can be traced to phonetic pressures, but these restrictions do not directly regulate phonological processes, and consequently, the constraints on voicing must be independently represented in the phonological system. Moreover, the design of phonological systems is relatively invariant across individuals despite large differences in the range of linguistic experience available to them. In the limiting case, core knowledge might emerge spontaneously in the absence of any model.

The propensity of humans to acquire knowledge systems of universal design suggests that those systems are innate. Innateness, however, is a term that I use with great trepidation, as many people consider a trait "innate" only if it is fully formed at birth, independent of experience – a possibility that is most likely false for many knowledge systems. But this common sense of "innateness" is not what I have in mind. Rather, "innateness," here, is used in a narrow technical sense to indicate systems that are acquired in the normal course of normal development without relying on representational learning mechanisms – mechanisms that compute new representations by manipulating existing ones, such as associationism, induction, abduction, analogical reasoning (Carey, 2009; Samuels, 2004; 2007). This does not mean that innate traits are fully formed at birth, nor does it require that their acquisition is independent of experience. For these reasons, innate traits could manifest some limited variation across individuals due to variation in their external and internal conditions. The failure of Zebra Finches that are reared in isolation to acquire a normal song is one such example (Fehér et al., 2009). But while core knowledge systems can be modulated by experience, they are not the product of representational learning. The propensity of Zebra Finches to converge spontaneously on a single song pattern and the spontaneous emergence of phonological patterns in human languages demonstrate this fact.

While the universality of design is by far the most significant hallmark of core knowledge, instinctive knowledge systems are also likely to manifest several additional characteristics. One such feature concerns the fit between the design and its function.[2] Generally speaking, one would expect core knowledge systems to fulfill a function that benefits their carrier and to exhibit a specific design that fits their function, the architecture of their carrier, and its environmental niche.

---

[2] What counts as "fit" critically depends on the forces shaping the system, but the nature of those forces is largely unknown. Like many aspects of the human body, the structure of a core knowledge system is likely to be determined by multiple forces operating in ontogeny and phylogeny, including natural selection (either directly targeting the system or some other correlated traits), genetic drift, architectural constraints, stochastic developmental processes, environmental and historic factors, and the dynamics of self-organization (e.g., Balaban, 2006; Chomsky, 2005; Cosmides & Tooby, 1994; Hauser & Spelke, 2004; Marcus, 2004; Pinker & Bloom, 1994). Because the precise contribution of these various factors is an open empirical question (cf., Dawkins, 1987; Fodor & Piattelli-Palmarini, 2010; Gould & Lewontin, 1979), it is also uncertain which design best fits the function of any specific system of core knowledge.

For example, if the phonological system were designed for communication, then, all other things being equal, designs that optimize communication and meet the constraints imposed by their phonetic channel would be more likely to occur. Although we do not necessarily expect to find any brain regions unique to a given domain (a consequence of the above-mentioned tendency of evolutionary tinkering to recycle existing neural circuits), the relevant network should be genetically regulated, and consequently, it should be relatively invariant across healthy individuals. And, given the role of many core systems as learning organs (Gallistel, 2007), one would also expect them to be active in early development. Early onset, to be sure, does not imply experience independence. For example, while some aspects of core phonology could manifest themselves in infancy, this system might be nonetheless triggered by phonetic experience, leading to some significant differences between the design of aural and manual (i.e., sign language) phonologies. Finally, the signature of core knowledge is seen not only in early development but also in knowledge acquired in later development. Many cultural inventions, such as the systems of mathematics, physics, and biology, are founded on the heels of the instinctive core knowledge of number, object, and motion. Although such cultural inventions are clearly different from the instinctive core knowledge, the systems are nonetheless linked.

Core knowledge systems have so far been documented in various areas of cognition, including knowledge of physics, number, biology, morality, and other people's minds (Carey, 2009; Hamlin et al., 2007; Keil, 1986; Onishi & Baillargeon, 2005). The quintessential argument for domain-specificity, however, is the one advanced by Noam Chomsky with respect to the syntactic component of the grammar (e.g., Chomsky, 1957; 1965; 1972; 1980). Informed by observation of universal principles of syntactic structure, Chomsky asserted that the human capacity for language reflects a specialized biological system – a language organ. The generality of such principles in speakers of all communities, irrespective of their culture and education, the documentation of spontaneous regenesis of syntactic organization in the linguistic systems invented *de novo* by children, their demonstrable function in communication, their implementation in specific brain networks that are genetically regulated, and the absence of any animal homologue to language further led Steven Pinker (1994) to characterize language as a uniquely human instinct, shaped by natural selection. In his words:

People know how to talk in more or less the sense that spiders know how to spin webs. Web-spinning was not invented by some unsung spider genius and does not depend on having had the right education or having and aptitude for architecture or the construction trades. Rather, spiders spin spider webs because they have spider brains, which give them the urge to spin and the competence to succeed. (Pinker, 1994: 18)

These discussions, however, rarely acknowledge that similar hallmarks hold for the phonological component of the grammar. Phonological patterning appears to be both universal and idiosyncratic – it exhibits structural characteristics that are common across languages, both signed and spoken, and distinct from those found in other communication systems sharing the same modality, such as music and gestures. But far from being an arbitrary quirk, phonological patterning is functionally adaptive. The discussion in Chapter 2 has offered several illustrations of the tendency of phonological systems to optimize constraints imposed by the phonetic channel. Moreover, phonological patterning itself, while certainly not necessary for communication, is nonetheless beneficial. Using evolutionary game theory, Martin Nowak and David Krakauer (1999) showed that if a communication system encodes each concept by a holistic signal, then, as the number of concepts increases, the signals become acoustically similar, and hence prone to confusion. Phonological patterning greatly reduces the risk of perceptual confusions. The adaptive value of phonological patterning might explain why a phonological mechanism could have become fixed in the human population. To the extent that the capacity for phonological patterning is genetically regulated, one would expect it to be generally available, and largely independent of linguistic experience. And indeed, like the spider's instinct to spin webs, people weave phonological patterning spontaneously, they extend phonological patterns to words that they have never heard before, and they regenerate phonological patterns when none exist in their language. Phonological patterning also manifests itself in early infancy, and recruits a well-defined brain network. Finally, a core system for phonology would also account for the obligatory recruitment of phonological principles in the design of the invented reading and writing systems and the strong link between reading ability and phonological competence.

3.3   Domain-general and non-algebraic alternatives

While there is much evidence to suggest that phonology is potentially a special human instinct, this conclusion is hardly inescapable. Even Noam Chomsky, the originator of the "language organ" hypothesis, typically reserves the terms "grammar" and "universal grammar" to the syntactic domain, implying that the computation of phonology is attained by a subsidiary sensory interface. Other linguists and psychologists have outright rejected the possibility that the phonological system is either algebraic or specialized. We consider the objections to each hypothesis – algebraic machinery and specialization – in turn.

3.3.1   Against algebraic phonology

The possibility that phonological generalizations are the product of an algebraic computational system faces two major challenges. One challenge is based on the continuity between the phonological system and the non-algebraic phonetic component, whereas a second underscores the success of non-algebraic connectionist systems in describing phonological patterns.

3.3.1.1 The phonology–phonetics continuity
As noted earlier, many phonological alternations recapitulate natural phonetic processes (e.g., Blevins, 2004). Phonetic processes, on their part, resemble phonological knowledge inasmuch as they are shaped by principles that vary across languages (e.g., Keating, 1985; Zsiga, 2000). The intimate links between phonology and phonetics would seem to blur the separation of these two domains. In view of such facts, some researchers have moved to erase the phonology–phonetics divide altogether, whereas others incorporate continuous analog phonetic information in the phonological grammar (e.g., Flemming, 2001; Kirchner, 2000; Steriade, 1997). Either move would render the design of the phonological grammar inconsistent with an algebraic system.

3.3.1.2 The success of connectionism
A second challenge to an algebraic phonology is computational. An influential research program initiated by David Rumelhart and Jay McClelland (1986) has sought to account for linguistic knowledge while eschewing most tenets of the Computational Theory of Mind. In their proposal, linguistic knowledge does not require a grammatical component separate from the lexicon, structure-sensitive algebraic rules, or syntactically complex representations. Knowledge and generalizations follow only from massive associations among unstructured atomic representations, acquired solely from linguistic experience.

The contrast between the connectionist computational framework advocated by Rumelhart and McClelland and the CTM can be plainly illustrated by their distinct approaches to the representation of geminates. Recall that, according to the CTM, semantically complex representations, such as /pp/, are syntactically complex, and consequently, knowing something about /p/ automatically transfers to the geminate /pp/. Not so in Rumelhart and McClelland's connectionist network. Here, a semantically complex /pp/ is not structured syntactically (see Figure 3.4). In the absence of syntactically structured representations, mental processes can only be guided by the associations among atomic labels (e.g., an association between the representation of /p/ and the atomic label that

Figure 3.4 The representation of semantic complexity using forms that are either syntactically complex (on the left) or simple (on the right)

stands for /pp/). Associative systems likewise eschew the representation of variables and relations among variables (e.g., identity). Such systems might learn about /pp/ or /dd/, specifically, but not about segment identity, generally (XX; where X stands for any segment; Marcus, 2001).

But despite eliminating many aspects of algebraic systems, these massive associative networks have been shown to account for complex syntactic dependencies, such as center-embedding (Elman, 1993) – dependencies that had been previously thought to require a complex class of grammar (Chomsky, 1957). And if associative systems can represent the syntactic structure of sentences – a structure comprising a potentially infinite number of lexical items – then these systems would surely suffice to capture phonological patterns of one's language, patterns which, by definition, include only a finite and small number of phonological primitives. An algebraic, phonological grammar is all but obsolete.

3.3.2 Against specialized phonology

The ability of general-purpose associative systems to capture complex linguistic knowledge casts doubt not only on the view of the grammar as an algebraic computational system but also on its specialization. The case for specialization has been traditionally motivated by three major arguments: the universal design of linguistic systems, its uniqueness, and the role of universal grammatical principles in language acquisition. But each of these arguments faces numerous challenges.

3.3.2.1 Typological vs. grammatical universals

One of the standard arguments for grammatical universals cites recurrent patterns in the distribution of linguistic structures across language, the so-called *typological universals*. Cross-linguistic regularities, so the argument goes, must be the product of principles that are represented in the language faculty of all individual speakers. Accordingly, the documentation of typological universals is suggestive of *grammatical universals*.

But linguists and psychologists have long known that those typological universals are statistical trends, not inviolable laws. For example, while many languages favor simple onsets (e.g., *ba*) over complex ones (e.g., *bla*), complex onsets are nonetheless attested in many languages; in fact, they are quite frequent. Moreover, many of the structures that are preferred across languages are ones that are phonetically natural (Stampe, 1973). The statistical nature of typological phonological universals, on the one hand, and their phonetic basis, on the other, have prompted some researchers to reject the existence of grammatical universals. To use the *ba* example above, the grammar, in this view, includes no principles that universally favor *ba* over *bla*. Rather, structures like *ba* are preferred because they are easier to produce and perceive. And since structures that are easier to perceive and produce are more likely to be accurately transmitted across generations of speakers and talkers, those structures will outnumber their more challenging counterparts. Typological universals, then, are emergent byproducts of cultural evolution, shaped by the properties of the phonetic channel and the laws of historical change, rather than a universal grammatical system (e.g., Blevins, 2004; Bybee, 2008; Evans & Levinson, 2009).

3.3.2.2 The role of experience in language acquisition
Additional key evidence for the specialization of the language system concerns its role in language acquisition. The argument, outlined by Noam Chomsky, asserts that language acquisition is unattainable given the impoverished linguistic experience available to the child. Chomsky's own solution for the poverty-of-the-stimulus conundrum was to move the burden of explanatory labor from experience to the child's innate mental structure: If the linguistic experience available to the child is insufficient to get the job (of language acquisition) done, then children must rely on special principles that they possess innately. These special principles correspond to universal grammar (UG) – a specialized language acquisition device, equipped with substantive linguistic constraints on the form of attainable grammars.

At the heart of Chomsky's poverty-of-the-stimulus argument is the assumption that the input available to the child is too impoverished to allow him or her to acquire certain aspects of language. But the success of simple associative systems in learning complex syntactic phenomena challenges this assumption. Far from being impoverished, linguistic experience would appear to present the child with all the information necessary to extract the syntactic structure of his or her language (e.g., Reali & Christiansen, 2005). And if simple associative systems can handle the acquisition of syntax – the quintessential challenge to language acquisition, according to the generative linguistic tradition – then surely, no specialized machinery is necessary for the case of phonology. Summarizing this commonly held sentiment, Peter MacNeilage (2008: 41) notes that

however much poverty of the stimulus exists for language in general, there is none of it in the domain of the structure of words, the unit of communication I am most concerned with. Infants hear all the words they expect to produce. Thus, the main proving ground for UG does not include phonology.

3.3.2.3 The contribution of domain-general mechanisms
A third major challenge to the specialization of the language system is presented by a large body of psychological research that underscores the role of domain-general mechanisms in language processing – mechanisms that are not specific to the processing of linguistic inputs.

One mechanism that has recently attracted much interest is statistical learning – the ability to track the statistical co-occurrence of events in one's experience. Countless studies demonstrate that humans are prodigious learners of statistical phonological patterns. After only two minutes of exposure to seemingly unstructured sequences as in *golabubidakupadotigolabutupiropadotibidaku*, 8-month-old infants detect the statistical contingency between adjacent syllables (e.g., *tu* is followed by *pi*), and consequently, they are less likely to attend to the familiar sequence *tupiro* compared to an unfamiliar sequence (a sequence whose syllables have never co-occurred in the familiarization sample, e.g., *dapiku*: Saffran et al., 1996). Such statistical learning, however, is not a specialized phonological ability – infants use a similar approach to track the co-occurrence of tones (Saffran, 2003b) and visual stimuli (Kirkham et al., 2002) – nor is it unique to humans: Cotton-top tamarins succeed in learning many (albeit not all) phonological patterns extracted by humans (Hauser et al., 2001; Newport et al., 2004; for a critique, see Yang, 2004).

Domain-general mechanisms have been likewise invoked to account for many other aspects of phonological knowledge. Consider, for example, the phenomenon of categorical perception – the identification of consonantal phonemes as members of distinct categories, separated by sharp perceptual boundaries (e.g., /b/ vs. /p/). While early reports of young infants' ability to perceive phonemes categorically (Eimas et al., 1971; Werker & Tees, 1984) were taken as evidence for a specialized speech-perception mechanism (Liberman & Mattingly, 1989), subsequent research has shown that categorical perception is highly sensitive to the statistical distribution of phonetic exemplars (Maye et al., 2002), and it is not unique to either speech or humans. Indeed, humans extend categorical perception to musical pitches (Burns & Ward, 1978; Trainor et al., 2002), whereas animals manifest categorical perception of elements of their own communication system (Wyttenbach et al., 1996) and, with training, they can even acquire categories of human speech sounds (Kuhl & Miller, 1975). Taken together, these observations suggest that phonological knowledge relies on mechanisms that are neither specialized for language nor unique to humans (but for critiques, see Pinker & Jackendoff, 2005; Trout, 2003).

3.3.2.4 *Shared organizational principles: phonology vs. music*
Not only does the acquisition of phonology appear to rely on mechanisms that are not specific to language, but the organization of the phonological system itself manifests numerous links with nonlinguistic systems, most notably music. While the discussion in Chapter 2 has noted numerous differences between the organization of musical and phonological structures, other observations underscore some commonalities.

Consider, for example, the links between linguistic tones and musical pitch. Many languages use tone to contrast among words' meanings. In Mandarin, the words for "mother" and "horse" are indicated by the same segments, *ma* – they contrast only on their linguistic tone: "mother" is marked by a high tone (the so called "first tone'), whereas "horse" is indicated by a falling-rising tone ("third tone"). Although linguistic tone is a linguistic phonological feature – tone contrasts among words in the same way that voicing contrasts *bill* and *pill* – several studies have shown that the representation of linguistic tones is linked to musical abilities. For example, speakers of tonal languages show greater propensity for absolute musical pitch than speakers of non-tonal languages (Deutsch et al., 2006), whereas musicians' brainstem responses are more sensitive to linguistic tone than non-musicians' (Wong et al., 2007).

Other studies have linked music and phonology in the representation of temporal structures. Just as musical listeners instinctively tap their foot to a perceptually fixed series of beats and contrast strong and weak beats, so do Spanish speakers, for example, use stress to distinguish between words (e.g., *bébe* 's/he drinks' vs. *bebé* 'baby,' Dupoux et al., 2008; Skoruppa et al., 2009). In both domains, people prefer to interpret longer and louder acoustic events as stress-bearing: Longer vowels tend to be heavy, or stress-bearing, whereas long, louder musical events are typically aligned with strong beats (Jackendoff & Lerdahl, 2006; Lerdahl & Jackendoff, 1983).

Phonological and musical representations likewise share many aspects of their rhythmical organization. In both domains, people group temporal events – either syllables and their constituents, or musical sounds – into hierarchically organized groups (Jackendoff & Lerdahl, 2006; Lerdahl & Jackendoff, 1983). Listeners note that English, for example, has the rhythm of a Morse code, whereas Spanish sounds like a machine gun (Lloyd James, cited in Ramus et al., 2003), and these rhythms allow young infants (Mehler et al., 1988; Nazzi et al., 1998; Ramus et al., 2000) and adults (Ramus & Mehler, 1999) to discriminate between different spoken languages. But as impressive as it may be, the ability to extract the rhythmical structure of speech is clearly shared with music. For example, Aniruddh Patel and colleagues (2006) observed that English and French music manifest distinct rhythmical properties (operationalized as the durational contrast between successive musical events), and those differences mirror the distinct rhythmical patterns of speech in the two languages (e.g., English exhibits a greater durational contrast than French in both music and speech). Remarkably, speakers of rhythmically distinct languages, such as English and Japanese, manifest different preferences even with respect to the grouping of musical tones (Iverson & Patel, 2008; Kusumoto & Moreton, 1997).[3] These numerous links between phonological and musical abilities

---

[3] Iverson and colleagues interpret these results as evidence that rhythmical organization is the product of a domain-general auditory mechanism, but these findings are open to alternative explanations. Because the cultural evolution of musical idioms is often shaped by vocal music

challenge the specialization of the phonological system. In fact, some precursors of rhythmical organization are even shared with our distant relatives. Like infants, cotton-top tamarin monkeys distinguish Japanese from French (Ramus et al., 2000) and Polish (Tincoff et al., 2005) – languages of different rhythmical class – although they might not rely on the same sources of information as infants.

To conclude, a large body of literature suggests that the design of phonological systems, their acquisition and processing relies on cognitive mechanisms that are not specific to language or humans. Summarizing this literature, an influential paper by Tecumseh Fitch, Marc Hauser, and Noam Chomsky (2005) has asserted that "much of phonology is likely part of the FLB [faculty of language broad – I.B.], not FLN [faculty of language narrow – I.B.]" – the subset of mechanisms that are unique to humans and language. In view of such evidence, why should the hypothesis of a phonological instinct receive any serious consideration?

3.4 Rebuttals and open questions

While the empirical observations reviewed in the previous sections (and summarized in 7) are indisputable, a closer look shows that they do not effectively refute the existence of a specialized algebraic system for phonology. Here, I wish to briefly outline some of these indeterminacies and suggest why an algebraic specialized system for phonology merits a closer look – a task I undertake in subsequent chapters.

(7) Some challenges to the hypothesis of an algebraic, specialized core system for phonology
   a. Against algebraic computation
     (i) Phonology–phonetics continuity: The similarity between phonological and phonetic processes favors a unified non-algebraic framework that incorporates analog, continuous detail in the phonological grammar.
     (ii) Phonological generalizations can be captured by (non-algebraic) associative machinery.
   b. Against domain-specificity
     (i) Typological phonological universals are statistical tendencies, shaped by properties of the phonetic channel and cultural evolution.
     (ii) Phonological knowledge can be gleaned from experience alone.
     (iii) Many aspects of phonological representation and processing rely on mechanisms and principles that are shared with nonlinguistic domains.

set to text, the musical idioms associated with different languages are likely to differ. Accordingly, the musical preferences of English and Japanese speakers might differ for reasons related to their distinct musical idioms, rather than their reliance on some generic "auditory" processor of rhythm.

3.4.1 Algebraic phonology reconsidered

Let us begin by revisiting the challenges pertaining to the computational properties of the phonological system. Specifically, consider the argument from the continuity of phonology and phonetics. Although the link between phonology and phonetics is undeniable (indeed, I take it as a defining feature of phonological knowledge), their conflation is quite a different matter. It does not logically follow from the phonology–phonetics link nor is it otherwise evident. Conflation, instead, is solely motivated by parsimony: all things being equal, a single-mechanism account is simpler than a double-mechanism alternative. Laudable as it is, however, parsimony is surely secondary to adequacy. The question at hand, then, is not whether a unified phonology–phonetics system is simpler, but rather whether it works: Can it capture the full extent of phonological knowledge and generalizations? Viewed in this way, the prospect for a unified phonology–phonetics system is not promising.

Here is the crux: Since phonetic knowledge is analog, a unified phonology–phonetics system must be likewise non-algebraic. For this reason, a unified phonology–phonetics mechanism would work only if the phonological system were, in fact, non-algebraic. While the success of connectionism in some areas of phonology would seem to suggest so, these conclusions might be premature. Most existing research has evaluated the phonological system by inspecting cases where the required generalizations are rather limited in scope. Decades of intense scrutiny, however, have revealed that associationist networks suffer from numerous in-principled limitations that render them too weak to handle many linguistic generalizations (e.g., Fodor & Pylyshyn, 1988; Marcus, 2001; Pinker & Prince, 1988). To the extent that these broad generalizations form part of phonological knowledge, then non-algebraic approaches – whether they are associationist or ones that conflate phonology and phonetics – are unlikely to prove adequate.

3.4.2 Why core phonology merits a closer look

Like the support for non-algebraic accounts of phonology, paucity of evidence also plagues many of the challenges for the specialization of the phonological system. At the heart of those challenges are correlations of various kinds. Specifically, cross-linguistic phonological regularities correlate with phonetic naturalness and familiarity. Similarly, several aspects of phonology parallel phenomena seen in nonlinguistic domains, such as music. But the problem with such correlations is that they are moot with respect to causation. The observation that putative cross-linguistic universals are phonetically functional and statistically frequent does not pinpoint the source of this correlation – whether those structures are preferred in phonological systems because of their functional advantages and familiarity, or whether it is phonological well-formedness that contributes to their prevalence across languages. In a similar vein, the congruence between phonological and musical principles is ambiguous. It does not tell us whether phonology borrows musical mechanisms, or whether it is rather music that adopts mechanisms that were in fact selected for some specific linguistic purpose (Pinker, 1997). The link between some key musical ingredients, such as rhythmical entrenchment (Schachner et al. 2009) and musical scales and chords (Bowling et al., 2010; Gill & Purves, 2009), and vocal learning and speech, respectively, hints at the latter possibility.

Beyond this empirical ambiguity, the a-priori denial of specialization leaves us with no explanation for several of the key observations outlined in Chapter 2. It fails to explain why all human languages – spoken or signed – exhibit a phonological system, why these systems share some of their characteristics irrespective of modality, why people spontaneously give birth to a phonological system even when they have no access to such a system in their own linguistic experience, and why the design of reading and writing – both fully developed systems and the ones spontaneously invented by children – recapitulates phonological principles.

To begin addressing these questions, it is necessary to gauge specialization from an interdisciplinary perspective. Such a research program must begin by identifying putative universal phonological principles and proceed to examine whether these principles are independently active in the minds and brains of individual speakers. Additional lines of research might investigate their neural implementation, evaluate their development in ontogeny and phylogeny by comparing our sound-patterning abilities with those of nonhumans, and document their interaction with reading ability and disability.

At the heart of specialization, however, is the universality of design. Until recently, this critical issue remained largely beyond the purview of psychological science. Most psychologists have exclusively concerned themselves with testing the adequacy of domain-general accounts of phonology. However, the phenomena informing such investigations are rather narrow in scope. For example, much research has examined the language- and species-specificity of processing of phonetic categories (Eimas et al., 1971; Kuhl & Miller, 1975; Kuhl et al., 1992; Werker & Tees, 1984). Other psychological models target restrictions on phoneme co-occurrence in speakers' own language (Gaskell et al., 1995; Harm & Seidenberg, 1999; McClelland & Patterson, 2002). But these interesting results may not necessarily scale up to a full account of phonology. The processing of phonetic categories does not address phonotactic knowledge, and the ability of associationist accounts to capture limited phonotactic generalizations in speakers' own language may not address the full extent of their phonological competence. The critical question is whether some of the typological patterns seen across languages might reflect grammatical phonological principles that are universal. This possibility remains mostly unexplored.

More recent developments in contemporary linguistics, however, have revitalized the age-old interest in language universals and their role in the grammar (Greenberg, 1966; Humboldt, 1997; Jakobson, 1968). Optimality Theory (OT), the dominant paradigm in formal phonology (Prince & Smolensky, 1993/2004), has been proposed to account for the close correspondence between the distribution of linguistic structures across languages and the linguistic processes attested in specific languages. According to this theory, both facts have a single source: a set of universal grammatical constraints that are active in the brains of all adult speakers. Although OT has made significant strides that have dramatically reshaped modern linguistics, these achievements have had little resonance in the conceptualization of the phonological system by most psychologists and neuroscientists. But a recent line of research in experimental phonology – the field that applies experimental methods in the testing of formal accounts of phonological knowledge – has offered support for several of OT's key predictions (e.g., Berent et al., 2007a; Moreton, 2008). These results, along with the many special phonological talents listed in the previous chapter, offer a new impetus for reevaluating the design of phonological systems.

3.5 A roadmap

At the center of this book is the hypothesis that phonology is an algebraic system of core knowledge. Subsequent chapters evaluate this hypothesis by adopting a fresh interdisciplinary perspective. Following the path paved by past research on the syntactic component (Crain & Nakayama, 1987; Fitch & Hauser, 2004; Gentner et al., 2006; Hauser et al., 2002; Lidz et al., 2003), the discussion integrates formal linguistic analysis with the experimental tools of cognitive psychology, neuroscience, genetics, and comparative research with nonhumans. To allow for a detailed, in-depth analysis, however, the cases discussed here are all drawn from the area of phonotactics – the phonological restrictions on phoneme co-occurrence. There also are several substantive reasons for the focus on phonotactics. Not only might phonotactics be formally distinct from utterance-level phonology (Hulst, 2009), but it also exhibits the clearest hallmarks of a specialized system. Unlike metrical structure and intonation, which bear parallels to metrical and pitch organization in music (Lerdahl & Jackendoff, 1983; Patel, 2008), phonotactic constraints have no obvious homologies in other sound systems. While phonotactic restrictions are a good candidate for specialized core knowledge, this possibility remains largely unexplored. To evaluate the phonotactic system, I examine its computational properties, the substantive universal constraints on its design, its developmental trajectory, neural implementation, evolutionary origins, and interaction with the cultural invention of reading and writing.

The organization of the book follows logically from the two components of my thesis – the claims that (a) the phonological grammar is an algebraic system and (b) the phonological grammar forms a system of core knowledge.

Chapters 4–6 evaluate the computational properties of the phonological system. Chapter 4 examines the representation of equivalence classes – classes that represent all category members alike in a manner that allows language-learners to extend generalizations to all class members, familiar and old. Chapter 5 further demonstrates that phonological generalizations extend not only to unfamiliar items but even to ones comprising elements that have never been experienced – a hallmark of across-the-board algebraic generalizations. Although such generalizations suggest that the phonological grammar is an algebraic system, the grammar alone is insufficient to account for the full range of phonological generalizations. Some linguistic generalizations appear to track the statistical properties of the lexicon. I conclude that a full account of phonological generalizations requires a dual-route approach (Pinker, 1999): It must include both an algebraic grammar and an associative learning mechanism that captures statistical knowledge.

The conclusion that some phonological generalizations are the product of an algebraic grammar sets the stage for examining the nature of grammatical phonological principles: Are these principles extracted from linguistic experience and shaped by the acoustic and articulatory interfaces, or is the phonological grammar constrained by universal principles that are specific to language?

The hypothesis of universal substantive constraints on the phonological grammar predicts the existence of grammatical phonological universals that are active in all synchronic grammars. This hypothesis is articulated and tested in the next part of the book. Chapter 6 reviews formal linguistic accounts that link typological universals to the structure of individual synchronic grammars, contrasts them with alternative diachronic and phonetic explanations, and considers how synchronic language universals can coexist with language-particular variation. Chapters 7–8 next demonstrate how the hypothesis of grammatical universals can be put to a direct experimental test.

What are the origins of such phonological universals in human ontogeny and phylogeny? Is speakers' knowledge of phonological universals present at birth? Are the broad principles of human phonological patterning shared with those found in the communication systems of other species? How is the phonological network implemented in the brain? And what is the link between the putative system of core phonology and the later developing "technologies" of reading and writing?

These questions are some of the topics addressed in the final part of this volume. Chapter 9 examines the role of universal grammatical restrictions in language development, with special focus on the precursors of grammatical restrictions in infancy; Chapter 10 examines the existence of nonhuman homologies to phonological competence and reviews existing proposals regarding the evolution of phonological instinct; and Chapter 11 evaluates neurological evidence from healthy speakers and probes for acquired and heritable congenital deficits that affect phonological processing. Finally, Chapter 12 investigates how phonological knowledge has shaped the design of writing systems, how it constrains skilled reading, and how it is implicated in dyslexia. Conclusions, caveats, and questions for future directions are considered in Chapter 13.

Part II

Algebraic phonology

4 How phonological categories are represented: the role of equivalence classes

> In earlier chapters, I suggested that the phonological grammar is an algebraic computational system. Phonological patterns, in this view, comprise abstract equivalence classes – categories whose members are all treated alike, regardless of whether they are familiar or novel. But on an alternative associationist account, phonological patterns bind chunks of phonological substance – the more likely two sound elements are to occur together, the more likely they are to form a chunk. Algebraic phonological categories are not represented by the human mind. To adjudicate between these two accounts, this chapter investigates the representation of two phonological primitives – syllables and the consonant/vowel distinction. If people represent these primitives as equivalence classes, then they should extend generalizations to any class member, irrespective of its statistical properties. The evidence emerging from a wide array of studies is consistent with this prediction.

4.1 What are phonological patterns made of?

Consider the phonological patterns in (1). In each line, the words share a pattern, and the patterns in the four lines are all different. The pattern in (1a) comprises a single unit; in (1d), it includes four units, and in (1b–c), the pattern has two units – either consonant- or vowel-initial. Our interest here concerns the nature of those units: What are the "beads" that form phonological necklaces, and what principles allow us to identify them?

(1) Some phonological patterns:
   a. cat, mop, big
   b. pardon, template, cartridge
   c. elbow, anchor, enter
   d. Mississippi, Massachusetts, Pennsylvania

Chapter 3 asserted that the phonological grammar is an algebraic system that encodes *equivalence classes* – abstract categories whose members are all treated alike. In this view, phonological classes such as "syllable" and "consonant" are abstract formal kinds, distinct from the set of exemplars that instantiate those categories. Like the algebraic category of integers, a class that is distinct from any subset of its members (e.g., the members 1, 3, 13, 333), our concept of a syllable is irreducible to specific syllable instances (e.g., *par*, *don*). Similarly, just as an "integer" is defined by formal conditions that are independent of the idiosyncratic properties of any given instance – whether it is large or small, frequent or rare – so are phonological classes defined according to structural conditions. For example, in order for *par* to qualify as the first syllable of *pardon*, it must include one – and only one – critical constituent called a nucleus (in English, this constituent typically corresponds to the vowel). If this and other conditions are met, *par* will be identified as a syllable; otherwise, it will not be. But crucially, these conditions consider only the grammatical structure of *par* – external factors, such as the familiarity with this unit, its ease of production and articulation are immaterial to this decision. Moreover, all instances that meet the same set of conditions will be considered equivalent. Just as 2 and 13,333 are each an equally good instance of an integer, so are the differences among specific syllable instances (e.g., *par* vs. *don*) ignored. Consequently, phonological knowledge can extend to any member of a class, familiar or novel. Phonological patterns (e.g., *pardon*), in this view, are made of abstract equivalence classes (e.g., syllable), and phonological units (e.g., *par*) are identified as such because they instantiate those abstract categories.

On an alternative, associative account, phonological units are considered chunks of phonological substance. The two units in *pardon* represent no abstract categories (e.g., syllable). Rather, *par* forms a unit because this sequence of phonemes is familiar (e.g., it forms part of *party, parking, parquet*). The principles defining these units, then, are not algebraic (i.e., the instantiation of equivalence classes), but associative: frequently occurring sounds form a unit – the stronger the frequency of co-occurrence, the more salient the unit. While these associative chunks will often coincide with algebraic phonological constituents, the convergence is accidental – categories such as "syllables," "consonant," and "vowel" are not effectively represented by the mind.

Chapters 4–5 evaluate the role of algebraic mechanisms in phonology. Chapter 4 examines the representation of phonological categories, whereas Chapter 5 evaluates the restrictions on their combinations. To gauge the role of equivalence classes, the present chapter offers an in-depth evaluation of two categories that are central to theories of phonology: syllables and the consonant/vowel distinction. In each case, we first review evidence that is consistent with the representation of the relevant category. We next examine whether such categories are in fact necessary to capture human performance. If the category forms an equivalence class, then generalizations concerning this category should extend to all its members alike, irrespective of their statistical properties. This prediction is assessed against experimental evidence.

4.2 The role of syllables

4.2.1 Some evidence consistent with representation of syllable-like units

Syllable-like units have been implicated in a wide range of phenomena. Before I consider whether those units are in fact syllables, let me briefly discuss some of the evidence for the encoding of syllable-like units. For the sake of clarity, my review of the findings is highly targeted, and by no means exhaustive.

One argument for syllable-like units concerns the restrictions on segment co-occurrence. The acceptability of any given segment sequence varies dramatically according to its syllabic position: A sequence like *lb* frequently occurs across English syllables (e.g., *elbow*), but it is disallowed at their onset (e.g., *lbow)*. Similarly, Dutch speakers restrict the co-occurrence of consonants and vowels depending on their syllabic position (Kager & Pater, 2012). Consonant clusters ending with a non-coronal consonant (e.g., *mk*, *mv*) cannot follow a long vowel if the consonants belong to the same syllable (e.g., *ba:mk*), but they are more acceptable across syllables (e.g., *ba:m.ver*).

Syllable structure constrains not only the co-occurrence of phonemes but also their phonetic realization. Recall (from Chapter 2) that English talkers aspirate voiceless stops at the beginning of a syllable (e.g., $t^h op$) but they typically do not do so at non-initial positions (e.g., *stop*, *at*). Interestingly, English listeners use aspiration as a cue for discerning syllable boundaries (Coetzee, 2011). The evidence comes from auditory strings in which the aspiration occurs at an illicit, non-initial position (e.g., $st^h a$). Such strings have no felicitous parse in English: If the aspirated $t^h a$ is parsed as syllable-initial ($s.t^h a$), then the preceding syllable *s* is illicit, whereas if it is parsed as syllable-medial, then the aspiration is likewise ill formed ($st^h a$.). To solve the conflict, English speakers posit a syllable boundary before the aspirated stop and perceptually "repair" the illicit initial syllable by inserting a vowel (e.g., $s.t^h a$ ➔ $sə.t^h a$). For this reason, monosyllables like $st^h a$ are judged to be disyllables (e.g., $sət^h a$). These findings demonstrate that the sensitivity to syllable-like units is quite strong – so much so that it promotes perceptual illusions (for additional such cases, see Dupoux et al., 1999; Dupoux et al., 2011).

Syllable structure also captures the phonological restrictions on morphological processes. Morphological operations, such as reduplication or truncation, are often restricted with respect to the size of their phonological outputs. In some languages, outputs must attain some minimal size; in others, it is the maximal size of the output that is restricted. But in all cases, size is defined not by the number of segments, but rather by prosodic units, such as syllables

(McCarthy & Prince, 1998). Hebrew, for example, forms verbs and adjectives by reduplicating nominal bases (Bat-El, 2006; see 2): The noun *kod* (code) gives rise to *kided* (he coded) and *vered* (rose) forms *vradrad* (pinkish). Note, however, that reduplication invariably yields disyllables, irrespective of whether the input is monosyllabic (e.g., *kod, fax*), disyllabic (e.g., *ve.red*) or trisyllabic (e.g., *te.leg.raf*). These observations suggest that Hebrew limits the maximal size of the reduplicative output to a disyllable.

(2) Hebrew reduplication yields forms that are maximally disyllabic
  a. *kod* (code)➔*kided* ('coded')
  b. *vered* (rose)➔*vradrad* ('pinkish')
  c. *fax* (fax)➔*fikses* ('sent a fax')
  d. *telegraf* (telegraph)➔*tilgref* ('sent a telegraph')

English nicknames, by contrast, are often monosyllabic, but those monosyllables are limited with respect to their minimal size (McCarthy & Prince, 1998; see 3). *Cynthia*, for instance, can give rise to *Cyn*, but not *Ci* (with a lax vowel, /sɪ/), whereas *Bea* (/bi/, for Beatriz, with a tense vowel) is perfectly fine. The reason is plain – monosyllabic nicknames must abide by the same restrictions applying to any other stressed English syllables. Such syllables must be "heavy" – they must minimally contain either a tense vowel (e.g., /bi/) or a lax vowel followed by a coda (e.g., /bɪn/). Since *Bea* and *Cyn* are both heavy, these nicknames are acceptable, whereas the "light" syllable *Ci* is disallowed.

(3) English nicknames minimally include a heavy syllable
  a. *Cynthia*➔ *Cyn, Cynth;* *Cy (/sɪ/)
  b. *Beatriz*➔ *Bea* (/bi/)
  c. *Alfred*➔ *Al, Alf,* *A

Syllable structure also affects the division of words into constituents. People naturally dissect words into chunks that coincide with syllables (e.g., *en-gine*, rather than *eng-ine*). So when an expletive is inserted, it is placed at the word's joints, marked by the boundary between syllables, rather than within them (McCarthy, 1982). Thus, the angry motorist from Chapter 2 exclaims *en-bloody-gine*, rather than *e-bloody-ngine* (see 4).

(4) Expletive affixation
  a. en-bloody-gine
  b. *e-bloody-ngine
  c. a.la-bloody-ba.ma
  d. *a.lab-bloody-a.ma
  e. fan-bloody-tas.tic
  f. *fanta-bloody-stic

In longer words, syllables (marked by σ) are grouped into larger constituents called metrical feet (indicated by Σ), so a larger prosodic joint occurs at the boundary between feet (see Figure 4.1). And, once again, these joints mark the insertion point for expletives. For example, *Alabama* and *fantastic* give rise to

Table 4.1 *An illustration of the materials in illusory conjunctions*
In 2/3 words, syllable boundary occurs after the second letter; in 3/2 words, it follows the third letter

	Color-syllable congruency	
Boundary location	Congruent	Incongruent
2/3 words	AN.vil	AN.Vil
3/2 words	LAR.va	LAr.va

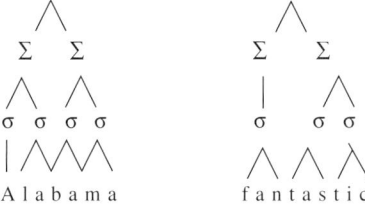

Figure 4.1 The prosodic structure of multisyllabic words

a.la-bloody-ba.ma and *fan-bloody-tas.tic*, but not **a.lab-bloody-a.ma* or **fanta-bloody-stic*.

People are sensitive to syllable structure not only when they are explicitly asked to segment words, but also implicitly, in word games (e.g., Treiman, 1986; Treiman & Kessler, 1995) and in tests that elicit judgments about segmented words (e.g., is *el-bow* a real word?; Levitt et al., 1991). In fact, syllable effects obtain even when the experimental task requires no attention to the word per se.

One ingenious demonstration of the indirect effect of syllable structure is presented by the phenomenon of illusory conjunction (e.g., Prinzmetal et al., 1986; Prinzmetal et al., 1991). In these experiments, people are briefly presented with disyllabic words, printed in two colors (illustrated here by different cases). In one condition (see Table 4.1), the boundary between colors coincides with the boundary between words (e.g., *AN.vil; LAR.va*) whereas in another, these boundaries are mismatched (e.g., *AN.Vil; LAr.va*). Participants are instructed to attend to one of the letters (e.g., the middle letter, *v*) and report its color (in our black-and-white example, this would be equivalent to detecting whether *v* appears in upper or lower case).

Here is the rationale: The distinction between a red *v* and green *v* relies on the conjunction of visual features – the visual features of a *v* and those of the color. But brief visual displays allow for only a partial encoding of such information. People might notice that the display had two colors (e.g., red and green), and that it included a *v*, but fail to register whether the *v* occurred in red, rather than green. To resolve this ambiguity, they might resort to additional, irrelevant features encoded with the display – and this is precisely where syllable structure might come into play. Recall from Chapter 2 that readers recode the phonological structure of printed words, even when they read them silently. If people divide printed words into syllable-sized units, then they might use this information to resolve the ambiguity regarding the conjunction of visual features. And since congruency between these various word attributes is preferable to incongruency, people will be biased to (incorrectly) align the visual boundary between colors with the boundary between syllables. Indeed, this is precisely what is reported (at least for printed words whose syllable boundaries are marked by letters that rarely co-occur together, such as *nv* in *an.vil* but not in *na.ive*, where the *ai* sequence occurs in *faith*, *rain*, *gain*, etc.; Prinzmetal et al., 1986; Prinzmetal et al., 1991).

4.2.2   What is represented: word chunks or an abstract syllabic category?

Why are people sensitive to the division of printed words into syllable-like units? And why do syllable-like units define the domain of phonotactic restrictions and constrain prosodic templates?

From an algebraic perspective, the answer is quite plain: People are sensitive to syllable-like units because they represent the syllable – an abstract category that treats all syllable instances alike. Because speakers represent syllables as an equivalence class, they can encode broad restrictions on syllable structure that apply to all members of the class alike, either existing instances or novel ones, irrespective of their familiarity. Accordingly, the requirement that English nicknames consist of minimally one heavy syllable would apply to any name – existing (*Douglas*, *Jennifer*) or novel, and even names with foreign phonemes. The Hebrew name *Chagit* (with the *ch* sounding like *Chanukkah*), for instance, will yield *Chag* or *Chaggy*, not *Cha*. In a similar vein, words are parsed into syllable-like units because those units are, in fact, represented as instances of syllables. But on an alternative account, syllables are not independently represented as an abstract category. The representation of *pencil*, for instance, specifies no syllables – people only encode word chunks (either chunks of letters, phonemes, or phonetic units) that happen to coincide with syllabic units.

The role of syllabic constituents has been challenged by both linguists and psychologists. Donca Steriade (1999), for instance, considers syllables to be artifacts of word structure. In her view, the dislike of sequences like *lba* would stem from the fact that words rarely begin with the *lb* sequence. Words, in turn, avoid initial *lb* sequences because such sequences are harder to perceive. But crucially, the dispreference of *lba* only concerns the linear arrangement of these consonants – not their syllabic role, specifically.

Psychologists have likewise questioned whether syllables truly exist. Consider speech production, for example. While numerous studies are consistent with the possibility that syllable-like units play a role in production, many of these observations are inconclusive. Some results are consistent with alternative explanations that do not postulate syllables (Shattuck-Hufnagel, 2011), and others strongly implicate syllable-like units that are phonetic articulatory plans (i.e., motor plans for executing the articulation of spoken words), rather than abstract algebraic constituents (e.g., Cholin, 2011; Laganaro & Alario, 2006). Such phonetic chunks could potentially explain also the preferences for syllable-like units in speech perception (e.g., Coetzee, 2011; Dupoux & Green, 1997; Dupoux et al., 1999).

Even when the role of articulatory units is minimized, in silent reading tasks, the role of syllable-like units is open to non-algebraic interpretations. In this view, the presumed effects of syllable structure are due to familiarity with word chunks, emerging not from the representation of syllables per se but from the statistical co-occurrence of letters and phonemes in words. Two statistical strategies might contribute to the emergence of syllable-like units. One strategy might track the absolute frequency of specific chunks. Thus, *pen* and *cil* are the chosen parts of *pencil* because those units each occur in many words, whereas alternatives such as *penc* or *ncil* are unfamiliar. A second strategy compares the frequency of co-occurrence within chunks (e.g., within *pen* and *cil*) to their co-occurrence across chunks (the frequency of *nc*). English speakers might notice that the *nc* letter-pair (a bigram) rarely occurs in English, whereas the preceding and following bigrams (*en* and *ci*) are each rather frequent. This so-called "bigram trough" will thus correctly signal the division of *pencil* into two chunks that happen to coincide with its syllables (e.g., Seidenberg, 1987).

But those linguistic and experimental challenges are not decisive. The observation that syllable edges often coincide with word edges does not establish why those particular sequences are preferred. As with any correlation, there is a chicken and egg problem: The preference for certain sequences could result either from their occurrence in word edges, specifically, or from the possibility that such word edges form part of better-formed syllables. Even if the word-edge analogy were firmly established, such a result would only indicate that word knowledge is encoded – it would not rule out the possibility that syllable structure is independently represented as well. The question at hand, however, is not whether word-edge effects can sway syllabification decisions, but rather, whether they alone are sufficient to account for word parsing. Word-level effects do not address this question.

Similar problems plague the psycholinguistic challenges. The possibility that phonetic chunks might mediate word production does not preclude the representation of syllables in phonology. Likewise, the correlation between the syllable-like units extracted in silent reading and the statistical structure of the language does not necessarily mean that people exploit this information. In fact, people demonstrably ignore relevant statistical information when it conflicts with the putative structure of their grammar (Becker et al., 2011a; Becker et al., 2011b). And even when people are sensitive to the statistical co-occurrence of word chunks (e.g., letters, bigrams), this does not show that the frequency of word chunks is, in and of itself, sufficient to account for the effects of syllable structure. In short, the evidence is amply consistent with the hypothesis that syllables form equivalence classes.

To support this possibility, let me consider two examples in greater detail. These examples are informative because they concern silent reading, so the role of articulatory or acoustic chunks is minimized. In both cases, however, there is evidence that people extract syllable-size units, and the units consulted by participants cannot be explained by the statistical co-occurrence of segments.

4.2.3 Dissociating syllables and their statistical correlates

4.2.3.1 Illusory conjunctions reconsidered

Consider, again, the evidence from illusory conjunctions. When asked to report the color of a medial letter in a briefly presented word, people are more likely to preserve syllable structure than to violate it. Thus, they tend to incorrectly report the *v* of *ANVil* (where the color boundary, indicated here by the case-alternation, is incongruent with syllable boundary) as consistent with the color of *il* (a preservation of syllable structure), whereas in *ANvil* (where case and syllable boundary coincide), they are unlikely to err, as errors would violate syllable structure (see Table 4.1). The word *larva* (syllabified as *lar.va*) exhibits the opposite pattern. Here, the middle letter is syllabified with the first unit, so the congruent *LARva* yields few (violation) errors whereas the tendency to preserve the *lar* unit elevates errors in the incongruent *LArva*.

Of interest is what drives these effects. Since the task calls for no articulatory responses, the units mediating performance are most likely phonological, rather than phonetic. These results, nonetheless, raise questions regarding the nature of this unit: Does the boundary between units follow from syllable structure or the familiarity with word chunks – the fact that the bigram (letter-pair) spanning the syllable boundary (e.g., *nv* in *anvil*) is far less frequent than its surrounding bigrams (e.g., *an* and *vi*)? If the preservation of syllable-like units is only informed by the bigram trough (the fact that *nv* is an infrequent bigram), then this effect should be eliminated when the statistical trough is eradicated – when the unit spanning the boundary is as frequent as the units on each side of the boundary. But results showed that illusory conjunctions persist irrespective of whether a bigram trough is present or absent (Rapp, 1992). Although subsequent experiments suggest that the effect of syllable structure can be greatly attenuated when the orthographic structure of the materials is set to strongly conflict with their syllable structure (Doignon & Zagar, 2005), this result shows only that bigram frequency can sway word parsing, not that it can subsume the syllable. Additional evidence against this possibility is presented by the effect of syllable frequency on printed word recognition.

4.2.3.2 The effect of syllable frequency

To recognize a printed word, we must locate a record of that word in our long-term memory storage for words – the mental lexicon. This allows us to retrieve various idiosyncratic facts associated with this word, such as its meaning and grammatical class (e.g., whether it is a noun or a verb). Although there is evidence that this process always results in some level of phonological recoding (see Chapter 2), the level of phonological detail depends on the orthography. Transparent orthographies, like Spanish, in which pronunciation is highly predictable from print, should allow far more detailed representation of phonological structure than orthographies plagued by irregularities (e.g., English). A series of experiments shows that Spanish readers do, in fact, divide printed words into their syllables, and they use this information for lexical access.

Consider, for example, the recognition of the Spanish word *foto* (photo). When Spanish readers encounter this word, they first divide it into syllables (*fo.to*). As soon as the initial syllable is extracted, readers immediately activate all Spanish words that begin with the syllable "fo" (e.g., *fo.to, fo.ca* 'seal,' *fo.co* 'center'), and search among them for the target word (*foto*). This search strategy is advantageous because it helps limit the cohort of relevant words: Rather than searching for *foto* among all Spanish words, one can focus on a smaller cohort of "relevant words" – those beginning with *fo*. But the actual ease of search depends on the size of the cohort (see Figure 4.2). While syllables like *fo* activate a relatively small cohort, others (e.g., the syllable *li*, in *li.la* 'purple') can activate a very large number of candidates (e.g., *li.so* 'straight,' *li.no* 'linen,' *li.mo* 'mud,' *li.cor* 'liquor,' *li.mon* 'lemon,' etc.), and this will make the target word harder to spot. Moreover, partially activated words are known to compete

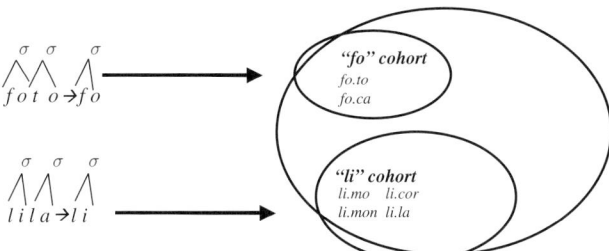

Figure 4.2 An illustration of the cohorts activated by the initial syllable of two Spanish words

with each other – the larger the cohort of activated words, the stronger the competition, and this competition exerts a cost on the recognition of each of those words, including the target. Since words beginning with a frequent syllable (e.g., *li.la*) activate larger cohorts of strongly competing words, they take longer to recognize (in a lexical decision task, e.g., "Is *lila* a word or a nonword?") and produce (in a naming task) compared to those beginning with an infrequent syllable (e.g., *fo.to*; Carreiras et al., 1993).

The initial-syllable effect presents an interesting twist to the algebraic proposal. The fact that people are sensitive to the frequency of the first syllable demonstrates that, in accessing the lexicon, they clearly consider the identity of specific instances (e.g., *fo* and *li*). But to the extent that the class of instances that they access is defined by their syllabic role (as the first syllable), then this effect would provide evidence for the representation of those instances as an equivalence class.

Before we can accept this conclusion, however, we must first rule out an alternative statistical explanation. According to this statistical alternative, the division of words into syllables is informed not by their algebraic structure, but rather by the statistical co-occurrence of their letters. Bigram troughs offer one such cue for the division of words into syllables. In this view, people do not encode the fact that *li* (in *lila*) is a syllable. Rather, they parse words into chunks depending on the frequency of their bigrams: Once they spot a bigram trough, they extract an internal boundary. It is indeed conceivable that the letter-bigram spanning the syllable boundary in *li.la* (e.g., *il*) is of lower frequency than the one spanning the boundary in *foto* (e.g., *ot*). This would make *lila* easier to parse than *foto* even if people did not represent syllable structure per se. Another possibility is that the division is informed by the frequency of the initial unit at non-initial position. Perhaps *li* of *lila* activates a larger cohort because *li* is a familiar orthographic unit that occurs in many words (e.g., *mo.li.no*; *lib.ro*) rather than as a syllable, or an initial syllable, specifically. This would make *li* "stand out" and activate many similar words. But crucially, the activation of the cohort, in this view, reflects not the encoding of syllable structure per se but rather the detection of a salient frequent word chunk.

Neither of these explanations, however, can account for the initial-syllable effect. To examine whether the initial-syllable effect is due to bigram trough, Markus Conrad and colleagues (2009) compared the effect of syllable frequency in the identification of two types of words. Participants in these experiments were presented with visually printed stimuli, and they were asked to indicate whether or not the printed string was a real word. In one condition, syllable boundary was marked by a strong bigram trough, as the letter-bigram spanning the syllable boundary (e.g., *il* in *li.la*) occurred, on average, in only 555 words per million – far less frequently than the bigrams occurring within a syllable (e.g., *li,la*; M=2,504 per million). In a second condition, syllables were not marked by a bigram trough, as the bigram across syllables was in fact *more* frequent than the one occurring within syllables (M=3,393 vs. M=1,618 per million, for the between- vs. within-syllable bigrams, respectively). Each such condition included words whose initial syllable was either frequent or infrequent, and these two classes were closely matched on numerous dimensions, including the frequency of their bigrams and their length and similarity to other Spanish words. Results showed that the disadvantage of the initial syllable emerged irrespective of whether the bigram trough was present or absent. In fact, the presence of the bigram had no measurable effect on the strength of the initial-syllable disadvantage. A second experiment demonstrated that the initial-syllable effect obtains even when words with frequent initial syllables were matched to their infrequent first-syllable counterparts on the frequency of all of their bigrams, including bigrams occurring within and between syllables. These results demonstrate that the effect of the initial syllable is inexplicable by the statistical co-occurrence of letters.

Summarizing, then, the linguistic and psycholinguistic evidence suggests that people encode the syllable structure of words, and that their sensitivity to syllable-size units cannot be subsumed by the statistical co-occurrence of letter chunks. While these results do not rule out all statistical accounts for the syllable (e.g., they do not address the role of phoneme and feature co-occurrence), the findings are certainly consistent with the possibility that syllables are encoded as equivalence classes. The role of phoneme and feature co-occurrence is further addressed in our subsequent discussion of consonants and vowels.

4.3 The dissociations between consonants and vowels

A second test for the role of equivalence classes is presented by the distinction between consonants (C) and vowels (V). One can hardly discuss phonology without talking about consonants and vowels. Many phonological theories use these categories to classify segments, describe syllable shapes (e.g., CVC vs. VCC, as in *pat* vs. *apt*), and distinguish between phonological processes that apply selectively to consonants and vowels. Building on such observations, Marina Nespor and colleagues have proposed that consonants and vowels are distinct categories that carry different grammatical roles (Nespor et al., 2003). In their proposal, consonants signal the idiosyncratic distinctions between words, whereas vowels are the chief carriers of grammatical information.

Consonants are indeed critical for word identification. Languages typically have more consonants than vowels, so if one were to misidentify a phoneme, the loss of information would be far greater if this missing phoneme was a consonant – absent the *t* in *cat*, for example, the *ca_* sequence would compete with many words (e.g., *cab, cad, cam, can, cap, car*), whereas a missing medial *a* would leave fewer options (e.g., *kit, cot, cut*). This, in turn, demonstrates that consonants carry more lexical information than vowels.

Given that consonants are critical for word identification, one would expect an adaptive phonological system to carefully protect its consonants, far more than it would preserve differences among vowels. In accord with this prediction, phonological processes tend to increase the salience of consonants (e.g., by banning adjacent consonants that are identical or similar, McCarthy, 1979; 1994), but they routinely obscure distinctions among vowels. English, for instance, conceals the identity of vowels by reducing them to schwas (ə) in unstressed positions. While the unreduced *atom* (/ætəm/) and *metal* (/mɛtəl/)/ clearly contrast on their initial vowels, many dialects of American English erase these distinctions in *metallic* and *atomic*, where these two vowels, now unstressed, are both realized as a schwa (/mətælɪk/ and /ətɑmɪk/; Kenstowicz, 1994). Although consonants are also subject to alternations, those alternations are far more restricted. For example, English assimilates the place of articulation of consonants, but these alternations are highly constrained. While the nasal *n* is likely to assimilate to *m* in the context of a labial (e.g., *green beans*➔ *greem beans*), this alternation does not occur in the context of a velar (e.g., *green gear*➔*greem grear*). Such restrictions are informative because they allow hearers to recover the identity of the original consonant (e.g., a perceived *greem* signals an intended *green*, not *Greek*; Gaskell & Marslen-Wilson, 1998; Gow, 2001). Accordingly, the assimilation of consonants does not fully obscure their identity. In contrast, vowel alternations, such as the reduction of unstressed English vowels, typically apply across the board, so given the schwa in /mətælɪk/ and /ətɑmɪk/, hearers have no way to determine whether the underlying vowel was a schwa, the vowel /æ/ (as in /ætəm/) or /ɛ/ (in /mɛtəl/). The small number of vowels and their susceptibility to alternation renders them far less informative with respect to lexical distinctions.

Vowels, however, often mark important grammatical distinctions. For example, vowels are the main carrier of stress (the dimension that contrasts *permít* and *pérmit*), and the location of stress, in turn, has been linked to the order of syntactic constituents. Consider, specifically, the order of the syntactic head and its complement. In the prepositional phrase *on the staircase* (e.g., *the cat sat on the staircase*), the head corresponds to the word "*on*," whereas "*the staircase*" is its complement. Languages like Dutch, for example, vary the order of these constituents (see 5): the syntactic head can occur either before the complement (e.g., *op* de trap 'on the staircase') or after it (e.g., de trap *op* 'the staircase on'). But because the syntactic complement tends to take the main prosodic stress (underlined), language learners can discern the syntactic function of words (as heads or complements) from their stress pattern: head-complement phrases tend to manifest a weak–strong pattern, whereas complement-head phrases exhibit the opposite stress. Similar links between syntactic structure and prosodic stress have also been noted across languages. Languages like French, in which the syntactic head occurs before the complement, tend to have a weak–strong stress pattern, whereas languages like Turkish, in which the head occurs after the complement, manifest the opposite stress sequence (Nespor et al., 2003). Thus, the capacity of vowels (but not consonants) to carry stress allows them to convey grammatical information.

(5) The link between prosodic stress and syntactic structure in Dutch (example from Nespor et al., 2003: 8). The syntactic head is italicized; main stress is underlined.
   a. Head-complement phrases (weak–strong stress pattern)
      op de trap
      'on the staircase'
   b. Complement-head phrases (strong–weak stress pattern)
      de trap *op*
      'the staircase on'

Further evidence for a distinction between consonants and vowels comes from their distinct roles in language processing. Recall that the perception of stop consonants (e.g., *b* vs. *p*) is categorical (see Chapter 2): When people are presented with instances of consonants that vary along a continuous acoustic dimension (e.g., voice onset time), they categorize the input as either *b* or *p*, with a sharp boundary between the two (Liberman & Mattingly, 1989). Vowels, in contrast, are perceptually organized around a "prototype": Certain members of the vowel category are perceived as exceptionally good exemplars that are more likely to stand for the category as a whole (Kuhl, 1991; Kuhl et al., 1992). Other important differences between consonants and vowels concern their susceptibility to rule- vs. statistical learning (Bonatti et al., 2005 ; Toro et al., 2008) and their dissociation in neurological disorders (Caramazza et al., 2000; Cubelli, 1991) – issues discussed later in this chapter.

Why do consonants and vowels differ with respect to their representation and processing characteristics? The distinction between consonants and vowels is readily explained by the hypothesis that people encode two equivalence classes – one for consonants, another for vowels – that afford broad generalizations to any class member, irrespective of its individual features or its co-occurrence with other members. But on an alternative explanation, the categorical distinction between consonants and vowels is only apparent. People encode no general categories for "consonants" and "vowels" – rather, they only encode specific instances (e.g., *b*, *o*) and their features (e.g., labial, round). The behavioral distinction between consonant and vowel phonemes emanates from the statistical properties of these instances in linguistic experience. Indeed, consonants are more numerous than vowels, and they tend to share more of their features with each other (Monaghan & Shillcock, 2003). The distinct behavioral patterns associated with consonants and vowels might only reflect these statistical distinctions – no actual categories exist.

The distinction between the algebraic and associative perspectives is rather subtle. Both accounts recognize the distinct effects of consonants and vowels on behavior. Likewise, neither view negates people's ability to differentiate among individual members of a class (e.g., *b* and *p*, members of the "consonant" class). The question, then, is not whether people can distinguish between individual class members, but rather, it is whether they can occasionally disregard their differences and apply certain processes to the class as a whole. To evaluate this question, we investigate whether the distinction between consonants and vowels can be explained by the statistical properties of consonant and vowel segments and their feature composition. We first examine whether generalizations learned over consonants and vowels can be subsumed by their associative statistical properties. Next, we move to examine dissociations – both natural and abnormal – that selectively affect one of those classes.

4.3.1 People encode CV skeleton

Our first test of the representation of consonants and vowels concerns their role in word frames. If members of the consonant and vowel categories are all treated alike, then two words sharing the same arrangement of consonants (C) and vowels (V) should be considered similar even if they share no common segments. For example, *bof* (a CVC sequence) would be more similar to *tep* (CVC) than to *ept* (VCC). These sequences of abstract placeholders for consonants and vowels are known as the CV skeleton (or frame).

Several studies have indeed shown that words that share the same CV arrangement are produced more readily (Costa & Sebastian-Gallés, 1998; but for conflicting results, see Roelofs & Meyer, 1998; and Schiller & Caramazza, 2002). For instance, Christine Sevald, Gary Dell and Jennifer Cole (1995) demonstrated that people can repeat word-pairs more rapidly when their initial syllable shares the same CV frame (*kem til-fer* or *kemp-tilf-ner*) compared to length-matched controls (*kem tilf-ner* or *kemp til-fer*). Similarly, speech errors often preserve the sequencing of consonants and vowels even when the precise identity of these segments is distorted. For example, *"did the glass crack"* is produced as *"did the grass clack"* – an error that maintains the CCVC structure of *glass* and *crack*, but alters their onset clusters (Shattuck-Hufnagel, 1992; see also MacNeilage, 1998; Stemberger, 1984).

While these results suggest that the production of speech is sensitive to the sequencing of consonants and vowels, it is not entirely clear whether they demonstrate the encoding of CV frames. Linguists have debated whether CV frames play a role in the grammar (see Clements & Keyser, 1983; McCarthy, 1979; 1981; cf., McCarthy & Prince, 1995). Moreover, words that share the same CV frame might be easier to produce for reasons that are specific to articulatory production. To gauge the role of CV structure in phonological representations, one might therefore inspect its effects in tasks that minimize articulatory demands.

The large literature on silent reading offers fertile ground to address this question. This literature suggests that readers assemble the phonological structure of printed words (for reviews, see Berent & Perfetti, 1995; Van Orden et al., 1990). If phonological representations specify a CV skeleton, then such effects might manifest even when printed words are read silently.

To test this prediction, Michal Marom and I have conducted a series of experiments that use a variant of the Stroop task (Berent & Marom, 2005; Marom & Berent, 2010). In a classical Stroop experiment (Stroop, 1935), participants are presented with letter strings printed in color. Their task is to name the color of the ink while ignoring the contents of the printed strings. But despite their best efforts, people typically cannot help reading. For this reason, it is harder for participants to perform the task when the letter string spells the name of an incongruent color (e.g., the word RED printed in green) compared to a neutral condition (e.g., XXX printed in green).

Like typical Stroop experiments, our experiments manipulated the congruency between the color name and the printed words. Unlike the typical Stroop experiments, however, the letter strings used in our experiments corresponded to novel words, and their congruency with the color name was defined in terms of their CV structure.

Consider, for example, the strings GOP and GROP, printed in red (see Table 4.2 for a monochrome illustration of the materials). The color name *red* has a CVC structure. Consequently, GOP, a CVC word, is congruent with the CV structure of the color, whereas GROP (a CCVC word) is incongruent. If readers represent the CV structure of those words, then color naming should be faster in the presence of a CV-congruent string compared to an incongruent one. The results were consistent with this prediction. For example, *red* was named more rapidly with CVC words compared to words with incongruent CCVC structure. Crucially, CVC words only facilitated the naming of CV-congruent colors. Thus, when the words were paired with the color *black* (/blæk/, a CCVC word), CVC words now produced slower color naming relative to CCVC controls.

Table 4.2 *An illustration of the materials in Marom & Berent (2009)* CV-congruent conditions are highlighted.

	Black *(CCVC)*	Red (CVC)
CCVC	GROP	TWUP
CVC	GOP	TUP
VCC	OSP	UPT

Subsequent experiments showed that these effects specifically depend on the sequencing of consonants and vowels. Here, we compared color naming for nonwords with two CV frames: CVC and VCC. Both frames were matched for length, but differed on the ordering of the C and V slots. Results showed that people were highly sensitive to the specific CV arrangement. For example, *red* was named significantly faster in the presence of CVC nonwords compared to length-matched VCC controls (see Figure 4.3; statistical significance may be gauged visually – since the error bars reflect confidence intervals for the difference between the means, the relevant means are statistically different if their error bars are non-overlapping). Moreover, the sensitivity to skeletal congruency obtained even when controlling for several statistical characteristics of these items using a step-wise regression analysis. These results rule out the possibility that the differential effect of CVC materials on *red* vs. *black* results from the statistical properties of the different sets of CVC items used in these two conditions.

Before closing, one additional aspect of the results is noteworthy. Although incongruent CV frames systematically impaired color naming across multiple experiments, one condition failed to exhibit this effect. Specifically, while incongruent CVC frames impaired color naming (with *black*), incongruent VCC frames did not exert any cost. Since VCC frames and CVC frames clearly differed when paired with *red*, the lack of a difference with *black* cannot result from the inability to differentiate consonant and vowel slots. Rather, this

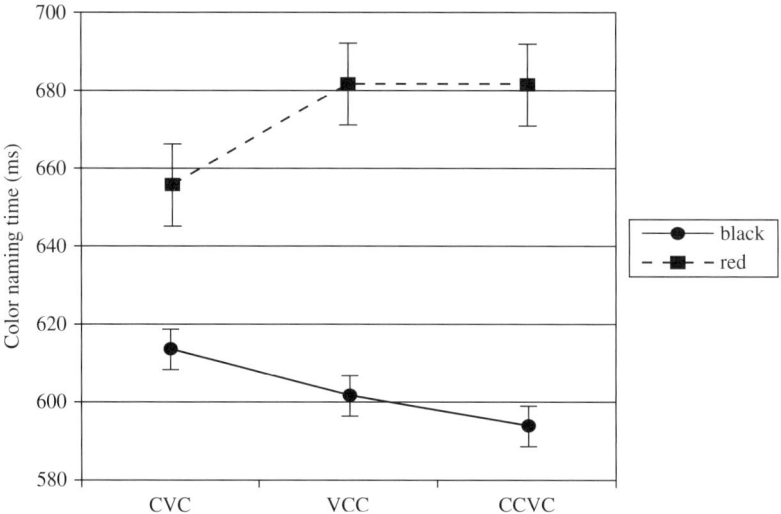

Figure 4.3 Color naming as a function of the CV-skeletal structure (from Marom & Berent, 2010, Experiments 1 & 3). Error bars are confidence intervals for the difference between the means.

divergence has to do with two factors. One is the gross similarity between the nonword and the color names: the VCC frame is more similar to *red* than to *black* (e.g., *red* and VCC words each include three phonemes), so VCC items might be easier to ignore when paired with the dissimilar name *black* than with more similar *red*. Gross similarity, however, cannot be the full story, as CVC items did interfere with *black* despite their salient differences in length and onset structure. So it appears that the potential for interference depends not only on the relationship between the nonword frame and the color name but also on the structure of the frame itself. Incongruent VCC frames might be easier to ignore because they are inherently ill formed – they lack an onset and they exhibit a coda cluster, two undesirable aspects of syllable structure across many languages (Prince & Smolensky, 1993/2004). Other Stroop-like experiments have shown that structures that are phonologically ill formed are ignored more easily than their better-formed counterparts (Berent et al., 2006; Berent et al., 2010). Readers' ability to ignore incongruency between VCC frames and *black* is consistent with this possibility. Crucially, these effects of well-formedness and congruency depend on the abstract arrangement of consonant and vowel categories, irrespective of the statistical co-occurrence of specific consonants and vowel segments. This finding suggests that consonants and vowels form distinct equivalence classes.

4.3.2   Consonants and vowels are favored by different learning mechanisms

Another demonstration of the functional distinction between consonants and vowels concerns their susceptibility to distinct mental processes. This rationale – the logic of double dissociations – is among the strongest and most elegant arguments for the separability of mental processes. If the encoding of two categories, A and B, can be selectively targeted by two mental processes, such that one process selects category A to the exclusion of B (a single dissociation of A and B), whereas another manifests the reverse pattern (targeting category B to the exclusion of A), then A and B form distinct mental categories. The two subsequent case studies follow this logic.

The first case concerns the susceptibility of consonants and vowels to distinct learning strategies. Recall that, according to a proposal put forward by Marina Nespor and colleagues (2003), consonants and vowels carry different roles in the language system – consonants signal lexical distinctions, whereas vowels typically convey grammatical information. These two roles, moreover, rely on distinct computational mechanisms. Because lexical information is largely unpredictable, its acquisition requires statistical learning – procedures that track the co-occurrence of specific instances (e.g., the syllable-size chunks *pen* and *cil* in *pencil*). The phonological grammar, by contrast, is algebraic, so grammatical phonological processes should apply to equivalence classes. To the extent that consonants and vowels serve lexical vs. grammatical roles, respectively, then statistical learning should favor consonants over vowels, whereas rule-learning should favor vowels over consonants. Two papers by Nespor and her colleagues demonstrate precisely this outcome (Bonatti et al., 2005).

One set of elegant experiments led by Luca Bonatti (Bonatti et al., 2005) compared the susceptibility of consonants and vowels to statistical learning. To this end, Bonatti and colleagues presented participants with "words" generated from three families. In some experiments, family members shared their consonants and differed on their vowels. For example, members of the p_r_g_ family included *puragi, puregy, poregi*; the b_d_k_ family included *biduka, bidoke, byduke*, etc. In other experiments, family members shared only their vowels, not consonants (e.g., _u_e_a, *kumepa, kuleta*), and the probabilities of these patterns were identical in consonant and vowel families. Of interest is whether consonant families lend themselves to statistical learning more readily than vowel families.

To examine this question, Bonatti and colleagues first familiarized participants with words generated from one type of family – either the consonant or vowel families – for a period of 14 minutes. To gauge learning, participants were next tested for their ability to distinguish "words" (familiar sequences that have been presented during the familiarization phrase, e.g., *puragi*) from part-words (novel sequences that strand across different words, e.g., *gibydo*, generated from the words *puragi* and **bydoka**). Results showed that participants were able to distinguish words from part-words only when these words were generated from consonant families, but not in the vowel-family condition. Subsequent research has shown that the greater salience of consonants to statistical learning is present already at 16 months of age (Havy & Nazzi, 2009).

A complementary demonstration by Juan Toro and colleagues (2008) suggests that rule-learning favors vowels over consonants. The hallmark of algebraic rules is their capacity to support free generalizations. Accordingly, people who have acquired a given rule (e.g., an ABA rule) should extend it to novel instances. Of interest is whether such generalizations are easier to extract for ABA sequences consisting of vowels (e.g., *tapena, tepane, badoka, bodeko*...) compared to consonants (e.g., **bitebo, nibeno, banube, batube**). To examine this question, Toro and colleagues first familiarized participants with a string of ABA "words" for a period of 10 minutes. In one condition, the rule targeted vowels (e.g., *tapena*) whereas in another, it concerned consonants (e.g., **bitebo**). Participants were next presented with novel test items: either items consistent with the ABA pattern or inconsistent ones. Results indicated that people generalized the ABA rule only for vowels, not consonants. For example, people trained on ABA vowel sequences (e.g., *tapena, tapona*) generalized to novel ABA vowels (e.g., *biduki*), whereas people trained on consonant sequences (e.g., **binebo, bitebo**) failed to generalize to the novel *pikepo*. Similar results were subsequently reported with 11-month-old infants (Pons & Toro, 2010). These findings suggest that consonants and vowels are favored by distinct learning mechanisms.

4.3.3 Consonants and vowels are selectively impaired in aphasia

The susceptibility of vowels vs. consonants to rule vs. statistical learning, respectively, shows that these two categories can doubly-dissociate in typical individuals. A report by Alfonso Caramazza and colleagues demonstrates such double-dissociations in neurological disorders as well (Caramazza et al., 2000).

The evidence comes from two patients with aphasia. Both patients exhibited errors in spoken word production, and their overall error rate was similar. In one patient (AS), however, the damage concerned mostly vowels (27 percent errors for vowels vs. 9 percent for consonants) whereas in another (IFA), the pattern was reversed (5 percent errors for vowels, 28 percent for consonants).

While this pattern is, of course, consistent with the view of consonants and vowels as distinct categories, such dissociations can also result from features that correlate with the consonant/vowel status, rather than the consonant/vowel distinction per se. Consider, for example, the role of sonority. Sonority is a scalar phonological property that varies continuously between consonants and vowels – least sonorous are stop consonants (e.g., *p*), intermediate are liquids (e.g., *l*), and most sonorous are vowels (for detailed discussion, see Chapter 8). Because consonants and vowels vary continuously along the sonority dimension, it is conceivable that the two impairments really reflect a deficit in encoding low- vs. high-sonority values, not consonants or vowels specifically. If so, then the "consonantal" deficit of IFA should be stronger for phonemes that are most consonant-like (e.g., stops), whereas the "vowel" deficit of AS should manifest itself more strongly for the less "consonantal" segments (high- sonority segments, such as liquids). But the patients' behavior did not correlate with the sonority value of consonants and vowels.

Another challenge to the view of consonants and vowels as distinct categories is presented by the computational simulations of Padraic Monaghan and Richard Shillcock (2003). These authors sought to capture the behavior of these two patients with a connectionist network equipped with no a-priori distinction between consonants and vowels. If the network could mimic the double-dissociation between consonants and vowels, then such a finding would challenge the need to encode these two classes as distinct primitives. The results, however, did not fully replicate the patients' behavior. Like the patients, damage to the model's "vowel-like" and "non-vowel-like" units yielded distinct patterns, but unlike the patients, the behavior of the damaged model varied continuously with the sonority of segments. While damage to the vowel units affected sonorants (nasals, liquids, and glides) more than obstruents (stops and fricatives), damage to consonantal units showed the opposite pattern (Knobel & Caramazza, 2007). The categorical distinction between consonants and vowels in the patients' behavior, on the one hand, and the failure of the connectionist network to fully capture those findings, on the other, suggest that the phonological mind encodes consonants and vowels as distinct equivalence classes.

4.4 Conclusions and caveats

What mechanism supports the human capacity for phonological patterning? Are phonological patterns assembled by an associative system that chains together chunks of phonological stuff, or do they require an algebraic system that operates on abstract equivalence classes?

To address this question, this chapter evaluated the role of equivalence classes with respect to two phonological categories: syllables and consonants vs. vowels. In both cases, people exhibited sensitivity to these classes, and their behavior was inexplicable by various statistical properties of phonological instances. Specifically, the sensitivity to syllable structure was not subsumed by word position, and it obtained even after controlling for the statistical co-occurrence of syllable-like units in the word. Similarly, the dissociation between consonants and vowels, evident in the encoding of word frames and their selective impairment in aphasia, was inexplicable by various statistical and phonological features that correlate with this distinction, including the similarity to existing words and the sonority of consonants and vowels.

While these results suggest that people encode equivalence classes, and apply generalizations to all class members alike, these observations leave open the crucial question of generalization. The hallmark of algebraic systems is that they allow learners to extend generalizations freely, not only to existing members of a class but also to novel, unattested ones. But most studies examine generalizations only to attested members of a category. The evidence from illusory conjunctions and the initial-syllable cost all concern sequences that function as syllables in participants' languages. Similarly, the distinction between consonants and vowels in CV frames and neurological dissociations is documented only for attested consonants and vowels. Although these results strongly suggest that the phonological mind ignores idiosyncratic features of specific consonants and vowel phonemes, they do not determine whether generalizations extend to novel members of a category. Only one case discussed in this chapter – the selective generalization of rules to novel vowels (Toro et al., 2008) – is potentially consistent with generalizations to unattested class members. But the inability of participants in these experiments to learn artificial rules over consonants is puzzling given that such productive rules are found in many natural languages (see Chapter 5).[1]

Summarizing, then, the findings reviewed in this chapter strongly suggest that the phonological mind encodes equivalence classes that support generalizations to all attested class members alike. But whether those classes also allow speakers to freely extend generalizations to novel instances – the quintessential hallmark of powerful algebraic mechanisms – remains unclear. The next chapter examines this question.

---

[1] Participants' failure to learn artificial rules over consonants is puzzling given the vast linguistic literature showing that such rules are readily learnable in natural languages. This puzzling divergence raises the possibility that the inability to learn artificial rules over consonants reflects not an in-principle property of consonants, but rather one that is specific to the materials used in these experiments. One immediate worry concerns the quality of the auditory materials. Because these stimuli were synthesized, it is possible that the learning failure could result from the quality of those speech sounds. While at first blush this explanation appears to be countered by the fact that consonants did allow for the extraction of statistical regularities, other factors might account for participants' success in this condition. An inspection of the materials reveals several cues that could have facilitated the discrimination of "words" and "part-words" in this condition. In Bonatti et al.'s experiments (2005), "words" invariably began with a labial consonant (p, b, or m) whereas "part-words" invariably began with non-labial segments. Accordingly, people could have discriminated words from nonwords even when the speech signal was overall unintelligible. Similar cues might have also allowed participants to differentiate words from part-words in Toro et al.'s experiments (2008). For example, Experiment 1 had velar-initial consonants occurring only in part-words, not words. Further research is necessary to address these problems.

5 How phonological patterns are assembled: the role of algebraic variables in phonology

> Chapter 5 further investigates the scope of phonological categories and the principles governing their combinations. Specifically, I examine whether phonological patterns encode relations among variables – the hallmark of powerful algebraic systems capable of generating discrete infinity. To this end, I systematically gauge the scope of phonological generalizations using a single case study taken from Hebrew. Hebrew manifests an interesting restriction on the location of identical consonants. It allows stems such as *simem* (with identical consonants at the right edge), but disallows forms like *sisem* (with identical consonant at the left edge). Remarkably, Hebrew speakers freely extend this generalization not only to any native consonant (thereby providing further evidence that consonants form an equivalence class) but also to novel ones, including novel consonants with novel features. Such generalizations, as I next show, are only attainable by computational devices that operate on variables. Accordingly, the documentation of such generalizations in Hebrew demonstrates that the phonological grammar is an algebraic system endowed with the capacity for across-the-board generalizations, comparable to syntactic generalizations. But while algebraic machinery is clearly necessary to account for phonological generalizations, further evidence suggests it is not sufficient. A full account of phonological generalizations thus requires a dual-route model, equipped with an algebraic grammar and an associative lexical system.

5.1 How do phonological categories combine to form patterns?

All patterns comprise building blocks, assembled according to some combinatorial principles. Chapter 4 suggested that the building blocks of phonological patterns are abstract categories that form equivalence classes. The equivalence of category members is significant because it supports broad generalizations of phonological knowledge. Specifically, if all instances of a category are treated alike, then knowledge concerning the category will automatically extend to new members. But while the findings reviewed so far strongly suggest that some phonological categories are equivalence classes, the classes we had considered were not only finite but also quite small. This limitation raises the question of whether the phonological grammar does, in fact, support open-ended generalizations, known as the capacity for "discrete infinity" (Chomsky, 1972).

In this chapter, we further address this question by investigating the principles governing the combination of phonological classes. All patterns are generated by conjoining a set of building blocks according to some combinatorial principles. As we next see, it is possible for those combinatorial principles to be quite powerful even if they operate on categories that are typically narrow in scope. Our question here is whether the principles of phonological combinations allow, in principle, for "discrete infinity."

To begin, let us distinguish between two types of combinatorial principles: concatenative and relational. Concatenative principles array a set of building blocks in a given order. Many phonological patterns are concatenative: A syllable, for example, concatenates a rhyme with an optional onset; a well-formed onset comprises an obstruent and a sonorant, and so on. To encode such functions, grammars require only the capacity to represent a small number of primitives (e.g., onset, rhyme, syllable) and sequence them in a particular manner. Such sequences, however, do not explicitly encode relations among the categories – the fact that the onset is "first" and rhyme "second" is implicit in the array. Likewise, a concatenative scheme can register that *bwa* comprises two labial consonants, but it does not explicitly represent the fact that these two features are identical.

Indeed, identity is a relation, and relations form a second combinatorial principle that is more powerful than concatenation. An identity function mandates that two occurrences of a category are instantiated by the same member. This requirement can apply, in principle, to any member of a category – familiar or novel. Likewise, identity functions can acquire multiple forms – they can either ban identical elements (*XX) or, conversely, enforce identity formation by means of reduplication (X➔XX). In all these expressions, identity is represented using a ***variable*** (X), akin to the algebraic variables familiar from mathematical expressions.

Variables (X) are abstract placeholders that stand for an entire category (e.g., the category of *any consonant*). Variables are critical for the representation of relations. First, by standing for entire categories, variables can support broad generalizations that apply to any member of the category, regardless of whether it is familiar or novel. Second, the use of variables can call a single category multiple times and ensure that all its occurrences are instantiated by the same member. Indeed, a ban on identity (e.g., *XX) does not merely require that the category X (e.g., *any consonant*) is instantiated (e.g., by the segment *p* or *t*). Rather, it mandates that the various instantiations of X are identical (e.g., *pp*, but not *pb* or *pt*) by binding those multiple occurrences. Binding guarantees that if *p* instantiates the first X category (in *XX), the same element *p* will occupy the second X slot as well. Note that the need for binding arises irrespective of the size of the category in question. Even if a relation applies to a category that is finite and small (e.g., the class of native labial segments in a given language), it is still necessary to ensure that all instantiations correspond to the same member. Accordingly, relations, such as identity, require the capacity to represent algebraic variables and to operate over those variables (Marcus, 1998; 2001; Pinker & Prince, 1988). Our question here is whether the phonological grammar makes use of this algebraic combinatorial capacity. We address this question by examining the representation of identity functions.

Identity functions, including restrictions on identity (*XX) and reduplication (X→XX), have been proposed at numerous levels of the phonological system, ranging from restrictions on identical features to segments and tones (Suzuki, 1998). But whether the phonological grammar does, in fact, represent identity remains unclear. A language that bans identical elements (e.g., *bb; *gg) might encode this fact as either a broad restriction on identity (*XX), or as a narrow ban on the co-occurrence of specific segments or features (e.g., *labial-labial; *velar-velar). Because many phonological classes are finite and small, the restrictions on the co-occurrence of identical members of the class could result not from powerful operations over variables, but rather from weaker restrictions on the concatenation of specific instances.

To adjudicate between these possibilities, one might inspect the scope of phonological generalizations. Because, by definition, identity concerns an operation over variables XX, and since a variable can stand for classes that are open-ended, identity restrictions can potentially generalize across the board, to any member of a category – familiar or novel. To the extent that people exhibit such generalizations, not only would this demonstrate that the phonological grammar encodes powerful operations over variables, but it would offer additional evidence that it has the capacity to encode phonological categories that are open-ended.

This chapter tests this prediction by means of an in-depth investigation of one famous case of identity restrictions – the restriction on identical root consonants in Semitic languages. After reviewing the constraint on identical root consonants, we assess the capacity of Hebrew speakers to generalize this restriction. We next gauge the scope of those generalizations in two steps. First, we examine whether the restriction on identity applies to all members of the class of native Hebrew consonants (section 5.3). Our main question, however, concerns the capacity of the grammar to extend such relational generalizations across the board, even to novel class members – phonemes that are nonnative to Hebrew. Section 5.4 examines this question.

5.2 A case study: the restriction on identical root consonants in Hebrew

Like many other Semitic languages, Hebrew constrains the structure of its base morphemes. It allows stems like *simem* ('he intoxicated (someone else)'), with identical consonants at the right edge of the stem, but disallows forms in which identical consonants occur at its beginning (e.g., *sisem*).

John McCarthy (1981; see also McCarthy, 1979) attributes this fact to a constraint on the structure of the consonantal root morpheme (hereafter "root"). Roots are units of word formation in Semitic languages. Like stems, a morphological unit familiar to us from English (e.g., *take*, in *retake*), roots convey the core meaning of a word. Unlike stems, however, roots in Semitic languages comprise only consonants – typically three consonants per root. The root *smm*, for instance, carries the core meaning of drug-related activities, such as 'he intoxicated (someone), he intoxicated (himself),' etc. Words, in this view, are generated by inserting the consonantal root into a word pattern – a prosodic template that includes vowels and affixes and specifies broad aspects of a word's meaning (e.g., active vs. reflexive actions, in verbs). Table 5.1 illustrates several words generated from two roots – *smm* and *ʃmr*.

Our interest here concerns the source of the *sisem–simem* asymmetry. McCarthy's account (1979; 1981) attributes this asymmetry to a series of grammatical constraints that conspire to ban identical consonants from occurring at the beginning of the root (e.g., **ssm*), but allow them at its end (e.g., *smm*). The details are provided in Box 5.1 (an alternative explanation, couched in reference to stems, is presented in Box 5.2).[1] While distinct proposals differ in several respects, two facts are common to virtually all linguistic accounts of the *sisem–simem* asymmetry. First, all accounts assume the *sisem–simem* asymmetry reflects a grammatical constraint on a base morpheme (e.g., the root *ssm*), rather than on the position of identical consonants in surface words. Indeed, the preference for *simem*-type

Table 5.1 *The structure of Hebrew words*

			Word pattern		
Root	$C_1 i C_2 e C_3$	Gloss	$hit C_1 a C_2 e C_3$*	Gloss	
smm	simem	He intoxicated someone	histamem	He intoxicated himself	
ʃmr	ʃamar	He guarded someone	hiʃtamer	He preserved himself	

* Another Hebrew rule metathesizes the *t* prefix and root-initial sibilants (e.g. hitsamem ➔ histamem).

[1] Although there is much linguistic and psycholinguistic evidence to favor stem-based accounts (see Box 5.2), in what follows, I discuss the *sisem–simem* asymmetry by appealing to the root. I do so for strictly expository reasons, as the root allows us to clearly see the pattern of consonants.

Box 5.1 A root-based account for the *sisem–simem* asymmetry by McCarthy (1981)

In a classic paper, John McCarthy (1981) showed how the *sisem–simem* asymmetry falls out from identity restrictions that specifically target the consonantal root. His account is couched within an autosegmental view of phonology, which posits separate "tiers" in phonological representations. McCarthy adopts the view that vowels and consonants belong to separate representational tiers, both of which are anchored to a CV skeleton. This skeleton, together with the vowel tier, determines the prosodic template of Semitic words, while root morphemes are represented on the consonant tier alone. The representation of the verb *simem* is provided in Figure 5.1.

Since the root is represented on a separate plane, the identical consonants in *smm* are effectively adjacent. Such representations, in turn, are banned by the Obligatory Contour Principle (OCP) – a principle that disfavors adjacent identical elements in phonological representations. Accordingly, roots with identical consonants (e.g., *smm*) cannot be stored as such in the lexicon. Instead, such roots are represented in a biconsonantal form, *sm* – the form *smm* results from a productive phonological process that reduplicates the biconsonantal root during word formation. To form a word, the root must be inserted in a word pattern in a left-to-right fashion. For a tri-consonantal root like *fmr*, this left-to-right alignment will fill all three root slots. But for *sm*, the alignment will leave an empty slot at the right edge. This empty slot will next be filled by spreading the *m* rightwards (indicated in the figure by the dotted line), and its association with two slots. This process yields the word *samam*. Crucially, the same process cannot yield *sasam*. Because association and spreading can only proceed from left to right in Semitic, reduplicated elements can emerge only at the right edge of the root. And if roots such as *ssm* cannot be obtained productively, and cannot be stored in the lexicon either (due to the OCP), then Hebrew (and other Semitic languages) can only exhibit roots like *smm*, not *ssm*.

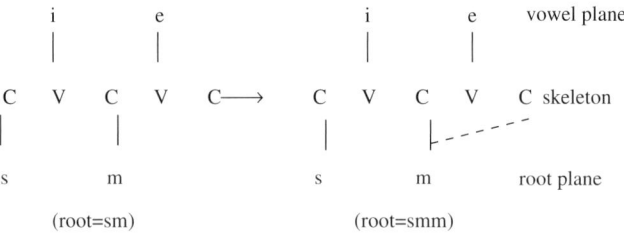

Figure 5.1 The formation of the root *smm* from *sm*

Box 5.2 A stem-based account for the *sisem–simem* asymmetry by Bat-El (2006)

While traditional linguistic accounts describe the structure of Semitic words by appealing to the root – a special morpheme that includes only consonants – subsequent linguistic analysis (Bat-El, 1994; 2003; Bat-El, 2004) and psycholinguistic evidence (e.g., Berent et al., 2007b) suggest that the structure of Semitic words can be better captured by appealing to morphological units that include both consonants and vowels (hereafter, "stems"). In what follows, I describe how the *sisem–simem* asymmetry can be captured in one such account, a proposal outlined by Outi Bat-El (2006).

In Bat-El's analysis, the representation of Hebrew words invariably encodes both consonants and vowel – no special "consonantal root" exists. The *sisem–simem* asymmetry must therefore reflect grammatical restrictions that operate over entire "stems." Her proposal represents verbs such as *simem* as the outcome of reduplication, and notes the identity relation (copy, c) between the consonant in the base *sim* (marked by $\{\}$) and its copy, the final consonant (e.g., $\{s_1im_2\}em_{2c}$). Crucially, however, reduplication is restricted to the right edge. Bat-El captures this fact by a Surface Correspondence by Position (SCorrP) constraint, which states that identical (but not similar) consonants are in a correspondence relation, provided that these corresponding segments are found at the right edges of their respective domains (the base and reduplicant, respectively). Consequently, in forms such as *simem*, the final consonant can be represented as a reduplicant (i.e., as $\{s_1im_2\}em_{2c}$). Moreover, such representations are also optimal – they are preferred to alternative candidates that do not encode the reduplicative relation (e.g., $\{s_1im_2em_3\}$, where the consonants are encoded as two non-corresponding consonants, as those candidates would violate a second constraint (Surface Correspondence by Identity, SCorrI), which prohibits identical consonants in a stem unless these consonants are correspondents.

In (1), I illustrate this proposal using an Optimality Theoretic tableau. The tableau lists the optimality of two candidates ([$\{S_1iM_2\}eM_{2c}$] and [$\{S_1iM_2eM_3\}$]) as outputs for SiMeM (the input, indicated at the top left corner). Optimality, here, is evaluated relative to two constraints, ranked in a left-to-right order (SCorrP, SCorrI). Because the first candidate, [$\{S_1iM_2\}eM_{2c}$], incurs a less severe violation of those constraints (here, no violation at all), it is the winning candidate, a fact indicated by the pointing hand.

(1) The representation of SiMeM

SiMeM	SCorrP	SCorrI
☞[{S₁iM₂}eM₂ᴄ]		
[{S₁iM₂eM₃}]		*!

Crucially, *simem*-type stems can escape the violation of SCorrI. Because identity resides at the right edge, a correspondence can be formed between these consonants, and consequently, no violation of SCorrI is incurred. In contrast, forms with initial identity, such as *sisem*, cannot be represented in this manner, as reduplication of the left edge (e.g., [S₁ᴄi{S₁eM₃}]) is disallowed. Left to their own devices, such forms will manifest identical consonants in the base, and as such, they will fatally violate the identity constraint (SCorrI; see 2). Summarizing, then, *simem*-type forms are well formed because they can be represented as reduplicants, whereas *sisem*-type forms fatally violate a constraint on identity; hence, such forms are banned.

(2) The representation of SiSeM

SiSeM	SCorrP	SCorrI
[S₁ᴄi{S₁eM₃}]	!*	
☞[{S₁iS₂eM₃}]		*

forms extends to entire morphological paradigms, regardless of whether the consonants are word-final (e.g., *simem* 'he intoxicated someone') or medial (*simamti* 'I intoxicated someone'). Second, all accounts assume that the restriction on the base morpheme specifically concerns identical consonants. And since identity is a relation among variables, the documentation of identity restrictions would suggest that the phonological grammar must be credited with the full algebraic power to encode variables and operate over variables.

Distributional linguistic evidence, however, is not in and of itself sufficient to show how such restrictions are mentally represented. The infrequency of forms such as *sasam* might be a historical fact that is not represented by speakers. Even if people did encode this restriction, and productively generalized it to novel items, such generalizations could potentially reflect not a productive restriction on identity but rather the statistical properties of linguistic experience. Indeed, because *ssm*-type forms are rare, novel forms such as *llk* are likely to include consonant combinations that occur infrequently at the right edge of the root. The dispreference for novel *llk*-type roots could therefore reflect only the infrequency of their consonant combinations, not their identity per se. Our question, then, is whether a grammatical restriction on identical consonants is active in synchronic grammars – a test of the broader capacity of the phonological mind to operate over variables. The next sections move to examine this question using experimental methods.

5.3 The restriction on identical consonants generalizes to native Hebrew consonants

Do Hebrew speakers represent a productive grammatical restriction on identical consonants? To address this question, we must gauge speakers' ability to generalize this regularity. Recall that operations over variables support generalizations that extend across the board to any member of a category – actual or potential, familiar or rare. So if Hebrew speakers restrict identical consonants – a constraint defined as a relation among variables – then they should be able to freely extend this restriction to novel forms. We evaluate this prediction in two steps. First, we demonstrate that Hebrew speakers generalize the *sisem–simem* asymmetry to novel consonant combinations that they have never encountered before. Moreover, people are demonstrably sensitive to the identity of consonants in *simem*-type forms, and they distinguish these items from no-identity controls (e.g., *pisem*) even when forms are equated for various statistical properties. Such generalizations are consistent with the possibility that speakers encode an abstract restriction on identity – a relation defined over variables. If this explanation is correct, then identity restrictions should generalize across the board, even to consonants that are unattested in Hebrew. The second part of this discussion (section 5.4) demonstrates just this.

5.3.1 *Identity restrictions generalize to novel roots*

To determine whether the restriction on identical consonants is productive, we can elicit participants' responses to novel words, generated from novel roots – combinations of three consonants that do not exist in Hebrew. If the grammar bans identical consonants root-initially, then novel *ssm*-type roots should be dispreferred to roots with either final identity (e.g., novel *smm*-type roots) or no-identity (e.g., *psm*-type) controls. These predictions are indeed borne out by the results of numerous experiments. For example, when Hebrew speakers are asked to rate novel words for their acceptability, they consider words generated from *ssm*-type roots as systematically less acceptable than *smm*- and *psm*-type controls (Berent & Shimron, 1997; Berent et al., 2001a). Similarly, when prompted to generate words from novel roots (e.g., *bg*), identical consonants are placed at the root's end (e.g., *bg*→*bgg*), but rarely at its beginning (e.g., *bg*→*bbg*, Berent et al., 2001a).

Rating and production procedures, however, both gauge knowledge offline, and offline procedures have a bad reputation among psychologists (e.g., Gibson & Fedorenko, 2010). Researchers worry that off-line experiments reflect conscious problem-solving strategies, rather than tacit linguistic knowledge. These concerns are unwarranted in the present case, as Hebrew speakers have no explicit knowledge of this restriction – when asked to justify their rating judgments, people are typically unable to explain why *ssm*-type forms are bad (Berent & Shimron, 1997). In fact, the dislike of *ssm*-type roots is also evident indirectly, in tasks using various online procedures.

Consider, for example, the lexical decision procedure. In this task, people are presented with word-like stimuli – either existing Hebrew words, or word-like stimuli that do not exist in Hebrew – and they are asked to quickly determine whether or not the stimulus is an existing Hebrew word. Our interest concerns the structure of these nonwords – these stimuli are generated from novel consonantal roots with root-initial identity (e.g., *ssm*), root-final identity (e.g., *smm*),[2] or no identical consonants (e.g., *psm*). If *ssm*-type roots are unacceptable, then novel words generated from such roots should appear less word-like. And indeed, people detect a nonword more readily (i.e., faster and more accurately) when it is generated from *ssm*-roots compared to *smm*- or *psm*-type controls (Berent et al., 2001b; Berent et al., 2004; Berent et al., 2007b). People's sensitivity to root structure in such tasks is important because it suggests that the restriction on root structure is active online, even when participants are not explicitly asked to attend to root structure.

In fact, Hebrew speakers cannot help but attend to the root despite explicit instructions to ignore it altogether. The evidence comes from Stroop-like experiments. In these experiments, participants are presented with printed words, displayed in various colors (e.g., *sisem* printed in red). Their task is to name the color of the letters (e.g., say "red") while ignoring the content of the printed word. But as it turns out, participants nonetheless process the printed stimulus. And since ill-formed *ssm*-type words are ill formed (i.e., less word-like), people can ignore them more easily. As a result, participants name the color (e.g., red) faster in the presence of words generated from *ssm*-type roots compared to *smm*- or *psm*-type controls (Berent et al., 2006). These findings demonstrate that Hebrew speakers generalize the *ssm*–*smm* asymmetry to novel forms.

We now turn to examine what mechanisms might support these generalizations. The results described in this section follow naturally from the view of the grammar as algebraic. In the case of Hebrew, specifically, the grammar might constrain the location of identical consonants relative to the base morpheme. Because categories such as "identity" and "base morpheme" are

---

[2] The root *smm* actually exists in Hebrew. Here I use this example for the sake of expository consistency, but the roots used in the experiment were all novel roots.

open-ended equivalence classes, these categories will support free generalizations, leading to the pattern of results reported here. But on an alternative, non-algebraic explanation, people do not encode such classes. We next consider this account. We first examine the possibility that people do not represent the base morpheme; the following section investigates whether people constrain identical consonants, specifically.

5.3.2 Speakers constrain the structure of a open-ended morphological classes

So far, we have shown that novel words generated from *ssm*-type roots are considered less word-like. One possibility is that this restriction reflects a constraint that bans the structure of base morphemes – roots or stems. But on an alternative explanation, people do not represent such morphological classes. Rather, they merely register the occurrence of consonants and vowels in **words**. For example, forms like *sisem* might be ill formed because no Hebrew word begins with an *s-i-s* combination.

The rich morphology of Hebrew allows one to adjudicate between these two possibilities. Hebrew allows a single root (e.g., *ssm*) to appear in numerous word patterns that dramatically alter the root's location in the word (see Table 5.2). The root *ssm*, for instance, can be presented either unaffixed (e.g., *sisem*), suffixed (e.g., *sisamti*) or even "sandwiched" between various affixes so that identical consonants do not coincide with word edges at all (e.g., *histasamti*). If the dislike of *sisem* reflects a restriction that applies to the edges of words, then forms such as *histasamti* should not be disliked, but if it concerns roots, then people might automatically "strip" the *ssm*-constituent of *histasamti*. Accordingly, *histasamti* (from *ssm*) should still be disliked relative to *histamamti* (from *smm*), even though neither word manifests identical consonants at their edges.

The results of numerous experiments confirm this prediction. In all cases, the dislike of *ssm*-type roots obtains regardless of the position of identical consonants in the word. In particular, highly opaque forms such as *histasamti* are still rated as less acceptable, they are considered less word-like (in lexical decision,

Table 5.2 *An illustration of various word classes, generated by inserting a root in various word patterns*

For viewing convenience, root consonants are indicated in upper case.

Root type	Class 1	Class 2	Class 3
SSM	SiSeM	maSSiMim	hiStaSaMtem
MSS	MiSeS	mamMSiSim	hitMaSaStem
PSM	PiSeM	maPSiMim	hitPaSaMtem

e.g., Berent et al., 2001b; Berent et al., 2004), and they are easier to ignore in a Stroop-like task (Berent et al., 2006). These findings clearly establish that the grammar restricts the structure of abstract morphological bases – either roots or stems – rather than surface words. They demonstrate that the grammar encodes open-ended morphological classes.

5.3.3 People constrain identity – not the co-occurrence of consonants or features

Now that we know people restrict the structure of morphological classes, we can return to our main question: What precisely is it that people restrict? One possibility is that they truly ban identity – an abstract relation defined over variables. However, not only are *ssm*-type roots ill formed, but they also manifest rare consonant sequences. Indeed, no Hebrew root has the sequence *ss* at its beginning. This observation opens up the door to an alternative associative account. In this view, the dislike of *ssm*-type forms reflects unfamiliarity, rather than root structure. Familiarity, here, should be defined in a rather nuanced manner. Since all roots presented in these experiments were novel, people cannot simply base their judgment on the familiarity with the root as a whole. Nonetheless, people could have gauged familiarity from the co-occurrence of root parts – phonemes or features. Consider, for example, a novel root such as *ssk*, as compared to *kss*. Although *ssk* itself is entirely novel, its *ss* part isn't – many Hebrew roots manifest this combination at their end (e.g., *mss* 'dissolve,' *bss* 'base,' *gss* 'die'), but no Hebrew root manifests *ss* at its beginning. So perhaps the dislike of *ssk* reflects only the rare occurrence of *ss* root-initially, not a ban on consonant identity per se.

But the experimental results suggest this possibility is unlikely, as people remain sensitive to the presence of identical consonants even when their statistical properties are controlled. The critical evidence, however, comes not from *ssm*-type roots, but rather from their *smm*-type counterparts. Because *ssm*-roots are highly ill formed and infrequent, it is virtually impossible to find any consonant combination that is as rare as *ss* root-initially. In contrast, *smm*-type roots are highly frequent, so these roots can be closely matched on their statistical properties to no-identity controls (e.g., *psm*). If people do not recognize identity per se, then these two forms, *smm* and *psm*, should be indistinguishable, but if they encode identity, then the structure of these two forms should be distinct. This latter prediction is indeed borne out in several experiments. Findings repeatedly show that people differentiate *smm*-type roots from frequency-matched *psm*-type controls in rating experiments. Compared to controls, *smm*-type roots are rated as more word-like (Berent et al., 2001a), and since the structure of such words is well formed, novel words generated from *smm*-type roots are harder to classify as nonexistent words (Berent et al., 2001b; Berent et al., 2004).

Together, these results suggest that people do not merely track the co-occurrence of consonants but specifically encode their identity.

Still, one escape hatch remains open for the associative explanation. The results discussed so far show that the distinction between *smm*- and *psm*-type roots is inexplicable by their consonant co-occurrence. It is conceivable, however, that people distinguish *smm*-type roots from controls by attending to the statistical co-occurrence of features. For example, forms like *smm* might be preferred because labial consonants frequently occur at the end of the root. Similarly, the dislike of forms like *ssm* might reflect not a ban on identical segments, but rather the fact that two coronal consonants rarely occur at the beginning of the root. It is indeed well known that Semitic languages constrain not only the co-occurrence of fully identical consonants but also the co-occurrence of nonidentical segments that share the same place of articulation, the so-called homorganic consonants (for contrasting explanations, see McCarthy, 1994; Pierrehumbert, 2001), and such bans have been widely documented in numerous languages (in Javanese: Yip, 1989; Russian: Padgett, 1995; Muna – an Australian language: Coetzee & Pater, 2008; Yamato Japanese: Kawahara et al., 2006, as well as English, French, Latin: Berkley, 2000). Considering Semitic, roots such as *skg*, with two velars, are dispreferred, and hence infrequent compared to roots whose nonidentical consonants manifest distinct places of articulation. Such observations open up the possibility that the dislike of *ssm*-type roots might be due to the rarity of their feature combinations, rather than their consonant identity.

But this possibility too is countered by the experimental evidence. The critical findings, once again, concern roots with identical consonants at their end (*ssm*-type roots are too rare to allow one to match their statistical properties with no-identity controls). To determine whether the responses to *smm*-type roots are informed by segment identity or feature co-occurrence, my colleagues, Vered Vaknin-Nusbaum and Joseph Shimron, and I have compared roots whose final consonants are either fully identical (e.g., *skk*), homorganic (e.g., *skg*) or heterorganic controls (e.g., *skf*). These roots were all matched for their consonant co-occurrence as well as the co-occurrence of their place of articulation feature. If the dislike of *ssm*-type roots in previous experiments is due only to the rarity of its feature combinations, then roots with identical features should not be dispreferred relative to controls, irrespective of whether they are identical (*skk*) or homorganic (e.g., *skg*).

The predictions of the algebraic ban on segment identity are markedly different. If the grammar encodes identity, then responses to roots with identical and nonidentical elements may differ despite their matched statistical structure. Moreover, if the restriction on identity distinguishes between consonant and feature identity (McCarthy, 1994), then responses to *skk*- and *skg*-type roots should further differ. Unlike the restriction on identical consonants, the restriction on identical features bans homorganic features irrespective of their location. Accordingly, homorganic *skg*-type roots should be ill formed relative to *skf*-type controls. Fully identical consonants (e.g., *skk*), by contrast, are represented as reduplicants, and consequently, these roots do not effectively manifest two occurrences of the same feature. Because *skk*-type roots do not violate the restriction on homorganicity, they should be preferred to their homorganic *skg*-type counterparts (see 3).

(3) Violations of the restrictions on consonant and feature identity

	Consonant-identity violation	Feature-identity violation
skk		
skg		*

The results of rating and lexical decision experiments support this latter prediction. Hebrew speakers rate homorganic roots as less acceptable than controls (Berent & Shimron, 2003) – a result that agrees with the behavior of Arabic speakers (Frisch & Zawaydeh, 2001). Words generated from the homorganic, *skg*-type roots were classified more rapidly as nonwords in the lexical decision task (Berent et al., 2004). Crucially, roots with homorganic consonants were less acceptable compared to roots with identical consonants, suggesting that fully similar *skk*-type roots are *better* formed than homorganic *skg*-type roots (Berent & Shimron, 2003; Berent et al., 2004). This finding is inconsistent with two classes of feature-based explanations for the *ssm–smm* asymmetry. Because *skk*- and *skg*-type roots were all matched for the co-occurrence of their place feature as well as their segment co-occurrence, this finding challenges statistical accounts couched in terms of segment and feature co-occurrence. The fact that roots whose consonants are identical (i.e., maximally similar, e.g., *skk*) are dispreferred to homorganic (i.e., partially identical, e.g., *skg*) roots further counters the possibility that people ban feature similarity (for converging conclusions based on computational simulations, see Coetzee & Pater, 2008). These results thus suggest that people encode two distinct bans on root structure – a ban on identical consonants root-initially, and a separate ban on homorganic features. Importantly, both bans specifically concern the identity of phonological elements.

5.3.4 Conclusion

The experimental results described in this section examined whether the phonological grammar is endowed with the capacity to operate over variables. As a case study, we investigated the restriction on identical consonants in Hebrew. Identity is a relation among variables, so a restriction on identity requires the ability to operate over variables. Our question was whether Hebrew speakers indeed encode such a constraint. The results from numerous experiments demonstrate that Hebrew speakers possess a productive restriction on the structure of the root, and that they generalize this restriction to novel roots, irrespective of their position in the word. Furthermore, the distinction between identical and nonidentical consonants cannot be captured by the frequency of consonant or feature co-occurrence. These results are consistent with the hypothesis that Hebrew speakers represent all consonants alike, and extend the restriction on consonant identity to any member of the class.

5.4 The restriction on identical consonants generalizes across the board

The ability of Hebrew speakers to generalize the restrictions on root structure to any Hebrew consonant is certainly consistent with the algebraic account of the grammar, generally, and specifically, with the possibility that the Hebrew grammar constrains consonant-identity using variables. Nonetheless, variables are not strictly required to account for these results. The hallmark of operations over variables is that they support across-the-board generalizations that extend to any member of a category – familiar or novel. But the results described so far show only that people treat alike all *existing* members of a phonological category – the category of *any consonant*; these findings do not demonstrate that people extend the relevant restrictions across the board to novel members of that category.

In what follows, we probe for such generalizations. But before we can evaluate whether people generalize phonological restrictions across the board, we must first define what, precisely, across-the-board generalizations are, and how one is to gauge them. Armed with the right yardstick, we move to examine whether Hebrew speakers generalize their phonological knowledge in this manner. Such generalizations, as we next demonstrate, require the encoding of operations over variables, as they are not attainable by various computational simulations that lack such operations. Accordingly, across-the-board generalizations will indicate that the phonological grammar represents identity relations by means of operations over variables.

5.4.1 Across-the-board generalizations: what they are and how to gauge them

Across-the-board generalizations are the hallmark of a powerful algebraic system, equipped with operations over variables. So to the extent that people broadly generalize their phonological knowledge, this would indicate that the phonological grammar possesses the full computational power of an algebraic system. While this prediction is straightforward, its evaluation is hardly trivial. The main obstacle is that we do not have a clear definition for the scope of generalizations and a method for their evaluation. What, precisely, is meant by saying that a generalization extends "across the board"? And how is one to gauge such generalizations empirically?

One way to approach this issue is by the method of brute force. One could train people on a given regularity, and next test whether they extend it to a large number of novel test exemplars – as many instances as one can possibly imagine. But brute force is neither practical nor adequate. Because the number of exemplars one could practically test is limited, such tests can never exhaust the theoretical set of "all possible exemplars." So even if people successfully generalized on all test items presented to them, this would not guarantee that they do possess the capacity to generalize to any potential member. Conversely, people might fail to generalize for a host of extraneous reasons: Certain strings might be hard to articulate or perceive, and consequently, generalization would fail for reasons that are unrelated to the phonological grammar. In short, a brute force approach might both underestimate the true scope of phonological generalizations and falsely overestimate them. A far more productive approach begins by offering a principled definition for the scope of generalizations. Such a definition provides a yardstick that allows one to systematically gauge the performance of language learners. The work of Gary Marcus (1998; 2001) provides us with this foundation.

Marcus captures the scope of generalizations by two factors. One is the learner's experience, namely, the set of exemplars of the desired generalization that the learner has encountered. A second, equally important, factor is the representation used by the learner to encode those exemplars.

To illustrate this proposal, let us consider a simple regularity concerning syllable reduplication, such as *ba*➔*baba*; and *ta*➔*tata*. Suppose that after hearing these two exemplars, a learner is presented with a novel test item, either *be, mu* or *gu* (see 4). Our question is whether the generalizations to these three test items are comparable. That is, is the generalization to *be* as easy as the generalization to *mu* and *gu*?

(4) An illustration of items presented in the training and testing phases of a simple learning experiment
    Training phase:
        ba➔baba
        ta➔tata
    Testing phase:
        gu➔?
        mu➔?
        be➔?

The answer, as it turns out, depends on how these exemplars are represented. If the learner represented the training instances as unanalyzed whole syllables, then generalization to all test items would be equally difficult, as neither would share any component of the familiar training items. To verify this conclusion, we can compare the representation of test items with the representation of

Identity restrictions generalize across the board 99

training items (the elements that comprise each such instance are indicated by a +, see 5). An inspection of these representations makes it plain that there is no overlap between the two sets.

(5) The scope of generalizations using a syllable-based representation: gu=mu=be

		Representational scheme				
		ba	ta	gu	mu	be
Training	ba	+				
	ta		+			
Generalization	gu			+		
	mu				+	
	be					+

Things would be quite different if the learner relied on a more detailed representation that encodes segments (see 6). Now the novel *be* shares a segment with the familiar *ba* (see the highlighted *b*), whereas the novel *gu* and *mu* do not share any component with the training exemplars. Consequently, the generalization to *be* is narrower in scope (i.e., easier) than to either *gu* or *mu*, which, in turn, do not differ.

(6) The scope of generalizations using a segment-based representation: (gu=mu)>be

		Representational scheme						
		b	t	g	m	a	e	i	u
Training	ba	+				+			
	ta		+			+			
Generalization	gu			+					+
	mu				+				+
	be	+							

Things would once again change if the level of representational detail increased even further to specify features. For the sake of simplicity, let us consider only features concerning the place of articulation of consonants (see 7 – the argument can be easily extended to other features too). This representation scheme still renders the generalization to *gu* more challenging than *be*, as the velar *g* shares no place feature with any of the training items. But now, *mu* overlaps with *ba*, as they both share the labial feature (highlighted), so generalization to *mu* is easier than to *gu*.

(7) The scope of generalizations using a feature-based representation: gu>(mu=be)

		Representational scheme		
		Labial	Velar	Coronal
Training	b (ba)	+		
	t (ta)			+
Generalization	g (gu)		+	
	m (mu)	+		
	b (be)	+		

Summarizing, then, the actual scope of a generalization depends not only on linguistic experience (i.e., training) but also on how this experience is represented by the learner. While coarse, syllable-based representations fail to reveal any overlap between training and test items, finer-grained representations would allow learners to discern some shared constituents, and this could render the generalization to certain test items easier than to others.

Building on these conclusions, Marcus (1998; 2001) captures the scope of generalizations by comparing the representation of a test item with the learner's representation of his or her experience (i.e., of training items). If the representation of a test item can be exhaustively described in terms of the constituents of training items, then this item can be said to fall *within* the learner's training space. In contrast, if a test item includes some untrained units, then this item falls *outside* the training space. In our previous examples, the syllable-based representation in (5) would render all three test items entirely outside the training space (see 8), the segmental representation in (6) renders *be* (but not *mu* or *gu*) within the training space, and the feature-based representation in (7) renders both *be* and *mu* as falling within the learner's training space (whereas *gu* lies outside it).

(8) The scope of generalization relative to the learner's training space, defined using syllable-, segment-, and feature-based representations. Items falling within the training space are circled.

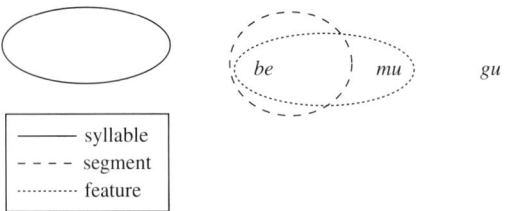

Armed with this definition, we can now proceed to precisely define the *scope of a phonological generalization*. Across-the-board generalizations are generalizations that extend irrespective of the position of test items relative to the training space. Accordingly, if phonological generalizations extend across the board, then they should apply even for test items that cannot be exhaustively described in terms of the representational constituents of training items, i.e., to items that fall outside the training space.

At first blush, it might appear that no phonological generalization would measure up to such stiff standards. Unlike syntactic categories, such as "nouns" and "verbs," categories that can stand for an infinite number of instances, phonological categories (e.g., the "English phonemes") are typically limited and small. Narrow phonological categories might attest to computational machinery that is inherently limited, incapable of "across-the-board" generalizations – the bread and butter of syntax, for instance.

But I believe this conclusion is premature. First, not all phonological categories are narrow. Although the class of phonemes and features attested in a language is small, other phonological categories, such as *any syllable* (the domain of prosodic processes) or *any base* (the domain of phonological reduplication), are potentially quite large. Second, the actual size of phonological classes such as "feature" and "segment" does not necessarily show that the phonological grammar lacks the in-principle capacity to generalize broadly. As shown above, it is certainly possible for small categories to manifest across-the-board generalizations (cf. 7). Moreover, there are good reasons to believe the small size of phonological categories might occur for reasons that are unrelated to the computational characteristics of the grammar. Indeed, the categories of *noun* and *feature* differ not only on size but also on their position in the language hierarchy. While *nouns* comprise words that have internal phonological patterning, *features* are linguistic atoms placed at the very bottom of the language hierarchy. Martin Nowak and David Krakauer have shown that this has direct bearing on perceptual discriminability: Large feature categories are inevitably harder to perceive, as it is harder to discriminate between their members. In contrast, nouns can be readily differentiated by their distinct phonological patterns (e.g., *god* vs. *dog*), so the addition of new noun instances to a category has little perceptual cost (Nowak & Krakauer, 1999). These considerations suggest that even if the phonological grammar were endowed with algebraic machinery that supports across-the-board generalizations, one would still expect phonological categories to maintain a small size. For these reasons, I believe the available evidence is fully consistent with the possibility that the phonological grammar might share with the syntactic component the capacity for discrete infinity – a capacity supported by the representation of variables and operations over variables.

One hint toward this possibility is presented by the systematicity of phonological borrowings – a process that occurs on a daily basis across many languages. Of particular interest are borrowings that introduce nonnative phonemes and features – elements that do not form part of the native phonological inventory. The borrowing of the *ch* (/x/) in *J. S. Bach* into English is one such example (an illustration suggested by Morris Halle; Pinker & Prince, 1988). Although forms such as *Bach* arguably fall outside the training space of native English speakers, people are known to extend their phonotactic knowledge to such items. Recall (from Chapter 2) that English requires the plural suffix to agree with the stem on voicing (cf. *dogz* vs. *cats*). Remarkably, English speakers extend this restriction to novel phonemes. Indeed, they pluralize *Bach* as *Bachs* /bɑxs/ (with two voiceless consonants; e.g., *The pianist played two Bachs and one Beethoven*) rather than *Bachz* (/bɑxz/ where the two consonants disagree on voicing) – a potential case of across-the-board generalizations. In what follows, I discuss a detailed experimental test of this possibility.

5.4.2 The restriction on identical consonants extends across the phonological space of Hebrew

To examine the scope of phonological generalizations systematically, let us move back to our Hebrew example. Recall that Hebrew constrains the location of identical consonants in its roots: it allows forms with identical consonants at their right edge (e.g., *smm*) but disallows them at the left edge (e.g., *ssm*). The results described so far demonstrate that Hebrew speakers generalize this restriction productively, irrespective of the statistical co-occurrence of consonants and features, and regardless of the position of the root in the word. While these findings are clearly consistent with an algebraic account, they do not strictly demonstrate that the restriction applies to any consonant.

The critically missing test concerns the scope of phonological generalization. If people constrain identity – an operation over variables – then they should extend this distinction across the board, even to items that fall outside their training space – items that cannot be exhaustively described in terms of the features experienced in the context of the relevant generalization. To test this prediction, one might examine whether Hebrew speakers extend their phonotactic knowledge to roots with nonnative phonemes. A series of experiments examined this question (Berent et al., 2002). These experiments generated roots with identical consonants that include one of the following phonemes: *ch, j, w*, and *th* (/tʃ/,/dʒ/, /w/, /θ/). None of these phonemes is native to Hebrew, but most adult Hebrew speakers are very familiar with these phonemes from borrowing (e.g., *check*) and from their knowledge of English as a second language. Of interest is whether Hebrew speakers constrain the structure of these roots.

As in previous experiments, the roots manifested either root-initial identity (e.g., *jjr*), or root-final identity (e.g., *jrr*) or no identity (e.g., *jkr*), and they were presented in a variety of word patterns that differ with respect to their morphological transparency (see 9). Because these foreign phonemes cannot be adequately transcribed in the Hebrew orthography, the words were printed using English letters (familiar to Hebrew speakers), and participants were asked to read each word aloud, so that the accuracy of their phonological encoding could be verified (to assure that participants did not assimilate the foreign phonemes into Hebrew phonemes, a possibility that would place these phonemes within the space of Hebrew phonemes). If Hebrew speakers extend the constraint on identical consonants beyond the space of Hebrew phonemes, then roots with initial identity would be reliably disfavored relative to root-final identity. To the extent the relevant restriction concerns the structure of the root, then this pattern should emerge irrespective of the position of the root in the word.

(9) Morphologically transparent and opaque words, formed by the conjugation of roots with nonnative phonemes.

Root type	Root	Word pattern	
		Transparent	Opaque
XXY	jjr	ja.jar.tem	hij.ta.jar.tem
YXX	rjj	ra.jaj.tem	hit.ra.jaj.tem
XYZ	jkr	ja.kar.tem	hij.ta.kar.tem

The results were consistent with these predictions. Roots with initial identical consonants were rated as significantly less acceptable than either final-identity or no-identity controls, and this result obtained irrespective of the location of the root in the word (see Figure 5.2). For example, people disliked root-initial identity even when the root *jjr* was presented in highly opaque words, such as *hijtajartem*.

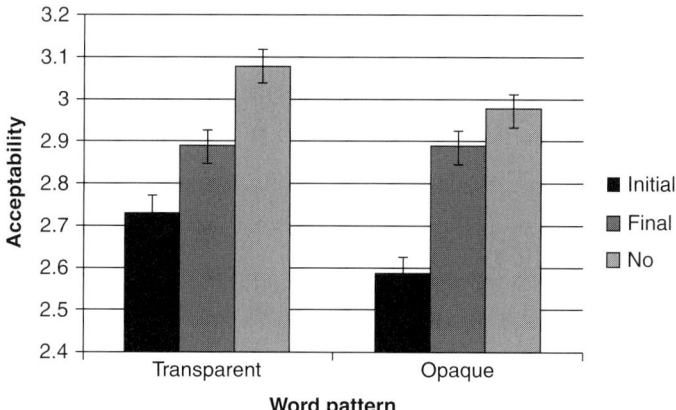

Figure 5.2 Rating result for novel roots generated from nonnative consonants (from Berent et al., 2002, Experiment 2). Error bars are confidence intervals constructed for the difference between the means.

Moreover, the dislike of root-initial identity obtained regardless of task demands. Specifically, the dislike for *jjr*-type forms was evident not only in explicit rating procedures, but also in tacit lexical decision experiments: because roots such as *jjr* are ill formed, participants were able more readily to determine that nonwords generated from such roots do not exist in Hebrew. These results suggest that Hebrew speakers generalize the restriction on identical consonants beyond the space of their native phonemes – a finding consistent with the possibility that the phonological grammar extends the restriction on identical consonants across the board, to items that fall beyond learners' training space.

Recall, however, that the definition of the training space intimately depends on the learner's representational scheme: An item that falls outside the training space of phonemes might be accommodated within the training space of features (see 8). It is indeed conceivable that some of the phonemes used in these experiments can be captured in terms of features that exist in Hebrew, and consequently, the generalization to such phonemes may not require Hebrew speakers to exceed their training space. For example, the phoneme /tʃ/ can be encoded as the combination of the features of /t/ and /ʃ/, phonemes that are each attested in Hebrew. Although it is doubtful that a feature-based account will correctly capture the restriction on Hebrew root structure (recall from section 5.3.3 that the restriction on identical consonants is inexplicable as a ban on identical features), it is nonetheless desirable to determine whether Hebrew speakers can extend identity restrictions even to phonemes that fall outside the space of Hebrew features.

One of the four phonemes used in these experiments, the coronal fricative /θ/, allows us to test this possibility. Coronal fricatives can be profitably captured in terms of the tongue tip constriction area – a feature that specifies the shape of the tongue's tip-blade on the cross-sectional dimension (Gafos, 1999). This feature takes three possible values. Two of those values – "narrow" (e.g., /s/, /z/) and "mid" (e.g., /ʃ/) are attested in native Hebrew phonemes, but the "wide" value, the place value of /θ/, is unattested in Hebrew, so this phoneme falls outside the training space of Hebrew, defined using either phonemes or features. And indeed, several observations suggest that *th* is particularly foreign for Hebrew speakers. First, *th* is never borrowed faithfully in loanword adaptation – it routinely assimilates into *t* (e.g., *termometer, termus, terapia*, for *thermometer, thermos, therapy*), whereas other foreign phonemes are borrowed faithfully (e.g., *check, job, walla*). Similarly, participants rate *th*-roots as less acceptable than roots with other foreign phonemes (Berent et al., 2002). These observations all suggest that *th* falls outside the Hebrew feature space. Remarkably, however, Hebrew speakers extended the restriction on identical consonants to *th*-type roots. Specifically, roots with initial identity (e.g., /θθk/) were significantly less acceptable than roots with final identity (e.g., /kθθ/) or controls (e.g., /θbk/), and this result obtained irrespective of the position of the root in the word

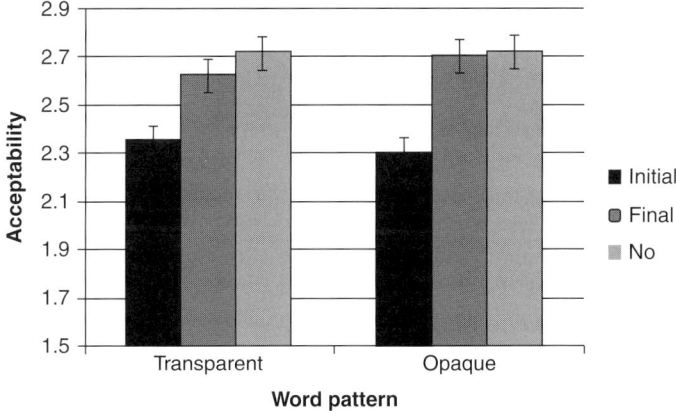

Figure 5.3 Rating result of novel roots generated from roots with the nonnative phoneme /θ/ (Data from Berent et al., 2002, Experiment 2). Error bars are confidence intervals constructed for the difference between the means.

(see Figure 5.3). Similarly, in the lexical decision task, novel words generated from the ill-formed θθ-initial roots (e.g., *θθk*) were classified more rapidly as nonwords relative to θθ-final (e.g., *kθθ*) controls. These results suggest that the restriction on identical consonants generalizes across the board.

5.4.3 Across-the-board generalizations require operations over variables

Let us summarize our conclusions so far. Our goal is to assess the computational properties of the phonological mind: whether the phonological grammar, specifically, has the capacity to encode algebraic relations that support across-the-board generalizations. Across-the-board generalizations naturally emanate from algebraic computational systems that operate over variables, so the demonstration of such generalizations could potentially show that the phonological mind possesses this computational machinery.

To determine whether the phonological grammar exhibits across-the-board generalizations, we systematically gauged generalizations of the identity function. We have shown that Hebrew speakers extend this restriction not only to any native Hebrew consonant, but also to nonnative consonants with nonnative features – items that arguably fall outside the representational space of the language. The capacity of Hebrew speakers to generalize identity restrictions across the board might suggest that the phonological grammar is algebraic.

Still, one link is critically missing from our argument. So far we have shown that (a) algebraic machinery that operates over variables supports across-the-board generalizations, and (b) the phonological grammar can exhibit such generalizations. However, it is not at all clear that across-the-board generalizations strictly *require* this algebraic machinery. Put differently, we have not shown that such generalizations cannot be attained in the absence of an algebraic grammar.

We now move to examine this question. To this end, we will inspect two distinct computational architectures that lack mechanisms to operate over variables. One case concerns a variety of connectionist architectures; a second example is offered by a constraint-induction model that uses a finite-state architecture. Although both architectures lack operations over variables, they have each been shown to capture many phonological generalizations. Of interest is whether these architectures can extend the identity relation beyond the training space.

5.4.3.1 Identity restrictions are unattainable by several popular connectionist networks

One of the strongest challenges to the view of phonology as an algebraic system is presented by associationist accounts of cognition. While all proponents of associationism reject the need for an algebraic architecture, their specific arguments differ in important details. The most common position outright discards the possibility that the mind exhibits across-the-board generalizations. In this view, phonological knowledge tracks only the statistical co-occurrence of specific instances. Other researchers, however, assert that associationist networks can effectively mimic the behavior of algebraic systems, including their ability to generalize across the board and to encode equivalence classes. Consider, for example, the representation of equivalence classes for consonants and vowels, discussed by Padraic Monaghan and Richard Shillcock (2003: 86):

> Apparently modular processing of vowels and consonants emerges from a neural network learning to represent phonemes in terms of phonological distinctive features.

Although these authors deny that such categories are hardwired, their view of them as "emergent" tacitly concedes that these entities do capture human behavior. Similar sentiments are expressed by James McClelland and Karalyn Patterson with respect to algebraic combinatorial principles – so-called "rules" (2002: 465):

> Characterizations of performance as 'rule-governed' are viewed as approximate descriptions of patterns of language use; no actual rules operate in the processing of language.

While these statements can be interpreted in multiple ways, one possible interpretation is that equivalence classes (e.g., "any consonant") and relations among those classes do effectively describe human behavior. But in order to manifest those capacities, a computational system does not have to be equipped with algebraic mechanisms "innately," in advance of learning.

These various objections to the algebraic hypothesis underscore two fundamental questions:
(a) Is the phonological grammar an algebraic system, endowed with the capacity to encode equivalence classes, variables, and operations over variables?
(b) Can the hallmarks of algebraic systems in (a) emerge in connectionist architectures as a result of learning the properties of specific instances?

While question (a) concerns descriptive adequacy – the ability of the model to accurately capture the properties of the phonological grammar, question (b) concerns essentially explanatory adequacy – namely, what allows a learner to acquire such generalizations in the course of language acquisition. The data presented so far suggest an affirmative answer to (a), but it is moot with respect to question (b). In particular, finding that the phonological grammar is algebraic leaves wide open the question of what properties must be specified in the grammar a priori, in advance of learning, in order for it to exhibit such generalizations. Does the capacity to operate over variables (i.e., rule-learning) have to be innately present, or could across-the-board generalizations emerge in connectionist systems that initially lacked algebraic machinery? Put differently, is the ability to learn algebraic rules (as opposed to learning any particular rule) learnable?

In a detailed set of analyses, Gary Marcus (2001; 1998) addresses this very question. As a case study, Marcus investigated the capacity of several popular connectionist networks to generalize the identity function (e.g., *a rose is a rose; a tulip is a ___?*). To this end, he first trained these networks on a set of items, and then examined their ability to generalize. His conclusions showed that these networks fall into two clear groups. One group of networks was able to properly generalize outside the training space, but these networks invariably implemented operations over variables. A second class of networks effectively lacked any means to encode variables. Although each of the networks in this second group was able to generalize the identity function to novel items, these generalizations were strictly limited in scope. In the absence of operations over variables, these models correctly generalized the identity function to novel items that shared features with training items (i.e., novel items within the training space), but they failed to properly generalize to items that fell outside the training space.

Strictly speaking, the failure of these networks to generalize outside the training space is an empirical problem, not a logical limitation. To see why, consider the case of a learner who is first trained on two instances (see 10). One instance maps 2 onto two instances of 2, and another maps 3 onto two instances of 3 (2→2,2; 3→3,3). Next, the learner is asked to extend this function to 5 (5→?). While 5→5,5, the identity relation, is one possible solution, it is by no means the only one that is logically warranted. Indeed, instances such as

108   How phonological patterns are assembled

2,2 and 3,3 can lead to numerous other inductive solutions. For example, since the two previous training outputs (2 and 3) differ from each other by one, learners could conclude that the answer to the test item should likewise differ from its predecessor output (3) by one, so the test item 5 should yield the output 4,4. Although the 5→4,4 generalization is certainly justifiable on logical grounds, it just happens to differ from the answer given by most humans (5→5,5). The failure of a network to exhibit this solution is therefore an empirical problem – an inability to capture the observed behavior of humans – rather than an inductive failure per se.

(10) Some inductive solutions to a simple regularity
   a. The induction problem
      Training:
      2→2, 2
      3→3, 3
      Test:
      5→?
   b. Generalization based on the identity function
      5→5, 5
   c. Generalization based on differences between successive output values (of each digit)
      Δ(output2– output1)=Δ (output3–output2)
      5→4 ,4

Marcus (2001; 1998) points out that the failure of such networks to generalize the identity function across the board stems from their training independence – the fact that their input nodes are independent from each other, and similarly, their output nodes are mutually independent. By definition, any item that falls outside the representational space of the network will activate a node that was not engaged during training. But because the operation of this node cannot be informed by knowledge acquired by the network as a whole, it is impossible to constrain the activation of that node to agree with that of trained nodes.

The example given in (11) illustrates this point. It describes a simple associative network that includes four nodes, trained to either turn on (a state of 1) or remain inactive (a state of 0). This network is trained on four items. An inspection of the representation of those training items shows that the state of each node is identical in the input and output – a case of an identity function. Note, however, that while each of the initial three nodes is always active in the input items (i.e., it has the value of 1), the fourth one invariably remains inactive (a state of 0). Of interest, here, is whether the network can overcome this bias and map the 1111 test item (with the fourth node on) to the 1111 output, as required by the identity function. But if learning for each input node is independent of the other input nodes (and similar independence holds among the output nodes), then the network will fail to infer the desired state of the fourth node from the activity of the other three. Marcus

(1998; 2001) demonstrates this problem in numerous connectionist networks (e.g., feed-forward networks, simple recurrent networks). Moreover, he shows that the failure to correctly map that fourth node is not merely due to its deprivation from training. Indeed, when the fourth node is first trained to stay on in some other function (unrelated to the identity mapping), the network still fails to generalize the identity function correctly. Marcus's results suggest that across-the-board generalization is not an emergent property of associative connectionist networks. It is important to keep in mind that "associationism" strictly refers to networks that lack the capacity to represent variables and operate on variables – it is not a synonym with connectionism generally. Indeed, Marcus goes to great lengths to show how variables can be incorporated in connectionist networks. Absent operations over variables, however, associationist connectionist networks will fail to generalize across the board, and this failure reflects a principled, unsurpassable limitation.

(11) The problem of training independence (from Marcus, 2001; 1998: 37)

Input nodes Training items:	Output nodes
1010	1010
0100	0100
1100	1100
1110	1110
0000	0000
Test item:	
1111	?

5.4.3.2 *Identity restrictions in Hayes and Wilson's Maxent model*

Like psychologists, linguists have also become increasingly interested in mechanisms that capture phonotactic knowledge by tracking the statistical co-occurrence of features (e.g., Adriaans & Kager, 2010; Albright, 2007; 2009; Cole, 2009; Coleman & Pierrehumbert, 1997; Frisch et al., 2004; Goldsmith, 2002). The possibility that such computational models can account for phonological competence, despite the elimination of variables, presents a second challenge to an algebraic account of the phonological grammar.

One influential exponent of this approach is the Maximum Entropy (Maxent) model, proposed by Hayes and Wilson (2008). This model represents phonotactic knowledge solely as induced restrictions on the co-occurrence of specific features. Despite this limitation, however, the model has been shown to account for numerous phonotactic generalizations, including generalizations to unattested structures. The success of this model would appear to suggest that, contrary to our conclusions so far, a full account of phonotactics might not require algebraic operations over variables.

To evaluate this possibility, my colleagues Colin Wilson, Gary Marcus, Doug Bemis, and I (Berent et al., 2012b) tested the ability of the Maxent model to generalize the restriction on identical Hebrew consonants. To this end, we first trained the model on a database including all productive trilateral Hebrew roots. Next, we evaluated the model's ability to generalize this knowledge to novel test items with identical consonants, either native Hebrew phonemes (e.g., *ssk* vs. *skk*) or foreign phonemes that are unattested in Hebrew (e.g., *θθk* vs. *kθθ* – Hebrew does not manifest the θ consonant).

Results showed that the Maxent model generalized quite well when test items comprised native Hebrew phonemes – items falling within this model's training space. But given roots with foreign phonemes and foreign features – items that fall outside its training space – the model's performance diverged from human behavior. While Hebrew speakers freely generalized the identity restriction to nonnative phonemes, the model failed to do so reliably. But once the model was revised so that it could represent segment identity, the revised model now generalized the identity restriction in a manner that matched the behavior of Hebrew speakers. The failure of the original Maxent model to generalize outside the training space in the absence of operations over variables, on the one hand, and the emergence of such generalizations in the revised model, equipped with such operations, on the other, establish that the capacity to extend phonological generalizations beyond the training space requires algebraic machinery, endowed with variables and operations over variables.

5.4.3.3 Conclusions

Hebrew speakers extend phonological generalizations across the board, to phonemes and features that fall beyond the representational space of their language. In contrast, computational models that lack operations over variables systematically fail to capture the full scope of human generalizations, and these limitations persist across different architectures, including various connectionist models and several variants of the Maxent model. Absent the capacity to operate over variables, these models generalized the identity function only to items falling within the training space, but failed to generalize beyond it. But once operations over variables were implemented, these models generalized across the board, to items falling either within the training space or outside it. The finding that humans extend the identity function across the board, and that such generalizations emerge only in models equipped with operations over variables, suggests that the phonological grammar is an algebraic system. This system is innately endowed with the capacity not only to form equivalence classes but also to represent abstract identity relations among such classes, akin to syntactic relations, such as recursion.

The vast scope of grammatical phonological generalizations stands in sharp contrast to the known limitations of many phonological categories. Earlier in this chapter, we noted that unlike the open-ended syntactic categories (*noun*), phonological categories such as *phoneme* are not only finite but also quite small. Recognizing this limitation on the building blocks of phonological patterns, many researchers assume that the phonological grammar (i.e., the combinatorial engine) must be likewise limited in its capacity to generalize. But as we pointed out, these arguments are inconclusive, as phonological and syntactic categories might differ for reasons related to the discriminability of their exemplars (Nowak & Krakauer, 1999). And indeed, our systematic investigation of phonological generalizations suggests that phonological principles are general in scope. So while phonology and syntax might differ in many ways (Bromberger & Halle, 1989), at their core they are nonetheless similar, inasmuch as they both share the algebraic machinery that allows, in principle, for discrete infinity.

5.5   Coda: on the role of lexical analogies

The results described so far strongly suggest that the phonological grammar encodes relations between classes using algebraic operations over variables that apply to any member of a class, existing or novel. For the identity rule, "a consonant is a consonant" – idiosyncratic differences among segments, familiarity, and phonetic aspects are ignored for the purpose of such generalizations. But these conclusions also leave us with a puzzle. While the evidence discussed in this chapter clearly demonstrates that the phonological mind is equipped with algebraic mechanisms that are blind to the idiosyncrasies of specific items, other evidence (reviewed in Chapter 3) shows that people are exquisitely sensitive to the statistical co-occurrence of instances in their linguistic experience (for review, see Pierrehumbert, 2001). For example, when given a novel verb (e.g., *spling*), English speakers are far more likely to assign it an irregular past tense if the verb is similar to familiar irregular verbs (e.g., *spling*→*splung*, analogized to *sting*→*stung*) compared to an unfamiliar test item (e.g., *nist*→*nust*; Prasada & Pinker, 1993). So it is patently clear that some linguistic generalizations are highly sensitive to statistical information. The dilemma, then, is what is the proper account of such generalizations.

Psychologists and linguists tend to take quite different positions on this question. Psychologists, on their part, have cited statistical effects as evidence against the existence of algebraic grammatical rules – a position I take to be untenable in view of the evidence cited in previous sections. Recognizing the role of algebraic generalizations, linguists, in contrast, have attempted to incorporate such generalizations into the grammar itself. Tacit in this approach is the assumption that the grammar is the sole home of linguistic productivity. Accordingly, the adequacy of a grammatical theory depends on its ability to fully account for *any* productive linguistic behavior.

All things being equal, an account that captures all linguistic generalizations using a single mechanism is certainly preferable to one requiring multiple mechanisms; however, there are nonetheless reasons to doubt that a single-route approach is feasible. More generally, it is unclear that all forms of linguistic productivity fall within the purview of the grammar. Because the grammar, as shown above, must be equipped with algebraic restrictions that are blind to the idiosyncratic properties of specific lexical items, it is difficult to see how this very same machinery might support generalizations that are tuned to lexical analogy.

But the problem gets even worse. It is conceivable that one could still uphold a single-mechanism grammatical account. But in order for this solution to work, phonological generalizations should fall into two complementary distributions: Some phenomena should respect only abstract algebraic structure (the ones falling within the purview of the grammar), whereas others might be guided by lexical analogy alone. Crucially, however, the two distributions should not overlap. Certainly, no single case can be *jointly* shaped by both algebraic and associative factors. But as we next show, such phenomena nonetheless exist, and their existence, I would suggest, requires that the burden of linguistic productivity be jointly shared by two sources – an algebraic phonological grammar and lexical analogies.

The evidence comes from one last twist on the *ssm–smm* asymmetry in Hebrew. Although *smm*-type roots are overwhelmingly far more numerous than their *ssm*-type counterparts, *ssm*-type roots are not entirely unattested. In fact, two counterexamples are quite productive and frequent, both beginning with the consonant *m* (e.g., *mimen* 'financed,' *mimesh* 'realized'). Of interest is whether the familiarity with such counterexamples has any bearing on the generalization of the identity restriction. To examine this question, one can compare generalizations to novel roots that resemble counterexamples (e.g., *mmk*, analogous to *mmn*) with non-analogous controls (e.g., *ggd*). If people only track algebraic relations among variables, then the acceptability of roots with initial identity should be independent of their similarity to counterexamples. But if generalizations are partly constrained by item-similarity, then novel roots that resemble counterexamples (e.g., *mmk*) should be better able to activate these exceptions from the mental lexicon (e.g., activate *mmn*), and consequently, those novel roots will be more acceptable than non-analogous controls.

Results suggest that the similarity to counterexamples does, in fact, modulate the acceptability of roots with initial identity (Berent et al., 2001a). Hebrew speakers rate words generated from novel roots that resemble familiar counterexamples (e.g., *mmg*, analogous to *mmn*) as acceptable as *gmm*-type controls.[3] This result obtained regardless of whether these roots were presented unaffixed (e.g., *mimeg*

---

[3] An apparent exception comes from the second word pattern, where *mmg*-type roots were still less acceptable than their *gmm*-type counterparts. Recall that conjugation of initial-identity roots in this pattern yields words with two consonants that are not separated by a vowel – forms that were

vs. *gimem*) or highly affixed (e.g., *hitmameg* vs. *hitgamem*), and irrespective of the precise method of the rating procedure (whether people rated each word in isolation, or directly compared *mmg*- and *gmm*-type items to each other).

The acceptability of *mmg*-type roots suggests that these roots activate counterexamples that are stored in the lexicon, and that the familiarity with such forms overshadows their grammatical unacceptability. Thus, not only is the restriction on root structure subject to counterexamples, but knowledge of these counterexamples is productive. The finding that the restriction on identical root consonants can give rise to two different types of generalizations – one guided only by algebraic structure and another led by lexical analogies – suggests that algebraic and associative mechanisms each play a distinct role in the phonological mind.

5.6 Conclusion

Chapters 4–5 examined the computational properties of the phonological system. We investigated whether the phonological mind is an algebraic system, or whether phonological generalizations only track the co-occurrence of specific instances in linguistic experience. To this end, we inspected the building blocks that define phonological patterns and the principles that govern their combinations. The findings demonstrated that the phonological grammar manifests all hallmarks of algebraic systems (see 12)). The evidence reviewed in Chapter 2 establishes that phonological systems operate on discrete elements, rather than on analog phonetic properties (Hayes, 1999). In Chapter 4, we demonstrated that categories such as *any consonant*, *any vowel*, and *syllable* are equivalence classes that support generalizations that apply equally to all class members. Moreover, in Chapter 5 we showed that the grammatical principles that operate on those classes have the capacity to reference a class by a variable, encode abstract relations among variables, and extend them across the board, even to novel phonemes and features. Taken together, these results establish that the phonological grammar is an algebraic system.

(12) Some features of algebraic systems
  a. Algebraic representations are discrete and combinatorial.
  b. Algebraic systems can represent abstract formal relations, such as identity (XX).
  c. Algebraic relations define equivalence classes that support across-the-board generalizations to any class member.

But while algebraic machinery is certainly necessary to capture phonological generalizations, it is not sufficient. A large body of psycholinguistic research

---

markedly less acceptable in all previous experiments (e.g., *mammigim*). The unacceptability of *mmd*-type roots in this pattern is likely due to the surface adjacency of the identical consonants in the word, rather than their root structure per se.

demonstrates that people are sensitive to various statistical aspects of their linguistic experience, and they productively generalize this knowledge to novel forms. The investigation of the restriction on root structure suggests that such lexical analogies counteract the effect of algebraic grammatical restrictions. Accordingly, the answer to our "*either-or*" question – whether the phonological system is algebraic *or* associative – is "both." Neither an algebraic grammar nor a statistical mechanism of lexical analogies is sufficient to capture the full range of phonological generalizations. Rather, the phonological mind is equipped with two mechanisms for linguistic productivity: an algebraic grammatical component and an associative lexicon. This account for phonological generalizations mirrors the dual-route architecture outlined by Steven Pinker (1999) informed by morphological evidence. While the discussion in the remainder of this book will focus on the grammar, it is important to keep in mind that the grammar is only one of the two engines of phonological productivity. Both sources appear to make independent contribution to phonological generalizations, and neither one can be subsumed by the other.

Beyond its theoretical significance, this conclusion also carries methodological implications. Computational accounts of phonology are typically judged by two yardsticks: their capacity to fully capture the statistical structure of the lexicon, and their ability to model productive linguistic generalizations – the greater the fit between the model and these facts, the more adequate the model (e.g., Adriaans & Kager, 2010; Albright, 2009; Coetzee & Pater, 2008; Hayes & Wilson, 2008). Underlying this approach is the tacit assumption that the grammar is the sole generator of linguistic forms – both attested lexical instances and potential ones – evident in productive generalizations. Consequently, an account of the grammar is solely entrusted with the vast task of explaining the distribution of all linguistic forms, and the adequacy of a theory of grammar is judged by its ability to capture these data. But if the grammar is not the sole source of productivity, then these criteria for evaluating computational models are skewed. Since some aspects of linguistic productivity originate from our associative memory system, then grammatical outputs will not converge with the fine-grained structure of the lexicon and experimental data. A dual-route theory that explicitly postulates two generative engines – an algebraic system and an associative memory one – is likely to offer a far better account of linguistic productivity.

Part III
Universal design: phonological universals and their role in individual grammars

6 Phonological universals: typological evidence and grammatical explanations

> This chapter begins to examine whether the phonological grammar is a system of core knowledge. Core knowledge systems manifest a unique, universal design that features a common set of representational primitives and combinatorial principles. Our question here is whether such design is characteristic of phonological systems. To address this question, I first show that across languages, certain structures are systematically preferred to others. I next demonstrate how such regularities are explained in one influential theory of universal grammar (Optimality Theory) and illustrate how grammatical universals can be reconciled with several challenges, including the diversity of phonological systems and their strong grounding in phonetic constraints. I show that diversity can result from numerous sources, ones that are either internal to the grammar or external – most notably, the phonetic characteristics of the language. Phonetic pressures, however, cannot subsume grammatical phonological principles. Indeed, licit phonological structures are not invariably ones that are phonetically optimal. Moreover, spoken languages share aspects of their design with signed languages despite their dramatic phonetic differences. I conclude that the linguistic evidence is amply consistent with the possibility that grammatical universals form part of the phonological mind. The next chapters proceed to test this hypothesis.

The previous chapters have demonstrated that the phonological grammar is an algebraic system – a system that encodes equivalence classes, represents their relations, and generalizes them across the board. While this algebraic machinery is quite powerful, it is not unique to phonology. Humans (and nonhuman species) exhibit broad algebraic generalizations in many areas, including the representation of number, navigation, and logical inferences. Although each such domain manifests distinct knowledge, these different "programs" all run on algebraic machinery. Accordingly, merely possessing algebraic machinery does not account for the various idiosyncratic properties of phonological systems outlined in Chapter 2. Algebraic machinery alone does not explain why every human language manifests a phonological system, why distinct phonological systems exhibit a common design, and why some design features are shared across modalities – spoken and signed. Algebraic machinery also cannot account for the spontaneous emergence of phonological systems in home signs and the Al-Sayyid Bedouin sign systems, nor does it explain why the properties of such systems differ from other means of auditory communication – human and nonhuman alike. Finally, the generic computational features of algebraic systems offer no explanation as to why the cultural invention of writing systems is based on phonological systems, why readers decode phonological structure in silent reading, and why dyslexia might be associated with phonological deficits. These various shared properties of phonological systems raise the possibility that the phonological grammar is not only algebraic but also specialized – it is a core system designed for the computation of phonological structure.

The key hallmark of a specialized pattern-maker is its unique design. If all phonological grammars were shaped by a single mind/brain system that is specialized for phonological computation, then one would expect languages to converge on a single design that includes a universal set of phonological primitives and a common set of principles that constrain their combinations. Such primitives and principles, moreover, should be specific to language and demonstrably distinct from domain-general constraints, such as the constraints on auditory perception, motor control, and generic properties related to communication. The second part of the book addresses this hypothesis. Chapter 6 describes some of the regularities in the distribution of phonological structures across languages and examines how these universals (and variation) can be captured by universal grammatical constraints. Subsequent chapters review psychological experiments that evaluate the role of grammatical universals, and assess their neural implementation and development in ontogeny and phylogeny.

Before we launch the discussion, a few warnings are in order. The material presented in this chapter is linguistically more complex than in other sections of this book. I make every effort to introduce new concepts in a gradual, systematic fashion, but some effort is required of nonlinguist readers. Professional phonologists, on the other hand, might find this introduction too basic, and wish for greater nuance and depth. While it is impossible to fully meet the goals of either group, I nonetheless hope to bridge the gap between the different disciplines, and in so doing, lay down the foundation for subsequent chapters. Readers can adjust their level of reading depending on their interests and needs.

6.1 Phonological universals in typology: primitives and combinatorial principles

To determine whether the phonological grammar is shaped by universal principles, one might begin by comparing the phonological forms that are attested in different languages. If all those patterns are the product of a common pattern-maker, then, all things being equal, diverse languages should exhibit shared phonological primitives and combinatorial principles. I emphasize "all things being equal" because they usually aren't. Grammars are hardly the sole determinant of phonological systems. Rather, language structure is shaped by multiple factors, including historical and social forces, and limitations imposed by human memory, auditory perception, and motor control. But while cross-linguistic regularities do not mirror grammatical factors alone, they nonetheless offer a reasonable starting point for the study of universal grammar.

Typological research estimates the distribution of linguistic structures across human languages by inspecting the statistical tendencies found in representative language samples. Such studies have revealed numerous regularities. A comprehensive exposition of the numerous typological universals observed in the literature goes beyond the scope of this chapter, but the examples listed below suffice to illustrate some important candidates of universal primitives (see 1) and combinatorial principles (see 2) in spoken languages.

Consider first the set of phonological primitives in (1). All phonological systems include phonemes comprising discrete phonological features that are distinct from their phonetic correlates (e.g., Keating, 1984). Recall, for example, that voicing, the feature that contrasts *bee* and *pea*, is phonetically realized by a continuous change in the onset of the vibration of the vocal cords, but at the phonological level, this acoustic continuum is represented categorically: Consonants are either voiced or voiceless, and intermediate distinctions are ignored by the phonological system despite the fact that people can readily encode them (e.g., Miller, 2001; Miller & Volaitis, 1989; Theodore & Miller, 2010). Features are organized hierarchically to form consonants and vowels – two distinct classes that contrast on their phonetic properties, their grammatical roles, and various characteristics of their acquisition and processing (Nespor et al., 2003). Consonants and vowels, in turn, give rise to syllables – prosodic units that are demonstrably distinct from lexical, morphological, and syntactic constituents. Indeed, syllable boundaries often diverge from word boundaries. Spanish speakers, for example, parse phonological strings into syllables that invariably begin with an onset, even when these segments belong to different words. Thus, *los otros* ('the others') is syllabified as *lo.so.tros*, such that the syllable *so* spans different words (Hayes, 2009: 257). This case clearly shows that a phonological unit – *so* – need not coincide with any word part.

While the documentation of primitives such as features, consonants and vowels, and syllables describes phonological systems at a very coarse level, the existence of these primitives is not trivial. Phonological systems that lack consonants and syllables are certainly conceivable, so the recurrence of those primitives across languages potentially speaks to constraints that are inherent to phonological systems.

(1) Universal primitives:
   a. Phonological features. All languages define their speech sounds in terms of a small set of features (e.g., Clements, 2005)
   b. Consonants and vowels. All languages have consonants and vowels (e.g., Blevins, 2006)
   c. Syllables. All languages have syllables (e.g., Hyman, 2008; 2011)

Languages also manifest regularities concerning the combination of phonological primitives. Consider, for example, the cross-linguistic preferences that govern the shape of syllables (see 2a). Languages differ greatly on the type of syllables that they allow. Using C and V to refer to consonants and vowels, we can describe some of the syllables attested across languages as CV, VC, and CVC (e.g., *ba*, *ab*, *lab*, respectively). But while each syllable type is tolerated by some human language, not all types are equally preferred. Citing Roman Jakobson, Alan Prince and Paul Smolensky note several syllable-structure preferences. Specifically, syllables including an onset (e.g., *ba*) are preferred to syllables that lack it (e.g., *a*): Every language allows syllables that begin with an onset, and no language disallows them. Thus, any language that tolerates the onsetless syllable *ab* will allow *ba*, but the reverse does not follow. Similarly, syllables that end with a coda (e.g., *lab*) are dispreferred to those that lack them (e.g., *la*): every language admits open syllables like *la*, but none requires syllables to end with a coda. Finally, syllables that begin with a complex onset, as in *bla*, are dispreferred to those with a simple onset (e.g., *ba*), and consequently, the presence of syllables like *bla* will imply the presence of syllables like *ba*.

(2) Universal combinatorial principles: some candidates
   a. *Syllable-structure asymmetries*:
      (i) Every language admits consonant-initial syllables (CV) and no language requires onsetless syllables (Jakobson, 1962; Prince & Smolensky, 1993/2004: 105).
      (ii) Every language admits open syllables and none requires codas (Jakobson, 1962 ; Prince & Smolensky, 1993/2004: 105).
      (iii) Any language allowing complex onsets and codas admits simple ones (Greenberg, 1978: Universal 3).
   b. *Syllable peak/margin hierarchies* (Prince & Smolensky, 1993/2004: Chapter 8; Smolensky, 2006):

(i) Margin hierarchy: Syllable margins (onsets and codas) favor segments of low sonority over segments of high sonority (x≻y indicates that x is preferred to y):
e.g., t≻ d≻ s≻ z≻ n≻ l≻ y≻a[1]

(ii) Peak hierarchy: Syllable peak favors segments of high sonority over segment of lower sonority:
e.g., a≻ y≻ l≻n≻ z≻ s≻ d≻ t

c. *Place of articulation hierarchy* (de Lacy, 2006; Prince & Smolensky, 1993/2004):
Coronal consonants (e.g., *d, t*) are preferred to labials (e.g., *b, p*), which, in turn, are preferred to dorsal segments (e.g., *k, g*):
e.g., *t,d* ≻ *p,b*≻*k,g*

These syllable-structure preferences illustrate an interesting property that holds true for many typological regularities. Given (minimally) two phonological variants that contrast on some phonological dimension (e.g., variant A and variant B), phonological processes often favor one of those variants over the other (A≻B, indicating a preference for A over B). Not only is A typically more frequent than B, but the infrequent B implies A: Every language that tolerates B also allows A. Moreover, the preference is asymmetrical: While B implies A, A does not imply B; that is, languages that allow the more frequent A variant do not necessarily allow the infrequent variant B (e.g., Hengeveld, 2006). The phenomenon of one variant asymmetrically implying another is known as an *implicational asymmetry*.

Implicational asymmetries govern not only syllable shape but also the likelihood of specific segments occupying certain syllable positions and arise from phonological alternations. Let us first consider the position of a segment in the syllable (see 2b). A syllable comprises three internal constituents – the onset and coda, which form the margins of the syllable, and the peak (or nucleus), which forms its core. In the syllable /bɪg/, the onset and coda correspond to the two consonants, *b* and *g*, respectively, whereas the peak is instantiated by the vowel /ɪ/. While margins – onsets and codas – are typically occupied by consonants whereas the peak often corresponds to a vowel, this generalization is not always true. The English word *apple* (/æpl/) demonstrates that the syllable can lack a vowel – here, it is the consonant *l* that occupies the peak of the final syllable. Other languages (e.g., Imdlawn Tashlhiyt Berber, Dell & Elmedlaoui, 1985) allow even obstruents to occupy this position (tFsi, 'untie, 3 feminine singular,' where F is the nucleus, Dell & Elmedlaoui, 1985).

On the face of it, the observation that a given segment (e.g., *l*) can appear at either the peak or margins (e.g., *l*ake, pi*ll* vs. *apple*) would seem to suggest that the behavior of the peak and margins is entirely lawless. But a closer inspection

---

[1] In what follows, I use the orthographic *y* to mark the glide /j/.

suggests some interesting regularities. The regularities concerning the syllabic role of segments are captured by their sonority (Clements, 1990) – a phonological property that roughly correlates with intensity (Parker, 2002; 2008): Highly sonorous segments, such as vowels, tend to be louder than less-sonorous consonants, such as stops. Sonority, as it turns out, is intimately linked to the function of a segment in the syllable. And while languages differ on whether they allow low-sonority consonants to function as peaks, all languages favor highly sonorous peaks (e.g., vowels) over low-sonority ones (e.g., stops). The margins, onsets and codas, show the opposite preference: Here, low-sonority segments are preferred to higher-sonority ones.

The sonority scale illustrates a typological regularity that is scalar, rather than categorical. Another important scalar regularity is the place of articulation hierarchy (see 2c). Across languages, coronals (e.g., *t, d, l*) are typically preferred to labials (e.g., *p, b, w*), which, in turn, are preferred to dorsals (e.g., *k, g*): Whenever dorsal and labial consonants are admitted in a language, so are coronals (Prince & Smolensky, 1993/2004: 219). Coronals are also preferred to labials in various phonological processes. Some phonological processes insert or delete segments in order to satisfy syllable-structure restrictions. For example, in a language that requires all syllables to have an onset (see 2a.i), vocalic-initial strings like *an* are ill formed, so the language might launch a phonological process that "repairs" such illicit inputs by inserting an initial consonant (e.g., *an*➔ *tan*). Similarly, a ban on codas (2a.ii) might promote a process that deletes the final consonant from the syllable (e.g., *bab*➔*ba*). Our interest here is whether certain places of articulation are more susceptible to those additions and deletions than others. If all places of articulations were equally preferred, then coronals and dorsals, for instance, would be equally likely to undergo insertion and deletion. This, however, is not the case in attested phonological systems. When an onset must be supplied, a coronal is much more likely to be inserted than a labial (*an*➔ *tan*, but not *an*➔*ban*, Smolensky, 2006: 39; for alternative explanations, see Blevins, 2008; Steriade, 2001). Conversely, when a coda must be deleted (e.g., in order to avoid a consonant cluster, as in *ab* +*da*➔*abda*), labials are more likely to be "sacrificed" relative to coronals (*ab* +*da*➔*ada*, not *aba*; Smolensky, 2006: 40).

The handful of examples listed in (1–2) provide only a brief illustration of the many generalizations reported in the literature. Such generalizations demonstrate that phonological systems manifest reliable preferences with respect to the type of primitives and principles they admit. These preferences, moreover, are not limited to the frequency of a variant across languages. Not only are certain elements systematically less likely to occur than others (both within and across languages), but they appear to be less natural, harder to perceive and produce, later to emerge in language acquisition, and more vulnerable to loss in aphasia (for review, see Rice, 2007). The question we address next is what is the source of those preferences.

6.2 Grammatical accounts for typological universals

Why are certain phonological structures systematically underrepresented across languages and disfavored in phonological alternations? And why do the same structures also appear to present greater demands for language acquisition, why are they harder to process, and why are they more vulnerable to loss in language deficits?

Several theories attribute those facts to universal constraints on the structure of the phonological grammar (e.g., Chomsky & Halle, 1968; Jakobson, 1968; Trubetzkoy, 1969). Optimality Theory (McCarthy & Prince, 1993; Prince & Smolensky, 1993/2004) is one such proposal. This theory is particularly attractive because it addresses both the regularities present across languages and their diversity. To capture the putative universality of phonological preferences, the theory asserts that the grammar includes a universal set of constraints on phonological well-formedness. These statements, known as *markedness constraints*, express bans on certain phonological structures. For example, the constraint NoCODA universally bans any syllable that ends with a consonant. Accordingly, closed syllables such as *at* (with the coda *t*) violate NoCODA, whereas open syllables such as *a* do not. Structures that violate a constraint are considered more marked: *at* (violating NoCODA), for instance, is more marked than *ta*. Markedness, then, is a relative notion: While *a* is less marked (i.e., better formed) than *at*, *a* is nonetheless worse-formed (i.e., more marked) than *ta* (which satisfies the additional constraint ONSET). And indeed, all constraints are violable, as marked syllables like *at* are amply attested.

The existence of such marked structures requires an explanation. Given that *at* and *pet* violate NoCODA, one would not expect such words to exist in the language. What, then, prevents words such as *pet* from extinction? And why do some languages (e.g., English) tolerate marked syllables like *pet* whereas others (e.g., Japanese) ban them? To address these questions, let us take another look at the Optimality Theoretic grammar.

The grammar, according to Optimality Theory, is a representational system: Its task is to construct the best-formed representation for an input. Consider, for example, the input *pet*. To find the best-formed representation for this input, the grammar will generate many potential representational candidates (i.e., outputs) for the input *pet*, evaluate them, and select the candidate that incurs the least severe violation of grammatical constraint: the optimal candidate (or "winner"). The number of such candidates, however, is not limited, nor must these candidates be faithful to the input. An input such as *pet* can be represented either faithfully as *pet*, or unfaithfully as *pe* (with the *t* deleted) or as *pe.ta* (where a vowel is appended, and the disyllabic output is syllabified). The tableau in (3) presents the input *pet*, lists some output candidates for this input, and indicates their status with respect to the NoCODA constraint. By convention, the input is listed on the top left corner, and the output candidates are listed below. As can be seen in the tableau, the faithful output *pet* violates NoCODA (a fact indicated by the asterisk next to it) whereas the two unfaithful candidates incur no such violation; hence, they are better formed.

So, if the grammar were only governed by markedness pressures, then marked structures such as *pet* could never win the competition – such inputs would invariably lose in favor of some less marked, unfaithful output (e.g., *pe*, *peta*). Accordingly, languages should have never exhibited marked forms such as *pet*.

(3) Some possible representations (i.e., output candidates) for the input *pet* and their markedness with respect to the NoCODA constraint

Input=*pet*	NoCODA
pet	*
pe	
pe.ta	

Since marked structures like *pet* are attested, they must be protected by some other grammatical force that counteracts markedness pressures. This other force, according to Optimality Theory, is *faithfulness constraints*. Faithfulness constraints ensure that the outputs generated by the grammar are identical to inputs. Thus, given the input *pet*, the faithful output is *pet*. Faithfulness to the input is enforced by means of various constraints. For example, the constraint PARSE bans the deletion of input segments. Specifically, PARSE requires all input segments to be parsed into a syllable position, as segments that are not assigned a syllable structure are effectively deleted from the phonetic form. For example, if the coda of *pet* remained unparsed, then the output would be *pe*, where the coda *t* is deleted. PARSE promotes faithfulness by banning such deletions. Another faithfulness constraint, FILL, will penalize unfaithful outputs like *pe.ta* – outputs where new segments (e.g., *a*) are added. The tableau in (4) lists those various candidates and indicates their status with respect to three (unranked) constraints – the markedness NoCODA constraint, as well as the two faithfulness restrictions, PARSE and FILL.

(4) Markedness and faithfulness violations by several outputs for the input *pet*

Input=*pet*	NoCODA	PARSE	FILL
pet	*		
pe		*	
pe.ta			*

Faithfulness constraints counteract markedness pressures, so they can potentially explain the existence of marked structures (e.g., *pet)*. At the same time, however, these new constraints also generate some additional complexity. Now that faithfulness is introduced, each of our three candidates violates a constraint, so no candidate is a "perfect" winner – free of any constraint violation. To decide between these various candidates, we must therefore consider how these various constraints are *ranked*. Generally speaking, the admissibility of marked forms depends on the balance of two forces: the markedness pressures against them, and the faithfulness forces that protect them from being surpassed by unfaithful, less marked candidates. If a given markedness constraint (e.g., NoCODA) is ranked below a faithfulness constraint (e.g., FILL), then the relevant marked output (e.g., an output with a coda) will be tolerated; if markedness outranks faithfulness, then that marked output will be banned. Crucially, languages can vary on this ranking. So while the two sets of constraints – markedness and faithfulness – are universally active in all grammars, languages can differ on how they rank those universal constraints. In this way, Optimality Theory captures cross-linguistic universals as well as the diversity found in human languages. In what follows, I illustrate this approach using three examples of markedness restrictions: the ban on codas, the preference for onsets, and the place of articulation hierarchy.

6.2.1   Codas are universally dispreferred, but they are sometimes tolerated

All grammars, according to Optimality Theory, include a universal markedness constraint that bans codas. Languages differ, however, on whether codas are tolerated. To illustrate this proposal, let us compare the admissibility of codas in two languages: English and Lardil – a Pama-Nyungan Australian language (Prince & Smolensky, 1993/2004: Chapter 7).

Let us consider English first. All things being equal, English would avoid codas. Indeed, given the word *coda*, English speakers would rather parse the middle consonant *d* as an onset (i.e., *co.da*, rather than *cod.a*). For the word *pet*, however, this solution is far more "expensive," as parsing the *t* as an onset (e.g., as *pe.t*) would yield an illicit syllable ("*t*," a syllable that lacks a nucleus), and this illicit syllable, in turn, will have to be "repaired" (e.g., by adding a vowel, *pet*➔ *pe.ta*). So ultimately, the re-parsing of the coda *t* as an onset would violate faithfulness restrictions. The existence of English words such as *pet* suggests that faithfulness constraints, such as FILL, outrank the ban on coda (NoCODA). The tableau in (5) indicates the ranking of these two constraints. The fact that the constraints are now ranked (rather than unranked, as in 4) is indicated by separating the constraint columns with a continuous line (note that in 4, the separating line is dotted). The ranking itself is indicated by the left-to-right ordering of the constraints – the higher-ranked FILL is listed left of NoCODA. The tableau also lists the input *pet* at the leftmost top corner, and two (of the many) possible candidates for this input (*pet* and *peta*). Note that each of these candidates violates one constraint (as before, constraint violation is indicated by an asterisk), but the violations differ in severity. In English, NoCODA is ranked below FILL, so the NoCODA violation of *pet* is less severe than the violation of FILL by *peta* (this fatal violation is indicated by!), and the faithful candidate *pet* prevails (this is indicated by the pointing finger).

(5) The tolerance of marked codas in English

Input=pet	FILL	NoCODA
☞ pet		*
pe.ta	*!	

Unlike English, Lardil places strict restrictions on codas like *pet̪*.[2] To maintain the *t̪* in the stem, short stems, such as *pet̪* ('to bite'; Wilkinson, 1988), are repaired by appending a vowel (e.g., *pet̪* ➔ *pe.t̪a*). The added nucleus opens up a new syllable, so the *t̪* can now be retained as an onset. The strict ban on Lardil codas is captured using the same constraints active in English, but in Lardil (see 6), NoCODA outranks FILL (note that NoCODA is now indicated left of FILL). Although the two candidates, *pet̪* and *pet̪a*, still violate the same constraints as the English *pet* and *peta*, the severity of the violation has changed. Since NoCODA now outranks FILL, the NoCODA violation by *pet̪* has become fatal, and this candidate loses in favor of the less marked (but unfaithful) *pet̪a*. In effect, the marked *pet̪* is automatically "repaired" by recoding it as the less marked *pet̪a*. Notice, however, that on this account, repair happens for purely grammatical reasons. Lardil speakers are not necessarily incapable of enunciating the sequence *pet̪*, nor are they incapable of hearing it. It is their grammar, rather than the audition and motor interfaces, that is responsible for the recoding of *pet̪* as *pet̪a* – a possibility that we further examine in subsequent chapters.

(6) The ban on codas in Lardil

pet̪	NoCODA	FILL
pet̪	*!	
☞ pe.t̪a		*

---

[2] Lardil bans codas unless their place of articulation is either shared with the following consonant or strictly coronal (excluding labials and velars even as secondary place of articulation). Because the final consonant in *pet* (a lamino-dental) is velarized, such codas are banned (Prince & Smolensky, 1993/2004).

To summarize, English and Lardil both share the universal dispreference for codas, as both grammars include a NoCODA constraint, but the languages differ on their tolerance of such forms due to their different ranking of NoCODA relative to faithfulness constraint FILL.

6.2.2 Onsets are universally preferred, but onset-less syllables are sometimes tolerated

Variation in constraint ranking can also explain universals and diversity concerning onsets. Spoken languages generally prefer syllables that have onsets to ones that lack them, but some languages, (e.g., English) tolerate onset-less syllables (e.g., the syllable /i/, e.g., *eBay*, *eel*), whereas others disallow onset-less syllables altogether. Optimality Theory attributes the universal preference of an onset to a markedness constraint that requires all syllables to have an onset – the constraint ONSET. Variation, in turn, is explained by the ranking of ONSET relative to faithfulness constraints.

Let us first examine languages in which onsets are mandatory. The requirement for a mandatory onset might be either absolute, or limited to certain morphological environments (e.g., at the juncture between morphemes). But once an onset is required, its slot will be filled even at the cost of violating faithfulness requirements. Axininca Campa (a language spoken in Peru, see 7) presents such a case – an example discussed in detail by Paul de Lacy. Axininca Campa forms words by attaching the vocalic suffix *i* to the root (e.g., i-N-√tʃʰik-i➔ intʃʰiki, "he will cut"; the root is indicated by √, and it is separated from the prefixes *i* and N). When the root ends with a vowel (e.g., *koma* 'paddle'), however, the complex form i-N-√koma-i would yield an onset-less syllable at the juncture between the root and suffix (* iŋ.ko.ma.i) – an environment in which onset-less syllables are disallowed in Axininca Campa. To repair such ill-formed outputs, the root's final vowel is separated from the vocalic suffix by appending a consonant (e.g., i-N-√koma-i➔iŋ.ko.ma.ti, 'he will paddle'; de Lacy, 2006: 88).

(7) The enforcement of an onset in Axininca Campa (examples from de Lacy, 2006: 88; √ indicates the root):
  a. Consonant-initial roots take the suffix *i* directly
     i-N-√tʃʰik-i➔ intʃʰiki, 'he will cut'
  b. Vowel-final roots are separated from the suffix by inserting a consonant
     i-N-√koma-i➔iŋkomati, 'he will paddle'
     i+N+√koma-ako-a:-i-ro➔iŋkomatakota:tiro, 'he will paddle for it again'

Unlike Axininca Campa, English freely allows onset-less syllables (e.g., *eat* /it/; *alter* /ɔl.tɚ/), though syllables with an onset are preferred. For this reason (along with the dislike of codas), *elate* is parsed as *e.late* (/ɪ.leɪt/), not *el.ate*, even though the syllables /ɪl/ (e.g., *ill*) and *ate* are otherwise allowed to occur. What needs to be explained, then, is why English tolerates onset-less syllables, such as /i/ (e.g., *either* /i.ðɚ/), and how it differs from Axininca Campa. The tableau in (8) captures these facts with the ranking of two constraints. The English tolerance of onset-less syllables reflects the ranking of the markedness constraint that demands an onset (ONSET) below faithfulness constraints such as FILL (a constraint that penalizes the insertion of material that is not present in the input). The obligatory onset in Axininca Campa is the result of the opposite ranking (for simplicity, we ignore the morphological restrictions on those required onsets). Here, the demand for an onset outranks faithfulness, so onset-less syllables can never emerge. In other words, the faithful onset-less candidate will always lose in favor of an unfaithful candidate that manifests an onset.

(8) Cross-linguistic variations in the enforcement of onset due to variation in constraint ranking
   a. Onset-less syllables tolerated (English)

/i/	FILL	ONSET
☞ /i/		*
/ti/	*!	

   b. Onsets are required (Axininca Campa)

/i/	ONSET	FILL
/i/	*!	
☞ /ti/		*

6.2.3 The place of articulation hierarchy

The markedness constraint ONSET demands that a syllable begin with a consonant. When this constraint is highly ranked, the missing onset must be supplied, and languages typically do so by calling upon some default segment. Our interest now turns to the nature of that segment. In the Axininca Campa example, the default segment is the coronal consonant *t* (e.g., i-N-koma-i→iŋ. ko.ma.ti). And indeed, across languages, coronal consonants are preferred to labials, and labials, in turn, are favored over dorsals (see 2c). Such preferences reveal another markedness constraint concerning the place of articulation of consonants. Like most other markedness constraints, the constraint is expressed as a negative statement – i.e., as a ban on marked consonants. But unlike the constraints reviewed so far, the place of articulation constraint is a scalar hierarchy, rather than a binary opposition. Highest (i.e., most marked) on this negative hierarchy is the ban on dorsals, which, in turn, is followed by the ban on labials; at the bottom is the ban against coronals (>> indicates constraint ranking, e.g., A>>B indicates that A outranks B). So when a language must come up with some onset consonant, many languages, including Axininca Campa, will choose the unmarked coronals over the more marked labials and dorsals.[3]

(9) The place of articulation markedness:
*dorsals (e.g., *k*, *g*) >>*labials (e.g., *p*, *w*) >> *coronals (e.g., *t*, *d*)

All things being equal, the coronals are preferred to labials. Nonetheless, the preference for the least marked place of articulation may be overridden by other constraints that are more highly ranked. Chamicuro (another aboriginal language spoken in Peru; de Lacy, 2006: 106–107) presents a case in point (see 10). Like Axininca Campa, Chamicuro separates adjacent vowels by inserting a default consonant. For example, while the prefix *a* typically precedes the stem directly (a+kamáni➔akamáni, 'we wash'), stems beginning with a vowel will be separated from the vowel prefix by inserting a consonant (e.g., a+i:la➔awi:la, 'our blood'). But somewhat surprisingly, the inserted segment is a labial (*w*), rather than the (typically less marked) coronal.

(10) Default consonant in Chamicuro
 a. Roots beginning with a consonant take the prefix *a* directly:
    a+kamáni➔akamáni, 'we wash'
 b. Roots beginning with a vowel are separated from the prefix by *w*:
    a+i:la➔awí:la, 'our blood'
    a+oʔti➔awóʔti, 'we give'
    a+eʃtihki➔aweʃtíhki, 'we tie up'

Paul De Lacy (2006: 106–107) explains this fact by calling attention to other constraints that determine the markedness of segments. And indeed, every segment comprises multiple features that can each be associated with distinct markedness constraints. While the place of articulation hierarchy will favor coronals, any given coronal segment also carries additional features, and these features could be independently penalized by other constraints. So, whether or not the coronal segment is ultimately selected will depend on the ranking of those constraints. In the case of Chamicuro, such constraints conspire to rule out

---

[3] The description of the place of articulation hierarchy here is somewhat simplified. Paul de Lacy (2006) demonstrates that the least marked consonants on the place of articulation hierarchy are glottal stops – coronals are only the second-best option. But because glottal stops are higher in sonority than coronal voiceless stops, and onsets generally disfavor sonorous consonants, Axininca Campa opts for coronal voiceless stop /t/ over the glottal /ʔ/ (de Lacy, 2006: 93). The preference for a low-sonority onset likewise bans the glottal approximant *h* in Chamicuro, discussed below (the remaining glottal, the stop ʔ, is banned by the demand for an approximant).

all available coronals: non-approximant coronal stops (e.g., *t*) are banned because Chamicuro requires the inserted consonant to agree with the following vowel on the approximant feature (a requirement enforced by the AGREE [approx] constraint), and the remaining non-approximant coronals (i.e., anterior palatal *y* and liquid *l*) are excluded by additional bans against anterior consonants (*[-anterior]) and liquids (*liquid). The winner, *w*, a labial, certainly does not carry the least marked place of articulation, but when all grammatical constraints are considered as a whole, *w* is nonetheless the least marked candidate available (see 11)).

(11) The choice of the default segment in Chamicuro (de Lacy, 2006: 106–107)

/a+i:la/	AGREE [approx]	*[-anterior]	*liquid	*labials
atila	*			
ayi:la		*		
ali:la			**!	
☞awi:la				*

To summarize, Optimality Theory captures both cross-linguistic universals and linguistic diversity via the ranking of a universal set of violable constraints. This proposal explains several of the typological regularities listed in the previous section. First, markedness accounts for the systematic underrepresentation of certain structures across languages. To the extent a structure violates universal grammatical constraints, it is less likely to be represented faithfully by the grammar. And since marked structures invariably incur a more severe constraint violation than unmarked counterparts, any language that tolerates marked structures is bound to tolerate fewer marked structures (Prince & Smolensky, 1993/2004, Chapter 9). So, in this account, the structure of the grammar gives rise to implicational asymmetries in typology. Finally, the theory captures the link between the multiple manifestations of markedness, both in the grammar and outside of it. Paul Smolensky (2006) shows how the various grammatical manifestations of markedness all result from a single source: Unmarked items incur a less severe violation of grammatical constraints. It is their potential to escape constraint violation that renders unmarked items "transparent" to phonological processes and more likely to serve as the default output of phonological processes. Conversely, marked structures are less likely to emerge as grammatical outputs, and consequently, they should also be harder to acquire and process, and less likely to be maintained in the face of language deficits, both acquired and congenital.

6.3 Non-grammatical explanations for language universals

While Optimality Theory attributes cross-linguistic regularities to universal grammatical constraints, other accounts have challenged this proposal. Some researchers worry that the stochastic, statistical nature of most cross-linguistic regularities is inconsistent with the existence of "language universals." To the extent one chooses to uphold the universal grammar hypothesis in the face of so many counterexamples, this hypothesis would be weakened to the point of becoming virtually impossible to falsify.

For example, Nicholas Evans and Stephen Levinson (2009: 429) note:

> Languages differ so fundamentally from one another at every level of description (sound, grammar, lexicon and meaning) that it is very hard to find any single structural property they share. The claims of Universal Grammar, we argue here, are empirically false, unfalsifiable, or misleading.

Even if different languages manifested similar preferences, it would not follow that those shared preferences originate from the design of their grammars. Indeed, structures that are preferred across languages are typically ones that are also easier to perceive and articulate. For example, in addition to being grammatically unmarked, the CV syllable – the syllable structure that is most prevalent across languages – also optimizes the simultaneous transmission of consonants and vowels (Mattingly, 1981; Wright, 2004). Accordingly, the cross-linguistic preference for CV syllables might result not from common grammatical principles but rather from generic properties of audition and motor control. These generic factors limit both the range of structures that can be produced and comprehended by humans at any given point in time and the transmission of those forms across generations. Structures that are difficult to perceive and produce will not be transmitted accurately, and consequently, their frequency will decline over time, both within languages and across them. The eradication of those "marked" structures, however, is not the product of any universal grammatical pressures. Rather, these facts are due to sources that are external to the grammar. The citations below all share this sentiment.

> Universal sound patterns must arise due to the universal constraints or tendencies of the human physiological mechanisms involved in speech production and perception. (Ohala, 1975: 289)

> All aspects of language structure reflect the principles of experience-and usage-dependence [sic]. (Bybee & McClelland, 2005: 385)

An important premise, related to this general approach, distinguishes Evolutionary Phonology from Generative Phonology (Chomsky and Halle 1968; Kenstowicz and Kisseberth 1979; Kenstowicz 1994; to appear), and Optimality Theory (Prince and Smolensky 1993; Kager 1999; McCarthy 2002). The premise is that principled extra-phonological explanations for sound patterns have priority over competing phonological explanations unless independent evidence demonstrates that a purely phonological account is warranted. (Blevins, 2006: 124)

The next sections address these challenges. At this point, we will not attempt to demonstrate that grammatical universals exist – doing so will require further evidence considered only in subsequent chapters. Rather, our goal here is to merely show that the hypothesis of universal grammar is in principle tenable in the face of those challenges. We first consider how grammatical universals can be reconciled with language diversity. Next, we examine whether phonological universals can be fully captured by functional forces that are external to the phonological grammar.

6.4 Why are phonological universals non-absolute?

Let us first consider the challenge from diversity. Although some authors assert that the hypothesis of phonological universals is countered by the known diversity of human language, grammatical universals are not incompatible with cross-linguistic diversity. In fact, even if grammatical universals existed, language diversity would still be expected for three reasons. First, language diversity is partly shaped by factors that are external to the grammars. A second reason for diversity stems from the formal properties of grammars. Finally, we consider the possibility that language diversity might result from variation in phonetic substance that triggers the expression of a putative invariant genotype in specific grammars – its phenotypes.

6.4.1 Non-grammatical sources for language diversity

Many researchers believe that the diversity of human languages is inconsistent with grammatical universals. In their view, language diversity indicates that the grammars of different languages vary in an unconstrained manner, a possibility that directly counters the hypothesis that grammatical constraints are universal. Underlying this line of reasoning is the assumption that linguistic diversity reflects variation in the internal organization of the grammar itself. This, however, is not the only possible explanation for the diversity of human languages. Indeed, languages could also diverge for reasons that are unrelated to the grammar. While such "external" non-grammatical forces can engender considerable diversity across language, they do not rule out the possibility that the internal design of the grammar is universal. Here, we consider several such non-grammatical sources.

The most basic external engine of diversity is the need for varied communication. Phonological forms are ultimately designed to differentiate between words, and this creates an inherent conflict between markedness and expressive power. If phonological forms abided only by markedness pressures, then languages would have theoretically converged on a very small set of unmarked phonological forms. Such languages might allow the syllable ʔa, and perhaps even ta (syllables with no codas and onsets comprising the least marked places of articulation), but no *blogs*, let alone a *constitution*. While such systems are phonologically well designed, they would make for awfully limited conversations. Increasing the range of lexical distinctions requires a larger set of phonological forms, and this necessarily introduces phonological markedness and diversity. Once we recognize that marked structures must occur, it is not surprising that languages will vary with respect to the type of structures that they admit. Social and historical factors that favor certain communities over others will give rise to the predominance of their languages. Similarly, functional reasons will render certain structures easier to perceive, produce, and remember, and the effect of such functional pressures might further vary according to the communication settings that are typical for distinct linguistic groups. While spoken syllables are readily perceived by talkers sitting in close proximity, hunters who communicate over large distances might be better served by whistles, and indeed the Pirahã, an Amazonian group of hunter-gatherers, use whistle speech as one of their means of linguistic communication (Everett, 2008).

Social, historical, and functional pressures skew the distribution of linguistic structures spoken across languages for reasons that are external to the grammar. But typological universals are not synonyms with grammatical universals. Grammatical universals are the properties of an internal language system in the mind and brain of individual speakers – the so-called I-language (Chomsky, 1972). Language typology, in contrast, surveys external phonological objects – it describes (at various levels of abstraction) the chunks of sound produced by people, and as such, typology concerns itself with E-phonology (external phonology). Because E-phonology is shaped by multiple forces that are external to the grammar, language typology offers only an indirect window into I-phonology. The absence of absolute universals in the typology therefore does not negate the existence of grammatical phonological universals.

6.4.2 *Grammatical mechanisms that support diversity*

A second reason for language diversity concerns the properties of I-language itself. Critiques of the universal grammar hypothesis tacitly assume that universal grammatical constraints could not possibly give rise to phonological diversity. Optimality Theory, however, asserts that grammatical constraints are violable, and consequently, universal grammatical constraints can nonetheless result in considerable variation. The discussion in section 6.2 has already mentioned two sources of diversity that are due to variation in constraint ranking. One case concerned variation in the ranking of markedness constraints relative to faithfulness constraints (see 12a) – the tolerance of codas in English (see 5), but not in Lardil (see 6), was one such case. A second source of language diversity stems from the ranking of distinct markedness constraints – the cross-linguistic diversity with respect to the choice of a default consonant illustrates that second case (see 12b).

(12) Diversity resulting from variation in constraint ranking
    a. Variation in the ranking of markedness relative to faithfulness constraints:
        Language 1: $M_A \gg F_A$
        Language 2: $F_A \gg M_A$
    b. Variation in the ranking of distinct markedness constraints:
        Language 1: $M_A \gg M_B$
        Language 2: $M_B \gg M_A$

A third source of variation is the conflation of constraints. Certain markedness restrictions are expressed as scalar hierarchies. Such hierarchies indicate that some structure A is more marked than some other structure B, which in turn is more marked than a third structure C, etc. (see 13a). The sonority and place of articulation hierarchies (2b) and (2c) were two such cases. Although, by hypothesis, markedness hierarchies can never be reversed (e.g., A can never become less marked than C), these hierarchies may nonetheless be subject to some cross-linguistic variation. While some languages might obey the full hierarchy, distinguishing A, B, and C (13a), others might fail to differentiate among certain regions of the hierarchies. These languages, for example, will treat both A and B as more marked than C, but fail to distinguish between A and B (see 13b.i). Other languages might rank A above both B and C, but treat the last two structures alike (see 13b.ii).

(13) Variation due to the conflation of markedness hierarchies
    a. A basic full hierarchy:
        *A $\gg$ *B $\gg$ *C
    b. Partially conflated hierarchies:
        (i) *{A, B} $\gg$ *C
        (ii) *A $\gg$ *{B, C}
    c. An impossible conflation of a markedness hierarchy:
        (i) *{A, C} $\gg$ *B

Paul de Lacy (2004; 2006; 2007) attributes such phenomena to the conflation of certain regions of the markedness hierarchy (e.g., a conflation of the *A $\gg$ *B region). Conflation, however, is not a magical panacea designed to eradicate any counter-evidence to markedness and render the markedness hypothesis unfalsifiable. Conflation can only erase markedness distinctions among adjacent levels on a hierarchy (e.g., A and B; B and C). Nonadjacent regions (e.g., A and C) cannot be conflated, and consequently, conflation (e.g., of A and C) can never reverse the markedness hierarchy (e.g., see 13c). In a similar vein, variation in constraint ranking does not render Optimality Theory unfalsifiable. While variation in constraint ranking and constraint conflation can promote considerable diversity, certain phenomena are never expected to occur. For example, because the ONSET constraint is a positive condition that requires onsets, variation in constraint ranking can never give rise to a language that bans onsets. The mechanisms of conflation and constraint ranking thus offer a principled, falsifiable explanation for the coexistence of certain cross-linguistic diversity along with grammatical universals.

6.4.3   Phonetic sources of diversity: the role of phonetic triggers

A third reason for phonological diversity is the variation in the phonetic substance of different languages (e.g., Boersma & Hamann, 2008; Hayes et al., 2004a; Steriade, 2001 and chapters therein). In this view, the grammatical constraints seen in individual grammars are not independent of experience. Rather, they are triggered by some critical phonetic cues that inform the configuration of the phonological grammar. While a required cue might be present in most languages – a fact that would render the relevant phonological constraint nearly universal – some languages might lack it, and consequently, the constraints encoded in such grammars might diverge from the ones found in most other languages.

Consider, for example, the preference for syllables to begin with an onset. Across languages, syllables that include an onset (e.g., *ba*) are far preferred to onset-less syllables (e.g., *ab*) – a fact attributed to a grammatical constraint that requires all syllables to have an onset. While Optimality Theory asserts that this constraint is universal, it remains agnostic on how it emerges in the individual grammar – whether its encoding in specific grammars is guided by innate linguistic knowledge, and whether the expression of such innate knowledge can be modified by experience (Smolensky & Legendre, 2006). The phonetic triggering hypothesis addresses this question. In this view, language learners encode phonological constraints in their grammars as a result of experience with specific phonetic triggers. The ONSET constraint, specifically, might be triggered by salient phonetic cues present at the consonant's right edge (Breen & Pensalfini, 2001; Wright, 2004). Because the transition between the consonant and the following vowel often carries salient phonetic cues, the syllabification of the consonant with the following vowel (e.g., *a.ba*) optimizes consonant identification relative to alternative onset-less parses (e.g., *ab.a*). Thus, the salience of consonant release might function as a trigger for the encoding of ONSET constraint in the phonological grammar. To the extent that most languages manifest this cue, the ONSET constraint will become prevalent.

But when a critical phonetic trigger is absent, language learners will fail to encode this constraint, and diversity will ensue.

Whether the specific ONSET constraint could, in fact, be absent in a human language (due to either phonetic variations or some other reasons) remains controversial (Breen & Pensalfini, 2001; cf. Berry, 1998; McCarthy & Prince, 1986; Nevins, 2009; Smith, 2005, for opposing views), so until shown otherwise, I will continue to refer to this and other grammatical constraints as universal. There is likewise no firm evidence that phonetic cues cause the encoding of phonological constraints – the best available evidence shows only a correlation between phonetic and phonological factors (see Box 6.1). Nonetheless phonetic triggering is a hypothesis of great significance that merits close evaluation. To the extent that this hypothesis is supported, however, it will not challenge the view of the phonological grammars as a system of core knowledge, shaped by innate phonological knowledge.

At its core, the hypothesis that humans possess innate, universal knowledge of phonology is a hypothesis about the human genotype. In contrast, the constraints attested in any particular grammar (e.g., ONSET, in the English grammar) are the phenotypic expression of that putative genotype. Phenotypic expressions, however, are known to vary depending on triggering conditions. The biologist Evan Balaban, for example, underscores this fact by demonstrating how the shape of genetically identical Achillea plants can change quite dramatically according to their geographic location (Balaban, 2006). Extending this reasoning to language, there is every reason to believe that the expression of putatively innate phonological knowledge in individual grammars (e.g., as the constraint ONSET) may be modulated by details of the phonetic experience available to learners of distinct languages. Triggering, however, does not amount to learning. Constraints such as ONSET, in this view, do not result from phonological induction informed by the preponderance of CV syllables. It is merely the phonetic properties of CV syllables (e.g., with the burst associated with the release of stops), not their frequency, that presumably trigger ONSET. No known mechanism of inferential learning explains how people make the leap from phonetic experience to phonological knowledge. The hypothesis of core phonological knowledge provides the missing link by proposing innate biases that are universally shared by all humans. To the extent grammars do manifest shared design, the postulation of such innate constraints remains necessary irrespective of whether this design is phonetically triggered.

To summarize this section, typological regularities (E-language) are shaped by many factors, so language diversity does not necessarily show diversity in I-language. Nonetheless, it is possible for grammars to vary even if their design is innately constrained. One source of diversity is variation in constraint ranking and conflation. In addition, constraints that are putatively universal could conceivably vary due to variation in phonetic triggers. Accordingly, the

Box 6.1 The effect of phonetic variation on phonology: evidence from tone (Zhang, 2004)

Jie Zhang (2004) offers an interesting illustration of the link between cross-linguistic variation in fine phonetic detail and the propensity of syllables to carry tones. Many languages use tone as a phonological feature that contrasts among words' meanings. For example, the Thai word /nǎ:/ (with a rising tone) means 'thick,' whereas /ná:/ (with a high tone) indicates 'aunt' or 'uncle.' While languages assign tone to different types of syllables, some syllables are more likely to carry tones than others. Long vowels (such as /a:/) are more likely to bear tone than shorter vowels (e.g., /a/). Similarly, syllables with longer sonorant codas (e.g., /tan/) are more potent tone bearers than shorter ones, ending with an obstruent (e.g., /naat/). Crucially, on Zhang's account, the effect of these two factors – vowel length and coda sonority – depends not on their discrete, digital phonemic representation (e.g., the binary distinction between tense or lax vowels), but rather on their analog phonetic realization, operationalized as their actual duration (e.g., in milliseconds).

Consider first the phonological distinction between syllables that bear tone freely and those that do not. Thai CVVO syllables (where C is a consonant, V is a vowel and O is an obstruent) manifest a long vowel, yet they are restricted in their tone-bearing capacity: they generally take High-Low (HL) and Low (L) tones, but rarely exhibit the High and Low-High tones. Closed Thai syllables, however, can nonetheless freely support all tones if they contain a sonorous coda, i.e., when their rhyme is sonorous. And indeed, the CVR (R=sonorant) and CVVR both support all five tones. Thus, in Thai, merely having a phonologically long vowel is insufficient to freely support all tones, whereas having a sonorant consonant does allow a syllable to carry all tones freely.

(14) The effect of vowel length and coda sonority on the capacity of a syllable to bear tone: Standard Thai vs. Navajo

	Is a phonologically long vowel sufficient to license all tones?	Is the sonority of the coda sufficient to license all tones?
Standard Thai	No	Yes
Navajo	Yes	No

Interestingly, Navajo presents the mirror image case (14). In Navajo, syllables with long vowels can freely carry all tones, whereas those that have merely a sonorant rhyme (in the CVR syllable) do not bear all tones freely. Thus, in Navajo, having a long vowel is both necessary and sufficient to allow a syllable to freely bear tones, whereas having a sonorant rhyme is not.

Why do these two languages dissociate in this manner? Why does a given phonological property (e.g., long vowel) license a syllable to carry all tones only in one of these languages, but not in another? Zhang suggests that the answer can be found in the phonetic properties of those two languages. Specifically, the distinct durations of syllables and their constituents – long vowels and rhymes (see Table 6.1). In Standard Thai, phonologically long vowels are quite short phonetically, and the overall duration of the rhyme in CVVO syllables is likewise short – far shorter than the rhymes of syllables that are closed by a sonorant (e.g., CVR) or open CV syllables. In contrast, Navajo vowels that are phonologically long are also phonetically long, and for this reason, vowel length is sufficient to allow syllables with phonologically long vowels to carry all tones freely, irrespective of the duration of the sonorant coda.

While the link between these phonetic properties and the phonological structure of these two languages is intriguing, the interpretation of these facts requires some caution. Because these observations only outline a correlation between phonetic and phonological facts, they cannot support any strong claims concerning causation: We cannot tell whether the grammatical constraints acquired by the language learner depend on allophonic variation, or some other third factor. Certainly, these observations do not demonstrate that speakers determine the well-formedness of syllables by inspecting their acoustic duration – a proposal that would obviate algebraic grammatical constraints. Nonetheless, the results do open up the possibility that phonetic factors might inform the ranking of grammatical constraints, and as such, contribute to cross-linguistic diversity.

Table 6.1 *Tone-bearing capacity of syllables in Standard Thai and Navajo as a function of the duration of the nucleus, coda, and rhyme (in ms)*

	Standard Thai			Navajo		
	Nucleus	Coda	Rhyme	Nucleus	Coda	Rhyme
CV	447		447	122		122
CVR	160	264	424	152	167	319
CVV				314		314
CVVR	308	213	521	298	160	458
CVO	144		144			
CVVO	315		315			

Note: Cells corresponding to syllables that are restricted tone-bearers are shaded; cells corresponding to syllables that freely carry all tones are clear. C=consonant; V=vowel; R=sonorant; data from Zhang, 2004.

diversity of phonological systems is not inconsistent with the hypothesis of universal grammatical constraints.

6.5 Algebraic, phonological universals are autonomous from phonetic pressures

While the results reviewed so far are all logically consistent with the possibility of universal phonological constraints, some of these observations might lead one to doubt the necessity of a phonological explanation. The phonetic triggering hypothesis, discussed in the previous section, already concedes that phonetic factors can possibly shape the phonological grammar in ontogeny by triggering the encoding of grammatical constraints and their ranking. But if phonetic factors constrain the organization of individual grammars, then why should we assume that innate grammatical universals play an independent role? Why cannot phonetic factors also carry the full burden of linguistic productivity – of explaining the human capacity for across-the-board linguistic generalizations? In other words, could phonetic factors subsume the putative effect of grammatical universals altogether?

The possibility that phonetics subsumes phonology can acquire multiple forms, ranging from denial of the phonological grammar altogether to the view of phonology as an "*emergent property*" of the phonetic system. Proponents of emergence rarely articulate how, precisely, emergence happens, but roughly speaking, this view assumes that a phonological grammar can spontaneously arise in human brains in the absence of any innate phonological biases per se. To the extent phonology is not innate (in the sense discussed in Chapter 3), it must therefore be learned by mechanisms that use experience as evidence (e.g., statistical learning, analogical reasoning, and the like).

These two rival positions – the hypothesis that phonetics can subsume the phonological grammar and the "grounded phonology" alternative (that an autonomous phonological grammar is only grounded in phonetics) – are listed in (15),[4] along with the third logical alternative – the possibility that the emergence of the phonological grammar in ontogeny is unaffected by phonetic substance. This could occur either because the phonology–phonetics link was only significant in phylogeny, or because phonological universals are truly arbitrary, bearing no relation to functional considerations.

(15) The contribution of phonetic cues and universal grammatical constraints to the organization of specific phonological grammars (e.g., the grammar of English) in ontogeny

---

[4] For the sake of simplicity, the causal chain in (15) captures only the interaction between universal grammar and phonetic cues. By focusing on phonetic triggers, however, I do not wish to suggest that they alone are sufficient to account for universal grammar.

a. *No grammatical phonological universals*:
   Phonetic cues➔ $G_{English}$
b. *Grammatical universals grounded in phonetics*:
   {Universal grammar+Phonetic triggers}➔ $G_{English}$
c. *Arbitrary grammatical universals*:
   Universal grammar➔ $G_{English}$

While I believe there is ample evidence to suggest that grammatical universals are grounded in phonetics (e.g., see Box 6.1), and that phonetics might well play a critical role in the development of the phonological grammar in both ontogeny and phylogeny, it is doubtful that the phonological grammar can emerge from the phonetic system spontaneously, in the absence of innate biases. Indeed, absent a specific computational account of how phonological grammars emerge, "emergence" per se hardly constitutes a scientific hypothesis. But beyond its vagueness, there are also some specific reasons to question this possibility.

One argument against emergence is presented by the distinct computational properties of phonetics and phonology. As shown in previous chapters, the phonological grammar is an algebraic system, capable of across-the-board generalizations, whereas phonetics relies on analog, continuous representations. Not only are these computational mechanisms quite distinct, but there is no evidence that algebraic generalizations can arise from an analog phonetic component. The known failure of operations over variables to emerge in associative systems (Marcus, 2001) gives some good reasons to doubt that phonetic emergence is even remotely trivial. Accordingly, the spontaneous emergence of phonology from phonetics remains highly speculative.

A second argument against the view of the phonological grammar as spontaneously emergent from phonetics is presented by the observation that some phonological constraints are phonetically unmotivated, whereas others are amodal – shared by both spoken and signed languages. The following sections lay out the evidence. These observations suggest that algebraic grammatical universals cannot be reduced to analog functional pressures. As such, grammatical universals must be autonomous (i.e., distinct) from the phonetic component.

6.5.1   Phonological constraints in spoken languages are autonomous from phonetic factors

The possibility that phonological constraints are autonomous from analog phonetic pressures is consistent with a large body of work (e.g., Kiparsky, 2008; Pierrehumbert, 1975; Zsiga, 2000). Earlier in the book, we reviewed in detail one relevant case from Egyptian Arabic (Hayes, 1999). We showed that this language disallows voiceless stop geminates (e.g., *appa*) despite the fact that they are easier to produce than their attested voiced counterparts (e.g., *abba*), and consequently, this ban must form part of the algebraic phonological component, rather than the analog phonetic system. Indeed, many phonetically plausible alternations fail to occur in phonological systems, whereas some alternations that are attested across languages lack phonetic motivation. An example suggested by Paul de Lacy and John Kingston (2006) further illustrates this point.

The specific case concerns the relationship between historical language change and grammatical universals. Many historical changes result in phonetic simplification: Structures that are phonetically challenging are replaced by ones that are easier to perceive and produce. For example, while the phoneme *t* frequently changes to *k* (e.g.,*tina➔kiŋa,'mother') in Luangiua and many other Austronesian languages (Blust, 2004)), the reverse (a change from *k* to *t*) is far less frequent. And indeed, several phonetic factors favor *k* over *t*. Because *k* is marked by a long voice onset time and high amplitude, it might be a particularly good exemplar of a stop consonant (specifically, of consonants with long voice onset time, Blevins, 2006). In addition, the velar place of articulation is less distinct, as its articulation and acoustics vary considerably depending on the following vowel (far more so than *t*; de Lacy & Kingston, 2006). For these reasons, *k* might form a good substitute for *t*. Recognizing these facts, several linguists have suggested that language change is the result of "innocent" phonetic confusions (Blevins, 2006; Ohala, 1990). In this view, *t* changes to *k* because hearers confuse the phonetic properties of *t* with *k*. And since diachronic change ultimately shapes the regularities found synchronically across languages, these observations open up the possibility that synchronic phonological processes are governed only by phonetic pressures, rather than by universal phonological principles.

De Lacy and Kingston demonstrate that this possibility is unlikely. If synchronic alternations resulted only from phonetic pressures – the very same forces that shape historic language change, then synchronic grammars would invariably have favored *k* over *t*. But this prediction is countered by several phonological processes. Consider, for example, consonant epenthesis – a phonological alternation that inserts a consonant in the input (e.g., to supply a missing onset, see section 6.2). Although phonological epenthesis is quite frequent, *k* is actually the least likely obstruent to undergo insertion. This fact is unexpected on functional grounds, but it is fully predicted by the markedness hierarchy of place of articulation, discussed earlier (see 9). Because velar consonants are the most marked on this scale, they should be least likely to be selected as the output of phonological processes.[5] The dissociation between

---

[5] In her reply to de Lacy & Kingston, Juliette Blevins (2006) points out that *k*-epenthesis does occur in Mongolian, Maru and Land Dayak, citing Vaux (2002) as a source. An inspection of the facts described in that source, however, does not make it clear whether the insertion of *k* is in fact

phonetic and phonological processes suggests that at least some phonological processes are autonomous from phonetic factors.

6.5.2 Phonological universals across modalities

Another way to dissociate the effect of putative grammatical universals from phonetic properties is to compare languages across modalities. Given that many phonological constraints are grounded in phonetic substance, we do not expect to find many phonological properties shared across modalities – there are indeed many important differences between spoken and signed phonologies, and such disparities should not be overlooked (Sandler & Lillo-Martin, 2006; van der Hulst, 2000). But to the extent some similarities are found, they are nonetheless significant, as they suggest that some aspects of phonology might be autonomous from the phonetic component.

The evidence presented below shows that some phonological primitives are shared not only across spoken languages but also across signed ones. Markedness, moreover, appears to constrain the phonological grammar in both modalities, and there are also some specific markedness constraints that might be shared across modalities.

6.5.2.1 Shared primitives

All phonological systems, signed and spoken, encode phonological features that are grouped hierarchically into larger binary units – either consonants or vowels, in spoken language, or hand location and movement, in sign languages – and these units, in turn, give rise to syllables (Brentari, 1998; Perlmutter, 1992; Sandler, 2008; van der Hulst, 2000). In both modalities, syllables are meaningless phonological constituents that are distinct from morphemes – the constituents of word formation – or entire words (Sandler & Lillo-Martin, 2006). Indeed, monomorphemic words can include either one syllable or two, whereas bimorphemic words can be either monosyllabic or disyllabic.

In the case of spoken language, the distinction between morphemes and syllables is patent: *box* and *rocks* are phonologically similar forms that consist of a single syllable, but *box* comprises a single morpheme whereas *rocks* has two – the base *rock* and plural suffix *s* (see Table 6.2).

In a similar fashion, one can also distinguish syllables and morphemes in sign languages. Just as spoken syllables are units of phonological structure that must include a nucleus (typically, a vowel – a highly sonorous segment), so signed

> preferred – that is, whether *k* is *more* likely to be inserted than other consonants. Rebecca Morely (2008), however, shows that Buryat (an Altaic language spoken in Siberia, near the Mongolian border) presents some clear cases of velar epenthesis (but for an alternative analysis, see de Lacy, 2006: 139–142 and de Lacy & Kingston, 2006). The status of *k*-epenthesis thus requires further research.

Table 6.2 *The distinction between syllable structure and morphological structure in spoken language*

		Morphemes	
		One	Two
Syllables	One	box	rocks
	Two	pencil	undo

syllables are phonological units that must minimally include one (and only one) sequential movement (multiple movements can occur in a single syllable only if they are produced simultaneously; Brentari, 1998). Because syllables are defined on purely phonological grounds, the number of syllables and morphemes can differ in manual signs just as they differ in spoken words (Figure 6.1). The signs for MOTHER and WIFE in American Sign Language are both monosyllabic (as they comprise a single movement) despite their difference in morphological complexity – MOTHER includes a single morpheme whereas WIFE results from the combination of two distinct morphemes – one for 'mother' and one for 'marry.' Conversely, MOTHER and FARM are both monomorphemic, but the former is monosyllabic (with one movement) whereas the other is disyllabic (with two movements). Moreover the distinction between syllables and morphemes forms part of signers' linguistic competence, as they can demonstrably generalize it to novel signs (Berent et al., forthcoming).

6.5.2.2 Markedness matters across modalities

Signed and spoken phonologies not only share their building blocks but also manifest some common restrictions on their combinations. Before moving to discuss specific markedness restrictions, it might be useful to first establish that markedness is generally operative across modalities. In both modalities, markedness is defined by the violation of grammatical constraints on well-formedess, and as such, markedness is first and foremost an attribute of grammatical structure (i.e., of I-language). But to the extent that unmarked structures are preferred, markedness is expected to shape not only grammatical phonological computations (aspects of I-language) but also the processing of language and its acquisition (properties of E-language). Those various consequences of markedness, as we next see, tend to converge across modalities.

Our previous discussion of spoken language has shown that marked elements are less likely to serve as the output of phonological alternations, such as neutralization, epenthesis, and assimilation. Marked elements are also more

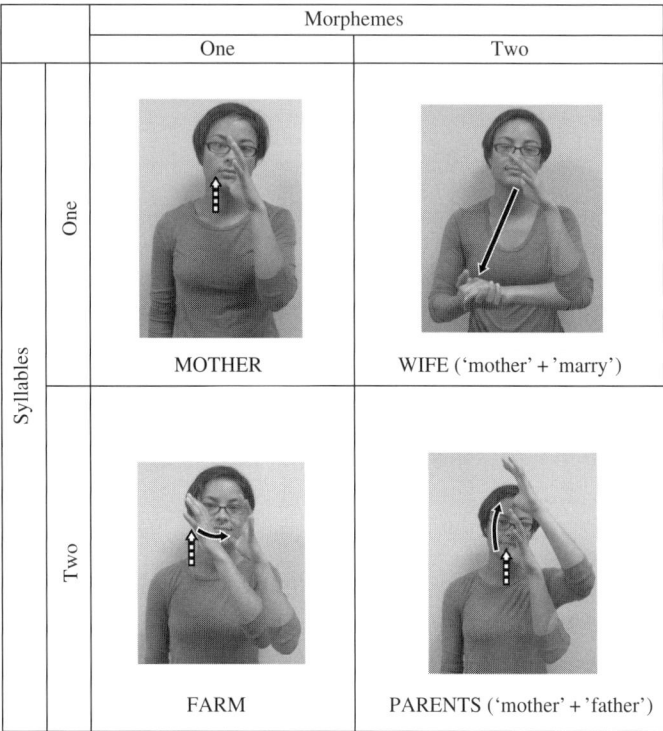

Figure 6.1 The distinction between syllable and morphological structure in American Sign Language

complex and harder to process – they appear to be less natural, infrequent (both within and across languages), harder to perceive and produce, later to be acquired, and vulnerable to loss in aphasia (Rice, 2007). Wendy Sandler and Diane Lillo-Martin (2006) note that these properties tend to converge in sign languages as well.

Consider, for example, the markedness of handshape. Across sign languages, certain handshapes appear to be unmarked, in the sense that they serve as a default (e.g., these feature values are the ones assigned to the non-dominant hand when it is involved in a sign that differs in shape from the dominant hand), and they are more likely to be inert to phonological restrictions (e.g., restrictions concerning the contact with the body or the other hand). Unmarked handshapes also tend to manifest the same extra-grammatical characteristics typical of unmarked elements in spoken language: Unmarked ASL handshapes are frequent in this language as well as in other sign languages (Israeli Sign Language and Sign Language of the Netherlands), they tend to be acquired earlier by children, and they are typically preserved in aphasia. These observations suggest that markedness might play a role in both modalities.

6.5.2.3 Some markedness constraints are general across modalities
Beyond their broad sensitivity to markedness, signed and spoken languages might also share some specific markedness constraints. Because sign language phonology has not been amply studied from the perspective of Optimality Theory, it is difficult to compare the role of markedness constraints across modalities. Nonetheless, some phenomena in sign languages are amenable to explanations that invoke markedness constraints shared with spoken languages.

Sonority restrictions on syllable structure
Earlier in this chapter we noted that the admissibility of a segment in a syllable depends on its sonority – an abstract phonological property that correlates with the phonetic salience of the segment. Sonorous segments (e.g., vowels) are preferred in the nucleus, whereas less sonorous ones (e.g., stop consonants) tend to occupy the margins. Interestingly, similar principles determine the role of manual elements in signed syllables. As in spoken languages, elements that are visually salient tend to be phonologically more sonorous. For example, hand movement is considered more sonorous than hand location. Moreover, sonorous elements are preferred as the syllable peak, whereas less sonorous ones are preferred as margins (Perlmutter, 1992; Sandler, 1993).

To illustrate the similar role of sonority in the two modalities, consider the phenomenon of cluster reduction. Recall that complex onsets as in *play* are more marked than simple onsets (e.g., *pay*). And indeed, many phonological systems – both mature and developing systems – allow only simple onsets. For example, in the course of acquiring their native English, young children often reduce complex onsets into simple onsets (e.g., Gnanadesikan, 2004; Ohala, 1999; Pater & Barlow, 2003). But interestingly, the reduction of complex onsets is not arbitrary. Rather, children systematically opt to preserve the onset member of lower sonority over its high-sonority counterpart (e.g., *play→pay*, not *lay*), indicating a preference for margins of low sonority (see 16a).

Sign languages manifest a similar phenomenon. Wendy Sandler (1993) arrays the sonority of sign-elements according to their amount of movement: At the top of the sonority scale are signs that include movement (M), movement is more sonorous than plain locations (L – with a sonority level of 2), and least sonorous are locations that contact the body ($L_{contact}$ – with a sonority level of 1). As in English child phonology, signed syllables ban complex margins and strongly favor monosyllables over disyllables. So when two monosyllabic words are combined to form a compound, it is often the case that only one syllable survives, and a choice must be made as to which of the various locations in the input will survive in the output (Liddell & Johnson, 1989). The example in (16b) illustrates a choice between two LML syllables (the subscripts indicate distinct locations and movements) whose location either manifests a contact (a sonority level 1, indicated below the sign in 16b) or has no contact (a sonority level of 2). Sandler shows that in such cases, it is the lower-sonority onset that is preferred to the higher-sonority one, a phenomenon that closely mirrors cluster reduction in spoken language.

(16) Reduction in spoken and signed language
    a. Reduction in English:
        play→pay
    b. Reduction in American Sign Language
        $L_1ML_{2(contact)}+L_3ML_4$ → $L_{1(contact)}ML_3$
        2 6 1   2 6 2   1   6 2

Phonological agreement

Another set of markedness restrictions that is potentially shared across modalities enforces agreement between adjacent phonological elements on the value of a shared feature. When two phonological elements are in close proximity and they share a feature, phonological processes often force these two elements to agree on the value of that feature. The case of English plural formation is an example close to home. Regular English plurals are typically generated by appending the voiced suffix /z/ to the base (e.g., /læb/+/z/ → /læbz/). Note that the two obstruents *b* and *z* both carry the voicing feature and agree on its value – both segments happen to be voiced. But when the base ends with a voiceless consonant (e.g., *cat*), the suffix (typically, /z/) now devoices (to /s/; Mester & Ito, 1989; Pinker & Prince, 1988). The fact that the suffix is altered (in violation of faithfulness constraints) to agree with the stem's final consonant suggests a higher-ranked markedness constraint that enforces agreement on voicing (Lombardi, 1999; for an alternative account, see McCarthy, forthcoming).

Feature agreement triggers similar alternations in sign language. For example, when two signs combine to form a compound, and their hand configurations mismatch on their orientation, the orientation feature can be selectively altered so that the two signs share the same feature (Sandler & Lillo-Martin, 2006: 156–157). The assimilation of phonological features suggests that agreement is a highly ranked constraint across modalities.

Identity avoidance

A third candidate for a shared markedness constraint concerns the restriction on identical phonological elements. In both modalities, identical phonological elements are distinguished from nonidentical elements, and their occurrence is highly restricted. The case of Semitic, where identical consonants are allowed only stem-finally (e.g., *simem*, but not *sisem*), is one such case (see Chapter 5). Similar restrictions on identity are also seen in English. Consider, again, the case of plural formation. While English plurals are typically formed by attaching the suffix /z/ to the base, stems ending with a sibilant (e.g., *bus*) take an additional schwa (e.g., *bus*es). Because the insertion of any phonological material that is absent in the base violates faithfulness restrictions, such additions must satisfy markedness pressures. Here, the schwa is appended to avoid adjacent identical consonants (e.g., /bʌss/). Identity avoidance has indeed been implicated at various levels of spoken language phonology, ranging from the avoidance of identical features and tones to identical syllables (Suzuki, 1998).

Diane Brentari points out that identity avoidance might also be at the root of one of the strongest phonological conspiracies in sign language – the preference for monosyllables. In her account, this preference reflects the application of identity avoidance to movement (Brentari, 1998). Because every movement defines the nucleus of a syllable, the avoidance of identical movements invariably favors monosyllabic outputs. Similar bans on identical elements could further explain the avoidance of identical places of articulation and identical selected fingers within the same morpheme (Sandler & Lillo-Martin, 2006: 223–225).

Summarizing, then, while signed and spoken phonology differ in many ways, they nonetheless appear to share some of their phonological primitives, they manifest markedness in similar ways, and they might even share a handful of specific constraints across modalities. The possibility that some phonological principles and constraints are amodal suggests that the phonological grammar, generally, and phonological markedness, specifically, are autonomous from the phonetic system.

6.6 Conclusion

Are phonological universals active in the grammars of all speakers? For many decades, this question has been addressed primarily by typological observations. While many regularities have been identified across languages, these patterns are mostly statistical trends, not absolute laws. The challenge presented by these data to phonological theory is to capture both the near universal regularities found across languages as well as their considerable diversity. In this chapter, we saw how Optimality Theory addresses both challenges. In this account, grammatical phonological constraints are universal. Grammatical constraints, however, are violable, conflatable, and potentially triggered by phonetic cues that vary across languages. These (and other) considerations explain why universal grammatical constraints do not result in absolute universals in external language.

But what is the nature of those linguistic universals? Do they reflect constraints that form part of an algebraic phonological grammar, or are they mere artifacts of the functional properties of spoken language? Previous chapters have presented computational challenges to the possibility that the algebraic phonological grammar can emerge from non-algebraic systems, such as the phonetic component. In this chapter, we further addressed this question by investigating whether universal phonological restrictions are autonomous from phonetic pressures. The evidence reviewed here suggests that phonological principles cannot be reduced to phonetic pressures. We saw that some phonetically plausible restrictions are unattested phonologically, whereas other phonologically attested restrictions are not phonetically optimal. In fact, some (albeit few) universal primitives and constraints might be shared across different phonetic modalities – in both spoken and signed languages. The demonstrable dissociation between phonological constraints and their phonetic basis, on the one hand, and the commonalities across modalities, on the other, suggests that the design of the phonological grammar is shaped by common universal principles that are distinct from phonetic knowledge.

Although these observations are consistent with the hypothesis of a specialized core system for phonology, they may be criticized as being simultaneously both too weak and too strong a test for the hypothesis. On the one hand, our search for phonetically arbitrary principles of phonological organization might be unnecessarily restrictive. Although the detection of such arbitrary principles would certainly implicate an abstract phonological cause, there is no reason to expect that had a core phonological system existed, its design should have been divorced from the properties of speech – the default medium of language in all hearing communities. In fact, some variation across modalities is expected. If grammatical constraints are the phenotypic expression of innate phonological knowledge, then variation in triggering conditions (e.g., input modality) is bound to yield variations in the resulting phonological system. So while the existence of amodal phonological constraints is certainly significant, the underlying expectation that core phonological knowledge be amodal presents too strong a test of the hypothesis.

At the same time, however, the inference of phonological universals solely from typological data is also a source of weakness. As we have repeatedly stressed, universal tendencies in the distribution of external objects are quite distinct from the putative grammatical sources that might shape them. Suggestive as it might be, linguistic evidence cannot, by itself, demonstrate that language universals form part of the synchronic phonological grammar. To address this shortcoming, we must complement the findings from typology and formal analysis with experimental investigation that examines the effect of putative universal grammatical constraints on the behavior of individual speakers. The following chapters undertake this task.

7 Phonological universals are mirrored in behavior: evidence from artificial language learning

> The view of phonology as a system of core knowledge predicts that the grammars of all speakers include a common set of universal constraints. This chapter illustrates the use of experimental methods to test this hypothesis. The experiments reviewed here compare participants' ability to learn structures that are universally marked to their ability to learn unmarked structures. As predicted, unmarked structures are learned more readily. Moreover, the advantage of unmarked structures obtains even when marked and unmarked structures are both absent in participants' language. While these results reveal a strong correlation between typological regularities and the behavior of individual speakers, they leave open questions regarding its source – whether the convergence results from universal grammatical constraints or from non-grammatical origins. These findings underscore some of the difficult challenges facing the experimental study of grammatical universals.

The view of phonology as a system of core knowledge predicts that all grammars converge on a common design, defined by a universal set of primitives and constraints. The evidence reviewed in the previous chapter supported this hypothesis. We saw that diverse languages, signed and spoken, manifest some strong structural regularities that systematically favor certain linguistic structures over others. We further showed how these cross-linguistic regularities (and variations) might emerge from the ranking of universal grammatical constraints. The universal grammar hypothesis, however, is not limited to an account of typological regularities. Indeed, the very same grammatical forces that shape the language typology are presumably active in the brains and minds of living breathing speakers. This strong hypothesis makes some interesting predictions that are amply amenable to experimental investigation. If this hypothesis is correct, then one would expect the behavior of individual speakers to mirror typological regularities: Unmarked phonological structures should be preferred to their marked counterparts. And if the preference for unmarked structures is universally active, then the preference for unmarked structures should be seen across all speakers, even if the structures under consideration are absent in their language.

This chapter begins to review some experimental results that are consistent with this prediction. The rationale guiding these investigations is the poverty-of-the-stimulus argument, outlined in Chapter 3 (see 1): Researchers first identify a typological generalization that favors a certain structural variant A to some counterpart B. They next proceed to select a population of naïve speakers whose native language (Language$_i$) lacks both A and B, and devise experimental tests that compare speakers' implicit preferences for these two structures. Of interest is whether speakers favor variant A to B despite no experience with either. If all speakers possess universal grammatical constraints that render structure A less marked than B, then participants in the experiment should manifest a similar preference – they should judge structure A as more acceptable than B, they should be more likely to represent it faithfully, and they should learn it more readily. To the extent one can rule out linguistic experience and other non-grammatical explanations for the results, then the observed preference for unmarked structures could suggest a broad grammatical constraint that is active in many languages, perhaps even universally.

(1) Experimental tests of grammatical universals (a test from the poverty of the stimulus)
Typology: A≻B
Language$_i$: ~~A, B~~
Test speaker of Language$_i$: is A≻B?

The experiments discussed in this chapter gauge the effect of markedness on the learnability of certain interactions among phonological elements. Phonological interactions typically involve segments that share a common feature, they result in changes that occur in a specific direction (e.g., from left to right), and they favor phonological outputs that mirror natural phonetic processes. As we shall next see, interactions conforming to such principles are also more readily learnable in artificial language experiments. The convergence between typological preferences and individual speakers' behavior is significant because it could suggest a common universal source that shapes them both. In what follows, we review the evidence for such a link and examine its origins.

The goal of this review is twofold. On the theoretical end, we aim to determine whether individual speakers' performance does in fact mirror the regularities seen in attested phonological systems. A second, methodological goal is to illustrate how one can use the tools of experimental psychology to systematically investigate phonological universals. Toward this end, we narrow the discussion to a handful of select studies that are reviewed in detail, leaving many other equally important studies unaddressed. A comprehensive review of the large literature on experimental phonology falls beyond the scope of this discussion.

7.1 Phonological interactions target segments that share features

In the previous chapter, we noted that phonological systems – both spoken and signed – frequently enforce agreement between elements that share a phonological feature. We illustrated this tendency with the alternation of the English plural suffix in *cats* (e.g., /kæt/+/z/➔/kæts/). In this example, the stem's final consonant *t* and the suffix *z* both carry the voicing feature, but they disagree on its value – the *t* is voiceless whereas the *z* is voiced. Since voicing disagreement is highly marked, phonological processes will frequently alter one of these elements to agree with the feature value of its neighbor. Crucially, feature alternations do not occur randomly: While alternations frequently target segments that disagree on a shared feature (e.g., the voiceless *t* and the voiced *z*), arbitrary interactions among consonants that do not share a feature (e.g., an interaction between the place of articulation of coronals and velar) are less frequent. Such regularities suggest a universal markedness constraint that enforces feature agreement (e.g., the constraint AGREE, Lombardi, 1999). If such a constraint is universally active, then interactions among adjacent segments that share a feature will be easier to learn than those that do not share features. In what follows, we consider several tests of this prediction.

7.1.1 Vowel height depends on the height of neighboring vowels, but not on consonants' voicing

Our first demonstration that phonological processes enforce feature agreement concerns the agreement of vowels with respect to their height – a case carefully argued by Elliott Moreton (2008). Moreton observes that in many languages, the height of any given vowel (H) depends on the height of its neighboring vowels (for a brief description of some of the main vowel features, see Box 7.1). Such systems, for example, favor forms like *tidi* (with two high vowels, HH) over *tidæ*, with one high vowel and one low vowel (see 2). In contrast, languages rarely condition vowel height (H) by the voice (V) of the following consonant (voice refers to properties related to glottis control, such as voicing, aspiration, or fortis/lenis). For example, few languages will require high vowels before voiced consonants (e.g., *tidu*) and ban them before voiceless ones (e.g., **titu*), so such interactions are considered "unnatural."
(2) Natural vs. unnatural restrictions on vowel height (Moreton, 2008)
 a. A natural restriction: vowels agree on height (HH language)
  tidi; *tidæ

b. An unnatural restriction: a vowel is high only if the following consonant is voiced (HV language)
   tidu, *titu

The asymmetry between HH and HV interactions is remarkable because the infrequent HV interaction is nonetheless expected on phonetic grounds. Since vowel height is signaled by the frequency of the first formant (F1), and since the first formant is modulated by the voice of the following consonant, there is ample phonetic reason for vowel height to interact with consonants' voice. Moreover, Moreton (2008) demonstrates that the phonetic cues for HH interactions (i.e., the change in a vowel's F1 as a function of the height of adjacent vowels) are not systematically stronger than the ones associated with HV

Box 7.1 A brief description of English vowels

To illustrate some of the main vowel features, here is a brief description of English vowels (following (Hayes, 2009). Vowels are modified by moving the tongue, jaw, and lips, and these modifications are captured by three classes of features: rounding, height, and backness (see Table 7.1). Rounding reflects the shape of the lips – rounded vowels (e.g., /u/ in *boot*) narrow the passage of air by rounding the lips, and they contrast with unrounded vowels (e.g., /i/ in *beat*). Another modification narrows or widens the shape of the vocal tract by changing the height of the tongue body in one of three positions – high (e.g., /i/ in *beat*), mid (e.g., ε, in *bet*), and low (e.g., /æ/ in *bat*). Finally, vowels can be modified by placing the body of the tongue towards either the front of the mouth (e.g., /i/ in *beat*) or its back (e.g., /u/ in *boot*).

Table 7.1 *English phonemes and diphthongs (following Hayes, 2009)*

					Back			
		Front unrounded		Central unrounded		unrounded		rounded	Diphthongs
Upper high	i	bead					u	boot	aɪ	bite
Lower high	ɪ	bid					ʊ	foot	aʊ	bout
									ɔɪ	boyd
Upper mid	eɪ	bayed, bait	ə	Abbot			oʊ	boat	Rhotacized upper mid central unrounded
Lower mid	ε	bed			ʌ	but	ɔ	bought	
Low	æ	bad			ɑ	Father			ɚ	Bert

interactions (e.g., the change in the vowel's F1 promoted by the voice of the following consonant). The fact that phonological processes rarely encode these phonetic interactions suggests that the phonological system does not follow phonetic cues directly. Rather, it is constrained by algebraic principles that enforce agreement among shared features. Because vowels, but not consonants, are specified for the height feature, vowel height depends only on other vowels, so height interactions among two vowels (HH) are phonologically more natural than interactions among vowel height and consonant voicing (HV). If such biases are universally active in all grammars, then one would expect them to constrain the behavior of individual speakers even when their own language does not give rise to those particular interactions.

To evaluate this possibility, Moreton (2008) next examined whether the typologically natural height interaction among vowels is indeed more readily learnable than the less-natural interaction among vowel height and consonant voicing. To minimize the possibility that such asymmetries might reflect knowledge that comes from speakers' native language, Moreton chose to study this question among speakers of English – a language in which neither of these processes is phonologically active. Moreton first familiarized two groups of participants with two artificial languages. Words in both languages were invariably disyllabic. In one language, the two vowels always agreed on height (e.g., *tiku*, with two high vowels, 2a); in another language, the first vowel was always high if the following consonant was voiced (e.g., *tidu*, see 2b). Each group of participants was exposed to only one of those languages. Participants were next presented with pairs of novel test items, and asked to make a forced choice as to which of the two pair-members formed part of the language they had previously studied. Results showed that participants exposed to the language with the natural interaction among vowels (HH) performed more accurately than participants exposed to the language with the unnatural vowel–consonant (HV) interactions. These results suggest that height interactions among vowels are more readily learnable than interactions among vowel-height and consonant voicing.

While these results are consistent with the possibility that phonological interactions are biased toward target shared features, vowel–vowel interactions could be preferred to vowel–consonant interactions because consonants and vowels form distinct phonological constituents (i.e., they belong to distinct phonological tiers; see Schane et al., 1974). Moreton (2008), however, points out that mere membership in a tier is insufficient to explain the height-height advantage, as his other findings suggest that HH interactions are learned more readily than interactions among vowel backness and height. Since backness and height engage members of a single tier (vowels), the advantage of the HH interaction must be specifically due to feature composition, rather than tier membership alone.

7.1.2 Vowel interactions target shared features

Further evidence that vowel interactions specifically depend on shared features is presented by Anne Pycha and colleagues (2003). These researchers compared the learnability of two types of rules. Both rules target vowels, but they differ in "naturalness" (see 3). The natural rule enforces stem-suffix agreement with respect to the feature [back]: Stems with a front vowel (e.g., CiC) require a suffix with a front vowel (-εk), whereas stems with a back vowel (e.g., CuC) require a suffix with a back vowel (-ʌk). The alternative, arbitrary rule likewise limits the co-occurrence of the same suffixes to stems of certain vowel categories, but those categories are now defined in an arbitrary manner, such that each category includes both front and back vowels.

(3) Natural vs. unnatural rules for vowel harmony (Pycha et al., 2003)
   a. A natural vowel harmony rule: the suffix agrees with the stem's vowel on [back]:
     (i) Front vowels take the front suffix -εk:
        CiC-εk
        CIC-εk
        CæC-εk
     (ii) Back vowels take the back suffix -ʌk:
        CuC-ʌk
        CʊC-ʌk
        CaC-ʌk
   b. An unnatural rule: the selection of the suffix depends on the stem's vowels (irrespective of feature agreement):
     (i) The vowels [i, æ, ʊ] take the front suffix -εk
     (ii) The vowels [ɪ, u, a] take the back suffix -ʌk

In the experiment, two groups of participants were first exposed to items instantiating one of the two rules, and they were next tested for their knowledge of the rule using both familiar test items and novel instances. Results showed that the natural rule was learned more readily than the unnatural one (for similar conclusions, see also Finley & Badecker, 2009). Thus, rules involving interactions among vowels with shared features are more readily learnable than rules involving arbitrary interactions among vowels that do not share a feature.

7.1.3 Consonant interactions target shared features

The tendency of phonological processes to target shared features also applies to consonants. The evidence, reported by Colin Wilson (2003), comes from interactions between stem and suffix. Wilson compared two types of interaction, captured by two competing rules (see 4). The natural rule concerns nasalization: Nasal stems (e.g., *dume*, including the nasal *m*) are paired with a nasal suffix (*na*), whereas non-nasal stems (e.g., *togo*, whose final syllable has a dorsal consonant, *g*) take the non-nasal suffix *la* (e.g., *togola*). This rule is considered natural because it enforces agreement between two consonants that share a feature (i.e., nasal). And indeed, this process frequently occurs in many languages (Walker, 1998). A second unnatural rule reverses the dependency between the stem and suffix, such that they no longer share a feature: Nasal stems now take the suffix *la*, whereas dorsal stems take the nasal suffix (*na*). Of interest is whether participants are better able to learn the natural rule than the unnatural one.

(4) Nasalization rules (Wilson, 2003)
 a. Natural rule: the suffix is [na] only if the stem ends with a nasal consonant:
  (i) Nasal stems take the nasal suffix [na]:
   dume**na**, binu**na**
  (ii) Dorsal stems take a non-nasal suffix [la]:
   toko**la**, dige**la**
 b. Unnatural rule: the suffix is [na] only if the final stem consonant is a dorsal consonant:
  (i) Nasal stems take the suffix [la]:
   dume**la**, binu**la**
  (ii) Dorsal stems take the suffix [na]:
   tuko**na**, dige**na**

To address this question, Wilson (2003) first exposed two groups of participants to words that instantiate one of the two rules. Next, participants were presented with test instances – either items that they had previously encountered during the familiarization phase or novel instances generated from the same set of segments. Participants were asked to determine whether each such item was "grammatical," that is, whether it was generated by the grammar of the language presented during familiarization. Results showed that participants who were familiarized with the natural alternation reliably favored grammatical over non-grammatical items, and they even generalized their knowledge to novel items. In contrast, no evidence for learning was obtained with the unnatural rule.

Taken as a whole, the studies reviewed in this section (Moreton, 2008; Pycha et al., 2003; Wilson, 2003) suggest that speakers favor phonological alternations among segments with shared features – either consonants or vowels. Accordingly, such interactions are learned more readily than arbitrary interactions among segments that do not share a feature.

7.2 Learners favor directional phonological changes

The tendency of phonological processes to target shared features imposes constraints on the segments that are likely to undergo phonological alternations. But in addition to delimiting the target of alternations, phonological processes must also determine their outcome. Consider, for example, an interaction between three vowels, specified for some binary feature F – either [+F] or [–F]. Given the discussion so far, such segments are likely to interact. Our question here concerns how the outcome of the interaction is decided: Which of the two feature values is the winner, and which one the loser?

Sara Finley and William Badecker (2008; 2010) outline several logically possible outcomes of feature interactions (see 5). One possibility is that the interaction is decided on directional grounds: The winner is invariably either the leftmost feature value or the rightmost one. A second type of system might a priori designate one dominant feature value (e.g., <+>) as the winner, irrespective of its position. Finley and Badecker observe that directional and dominant systems are both attested across languages. But there is also a third way to resolve the competition. In this third system, the outcome is determined by a majority rule: If most features are <+>, then this feature wins; if most features are <–>, then all features become a <–>. Although such a system is logically possible, Finley and Badecker note that it is unattested in human languages (possibly, a consequence of broader bias against phonological processes that operate by counting elements; McCarthy & Prince, 1995). The question is whether the absence of such systems reflects an active bias on the part of language learners.

(5) Some possible outcomes of interactions among vowels
    a. Directional systems: spread in a single direction:
      (i) Spread the leftmost vowel rightwards:
        + – – →+ + +
        – + + →– – –
      (ii) Spread the rightmost vowel leftwards:
        + + – →– – –
        – – + →+ + +
    b. Dominant system: spread a dominant feature (e.g., +), irrespective of direction:
      – + – →+ + +
    c. Majority rule: spread the majority feature:
      + – + →+ + +
      – – + →– – –

To address this question, Finley and Badecker compared the learnability of directional and majority rule systems. Their experiments first familiarized a group of English speakers with one of two types of interactions among vowels (also known as "vowel harmony" systems, see Table 7.2). These systems each manifested an alternation between front vowels [i, e] and round/back vowels [o, u], but they differed on the direction of alternation: One alternation spread the features of the left vowel rightwards, whereas the other spread the features of the right vowel leftwards. Regardless of direction, however, these alternations were designed to also enforce the majority feature, such that the feature value shared by

Table 7.2 *The design of Finley and Badecker's experiments (Finley & Badecker, 2008; 2010).*

		Spread left	Spread right
Familiarization		pu mi te➔pi**m**ite + – –➔– – –	ku ko pe➔ku**k**op**o** + + –➔+ + +
Test	Same direction / majority rule	nu pi ki➔<u>ni**p**iki</u>/ nu**p**uk**o** + – –➔<u>– – –</u>/+++	ni pi ku➔<u>ni**p**iki</u>/nu**p**uk**u** – – +➔<u>– – –</u>/+++
	Same direction / majority rule	ku ko pe➔<u>ki**k**epe</u>, ku**k**op**o** ++–➔<u>– – –</u>/+++	pu m ite➔pu**m**ut**o** /pi**m**ite +– –➔+++/– – –

Phonological changes to the output are indicated in bold, and test options that are consistent with familiarized alternation are underlined; + indicates a round vowel; – indicates an unrounded vowel.

two of the three input vowels always happened to win (e.g. + + –➔+ + +). Accordingly, the evidence presented to participants was ambiguous between two structural descriptions: either a directional rule (spread left or right, as in (5a) or a majority rule that enforces the feature most common in the input (as in 5c). Of interest is how participants interpret this ambiguous input – whether they infer a directional rule (the one attested in human languages), or a majority rule (a rule that is unattested in phonological systems).

To examine this question, Finley and Badecker next compared participants' ability to generalize the rule to two types of novel test items. In each test trial, participants were presented with a "disharmonious" input – vowels that disagreed on both their back and rounding features. For example, the input *nu pi ki* consists of one round/back vowel followed by two unrounded/front ones (i.e. + – –). This input was paired with two alternations that were both harmonious – either three unrounded/front vowels (e.g., *nipiki*) or three round/back ones (e.g., *nupuko*). Participants were asked to determine which of these two alternations was more consistent with the language they had heard. In one condition, the consistent option matched the familiarization items with respect to both the direction of assimilation and adherence to the majority rule whereas the other option was inconsistent with training items on both counts. For example, the *nu pi ki*➔ *nipiki* alternation spreads the unrounded/front feature leftwards, and this feature is also shared by two of the three vowels in the input. Accordingly, this alternation could be interpreted as either a directional rule or a majority rule. In another condition, the direction of change was pitted against the majority rule. The alternation *ku ko pe*➔*kikepe* illustrates this condition. Here, the unrounded/front feature still spreads leftwards, but this feature now corresponds to the minority feature in the input. Such instances force participants to choose between the two rules – they must opt for items that *either* maintain the directional change of familiarization items *or* follow the majority rule. If people are biased to acquire a directional rule, and they further generalize this rule to novel items, then they should favor the same directional change even when the majority rule is violated. To control for the possibility that participants might be a priori biased toward certain outputs for reasons unrelated to learning in the experiment, Finley and Badecker compared the performance of these two groups to a control group that was exposed only to isolated syllables (e.g., *ku ko pe*) but not to the concatenated harmonious output (e.g., *kukope*). To further ensure that participants' performance is not due to an inability to perceive vowels, these authors further tested their participants for their ability to discriminate between these options and excluded individuals who were unable to do so.

Results showed that participants interpreted ambiguous alternations as consistent with a directional rule, rather than the majority rule. Participants were more likely to choose items that maintained the familiar direction of assimilation over those that followed the majority rule. Moreover, participants in the experimental conditions (those familiarized with the assimilation rule prior to testing) were less likely to favor the majority-rule output compared to participants in the control condition (those presented with testing without any familiarization), indicating that the preference for the directional alternation was the result of learning, rather than the inherent properties of specific outputs. These findings suggest that learners are biased to infer directional phonological rules over majority rules.

7.3   Learners favor phonetically grounded interactions

The discussion so far has identified two factors affecting the likelihood that segments undergo a phonological alternation. We saw that phonological alternations typically target segments that share a feature, and that they favor directional alternations to majority-rule changes. Nonetheless, not all single-feature, directional changes are equally likely to occur in typology. Rather, phonological alternations tend to recapitulate phonetic changes that occur naturally (Jun, 2004; Ohala, 1975; Stampe, 1973; Steriade, 2001). The possibility that phonetic factors might constrain the range of possible phonological alternations has been the target of several recent experimental investigations (e.g., Becker et al., 2011; Coetzee & Pretorius, 2010; Hayes et al., 2009; Moreton, 2008; Zuraw, 2007). Here, we will illustrate this program with Colin Wilson's pioneering study of velar palatalization.

When a velar consonant (e.g., *k*) is followed by a high front vowel like [i], the front of the tongue is raised toward the hard palate, so its place of articulation becomes palatal-like. For example, the place of articulation of *k* is closer to the palate in *key* compared to *car* (Ladefoged, 1975). In English, the two *k* variants in *key* and *car* are instances of the same phoneme, so this subtle allophonic change is difficult for many English speakers to detect. In other cases (e.g., diachronic

Table 7.3 *The design of Wilson's (2006) palatalization experiment*

			Test	
			Familiar	Novel
Familiarization	High vowel (natural)	ki→chi	ki→?	ke→?
		gi→chi	gi→?	ge→?
	Mid vowel (unnatural)	ke→che	ke→?	ki→?
		ge→che	ge→?	gi→?

processes in Slavic, Indo-Iranian, Bantu languages, Guion, 1996), however, palatalization replaces the stop velar consonant with another phoneme of a different place of articulation – the coronal affricate *tʃ* (e.g., *ki*→ *tʃi*). This phonological process nonetheless mirrors phonetic palatalization inasmuch as it is more likely to occur in the context of a high front vowel [i] compared to mid and low vowels (e.g., ɛ, æ), and it renders velar stops (e.g., *ki*) acoustically more similar to palatoalveolar affricates (e.g., *chi*, Ohala, 1989). And indeed, when syllables like [ki] are masked by noise, they are more likely to be confused with the affricate *chi* compared to velars followed by non-high vowels (e.g., *che*, Guion, 1996; 1998).

Given that the *ki*→*chi* alternation is phonetically more natural than *ka*→*cha*, one wonders whether language learners might be biased toward such processes. If people favor phonetically natural alternations, then such alternations should be learned more readily. Moreover, if marked (phonetically unnatural) structures asymmetrically imply unmarked ones, then learners who master the less preferred alternation (*ke*→*che*) will generalize their knowledge toward the better-formed *ki*→*chi* change, whereas the reverse generalization (from the unmarked condition to the marked one) should not occur.

To test this hypothesis, Colin Wilson (2006) compared the learnability of the natural (*ki*→*chi*) and less natural (*ke*→*che*) alternations. His experiments first presented English-speaking participants with a language game in which velars (either voiced or voiceless) are palatalized (see Table 7.3). One group of participants was familiarized with palatalization in the natural context of the high vowel [i]; a second group was familiarized with a less natural change in the context of the mid vowel [e]. In the second phase of the experiment, participants were asked to generate the appropriate form to items with a familiar velar-vowel combination (e.g., *ki*, for the high-vowel group; or *ke* for the mid-vowel group) or novel sequences (e.g., *ke* for the high-vowel group; *ki* for the mid-vowel group). Assuming that participants can readily discriminate both *ki* from *chi* and *ke* from *che*, their performance on the two conditions should not differ on perceptual grounds. But if rule-learning is biased toward the phonetically natural *ki*→*chi* alternation, then learners who mastered the less natural (i.e., marked) *ke➔che* process should readily extend it to the more natural (unmarked) *ki➔chi* case, whereas generalization from the natural *ki➔chi* to the unnatural *ke➔che* alternation should be less likely.

Wilson's (2006) results are consistent with this prediction. Not only did participants familiarized with the unnatural *ke➔ che* alternation generalize to the novel *ki➔ chi* change, but the rate of palatalization for such novel instances did not differ from familiar *ke➔ che* sequences. Moreover, the generalization from the unnatural to the natural alternation is not simply due to an across-the-board tendency to form novel phonological alternations, as participants in the natural high-vowel group did not apply the unnatural *ke➔ che* pattern. These findings are in line with the hypothesis that learners are biased toward phonetically natural changes.

7.4 Discussion

The studies reviewed in this section all suggest that people systematically prefer certain phonological alternations to others even when these processes are unattested in their native language. In particular, learners favor interactions that target shared features, they prefer directional alternations to majority rules, and they are inclined to learn alternations that are phonetically grounded. Not only are learners biased with respect to the type of generalizations that they extract, but their biases mirror the regularities seen in existing phonological systems. The convergence between the biases of individual learners and typological regularities is significant because it opens up the possibility that the typology and individual learning preferences have a common source. But whether this source is in fact shared, and whether it is grammatical is not entirely certain from these findings.

Demonstrating the effect of universal grammatical constraints on human performance critically hinges on evidence for both convergence and divergence (see 6). The argument from convergence asserts that the performance of individual speakers converges with regularities observed across languages. But in order to establish that this convergence is due to universal grammatical constraints, it is also necessary to rule out alternative explanations. To do so, one must show that the pattern of human performance diverges from the outcomes predicted by various extra-grammatical sources, including the properties of participants' native language, generic computational biases, and biases that are phonetic rather than phonological in nature.

(6) Experimental evidence for universal grammatical constraints
   a. *Convergence*: the preferences of individual speakers converge with typological regularities

b. *Divergence*: the preferences of individual speakers diverge with the outcomes predicted by sources external to universal grammar:
   (i) Properties of speakers' native language
   (ii) Generic computational biases
   (iii) Phonetic restrictions

While all studies reviewed here convincingly demonstrate convergence, non-grammatical sources are quite difficult to rule out. In what follows, we briefly consider some of these non-grammatical explanations. Our goal here is not to criticize the particular studies considered in this chapter – many authors are quite aware of those challenges, and, for this reason, some researchers do not even attribute their findings to universal grammar. It is nonetheless important to outline those challenges in order to motivate future attempts to address these limitations.

7.4.1 *The role of linguistic experience*

One concern that immediately comes to mind is that participants' learning biases reflect constraints that originate not from grammatical universals but rather from properties of their native language, in this case, English. While the focus on learning an artificial language intends to minimize the effect of native language experience, it does not, in and of itself, rule it out.

One aspect of relevant native-language experience concerns the phonetic and phonological processes in participants' language. This concern is particularly urgent with respect to nasalization and palatalization – processes that are vigorously operative in English. Indeed, English systematically nasalizes vowels before nasal consonants (cf. [bid] *bead* and [bĩn] *bean*; Ladefoged, 1975). Some dialects of English also palatalize *s* as *ʃ* – both lexically (cf. *press–pressure*) and productively, before /j/ (e.g., *this year* [ðɪs.jir] ~ [ðɪʃ.jir]; Zsiga, 2000). But even when the relevant process is utterly unattested in participants' native language phonology, their preferences could be nonetheless guided by the statistical properties of their lexicon.

Consider, for example, the contingency of vowel height on the height of neighboring vowels, but not on consonant voicing (Moreton, 2008). If English words were more likely to agree on vowel height than on consonant voicing, then people's bias toward vowel-height agreement could reflect the familiarity with such outputs, rather than a grammatical preference, universal or otherwise. Being well aware of this possibility, Moreton (2008) goes to great lengths to gauge the statistical properties of the English lexicon, but the evidence is mixed. As Moreton notes, a preference for a given pattern could depend not only on the statistical support for this pattern (i.e., conforming patterns) but also on statistical evidence against it (i.e., nonconforming patterns). While the *support* for the preferred high-high pattern (as indexed by the frequency-weighted ratio of the words conforming to this pattern relative to the frequency expected by chance) is comparable to the high-voice pattern (1.02 vs. 0.97, respectively), the evidence *against* the high-high pattern (indexed by the frequency-weighted occurrence of nonconforming instances, 3,133,331) is actually weaker than the evidence against the high-voice pattern (4,902,618). It is, of course, possible that the statistical structure of the English lexicon is itself a consequence of phonological or phonetic biases. But in the absence of appropriate controls, one cannot rule out the possibility that the advantage of natural patterns reflects their greater frequency in participants' linguistic experience.

7.4.2 The role of complexity

Beyond the biases originating from learners' native language, certain phonological interactions might be favored for reasons related to their overall complexity. Consider, for example, the shared-feature advantage. Phonological alternations that share a feature (e.g., height) are arguably simpler to encode and remember compared to ones involving two features (e.g., height-voice). The shared-feature advantage could therefore emerge not from a universal grammatical bias on learning, but rather from the sheer simplicity of this alternation.

Computational complexity could likewise explain the dispreference of majority rules. Majority rules (see 7) indeed exact heavier demands on learning: Learners must compute the number of shared feature values, compare them to non-shared features, and determine which of the two sets is larger. Directional rules, by contrast, require only that the learner map the feature at one edge of the input (either the left or right edge) to the output. Because majority rules are arguably more complex than directional rules, the advantage of the latter is predicted on grounds of computational simplicity alone.

(7) Majority vs. directional rules
 a. Majority rule: spread the majority feature:
   + − + → + + +
   − − + → − − −
 b. Directional rule: spread in a single direction:
   + − − →+ + +
   + + − →− − −

An interesting follow-up experiment by Finley and Badecker (2010) addresses this possibility by showing that the preference for directional rules is less robust when participants learn alternations consisting of geometric figures. These findings, however, do not necessary mean that the directional biases on the processing of linguistic stimuli are guided by universal grammar. As these authors note, the reduced reliance on a directional rule with geometrical shapes might be due to their mode of presentation – while speech stimuli unfold in time continuously, visual sequences were presented sequentially. The selective advantage of the directional rule with speech stimuli might therefore result from the conjunction of computational complexity and domain-general biases on the allocation of attention, rather than a specialized linguistic constraint. These correlations between grammatical and extra-grammatical preferences should not be automatically interpreted as evidence for a domain-general origin. A simplicity bias, for example, could well have affected the configuration of a specialized grammatical system in phylogeny, so merely showing that unmarked alternations are simpler does not necessarily demonstrate that their advantage originates from a domain-general system. But because this explanation cannot be ruled out, simple-rule biases are ambiguous evidence for universal grammar.

7.4.3   The role of phonetic factors

A third, extra-grammatical constraint on learning is presented by phonetic factors. According to the core phonology hypothesis, phonetically natural alternations are preferred because the phonological grammar is biased (in either ontogeny or phylogeny) to encode naturally occurring phonetic alternations. While phonological constraints are phonetically grounded, they are nonetheless autonomous, inasmuch as they are separate from the phonetic system. It is these grammatical constraints, then, rather than the phonetic system, that are responsible for the preference for phonologically natural processes.

Speakers' bias toward phonetically natural alternations is certainly in line with this possibility, but the findings are also consistent with an alternative explanation. In this view, the advantage of phonetically natural structures in learning experiments is utterly unrelated to the phonological grammar. Such structures are preferred not for their phonological well-formedness, but only because their phonetic properties are more robust and easier to encode (Ohala, 1989; see also Gow, 2001, for example, for an experimental demonstration).

Unfortunately, many studies in the literature automatically attribute the disadvantage of marked structures to universal grammatical sources, ignoring phonetic (and statistical) explanations for their findings. Unlike those studies, several of the authors cited in this chapter are well aware of these alternative phonetic explanations. In an attempt to address these concerns, Finley and Badecker (2008; 2010) ensured that their auditory experimental stimuli were all perceptible. Similarly, Moreton (2008) has conducted a broad meta-analysis designed to show that the height-height and height-voice interactions are comparable for the robustness of their articulatory cues. While such steps are surely welcome, showing that marked structures are perceptible and robust does not guarantee that their phonetic encoding is as easy and as precise as the encoding of less-marked structures. The dissociation of markedness constraints from phonetic pressures thus remains an outstanding challenge.

To summarize, the experiments reviewed in this chapter present compelling evidence that link typological regularities and the learning preferences of individual learners. The results of these groundbreaking studies suggest that speakers possess broad restrictions that shape learning. But whether those restrictions are universal and grammatical is difficult to ascertain. To demonstrate that speakers are equipped with universal grammatical constraints, one must rule out the possibility that the relevant preferences reflect linguistic experience, phonetic attributes of the experimental stimuli, and domain-general restrictions. Although many of the studies reviewed here attempted to address at least some of these concerns, none has been able to rule them out. Evaluating these multiple concerns within the scope of a single study is indeed extremely difficult. To address these challenges, one might therefore adopt a complementary approach. Rather than examining the role of universals in multiple cases, one might conduct an in-depth analysis of a single case study. The following chapter takes this perspective.

8 Phonological universals are core knowledge: evidence from sonority restrictions

> This chapter examines the representation of grammatical phonological universals by pursuing an in-depth analysis of a single case study – the sonority restrictions on complex onsets (e.g., *bl* in *block*). Across languages, syllables like *block* are reliably favored over syllables like *lbock*. Of interest is whether these preferences reflect universal grammatical restrictions. We address this question in two steps. We first show that sonority restrictions are plausible candidates for a grammatical universal – they are amply evident in productive phonological processes and supported by typological data. We next proceed to examine whether this putative universal constrains people's behavior in psychological experiments. Results from numerous experiments demonstrate that people are sensitive to sonority restrictions concerning onsets that they have never heard before. Sonority preferences, moreover, cannot be explained by several non-grammatical sources, including the phonetic properties of the experimental materials and their similarity to onsets that are attested in participants' languages. By elimination, then, I conclude that people's behavior is shaped by grammatical restrictions on sonority, and these restrictions extend broadly, perhaps universally, even to structures that are unattested in a speaker's language. These conclusions suggest that universal grammatical restrictions might be active in the brains and minds of individual speakers.

8.1 Grammatical universals and experimental results: correlation or causation?

The results described in the previous chapter suggest that people's performance in psychological experiments mirrors putative universal grammatical constraints: Unmarked phonological structures – those that are systematically preferred across languages – are the ones learned more readily by individual speakers. While the correlation between human performance and grammatical constraints is suggestive, correlation is not evidence for causation. And indeed, the agreement between the typological and behavioral data could be due to various sources external to the grammar. The learning advantage of unmarked structures might reflect not existing universal knowledge that favors unmarked structures, but rather the fact that such structures are independently easier to process. The question then remains whether grammatical principles are, in fact, universally active in the brains of all speakers. More generally, our question is whether grammatical principles form part of a phonological system of core knowledge.

In what follows, we present some results that are strongly suggestive of this possibility. To outline the evidence in sufficient detail, the discussion is limited to a single case concerning the sonority restrictions on onset clusters. We will first present some linguistic evidence suggesting that sonority restrictions are a likely candidate for a universal markedness constraint. We will next demonstrate that universal sonority restrictions shape human behavior in psychological experiments.

8.2 Sonority restrictions are active in spoken languages: linguistic and typological evidence

Practically every known language constrains the co-occurrence of segments in the syllable. English, for example, allows syllables like *bla*, but bans syllables like *lft*. Such restrictions have been attributed to sonority. Sonority is an abstract phonological property that correlates with the intensity of segments (for phonetic evidence, see Parker, 2002; 2008). Most sonorous are vowels, followed by glides, liquids, nasals, and obstruents (the class of stops and fricatives; see 1a). Languages, however, are also known to make additional, finer-grained distinctions in sonority. Some languages overtly differentiate between the sonority levels of obstruents, treating fricatives as more sonorous than stops. In other languages, sonority distinctions are sensitive to voicing, such that voiced obstruents are more sonorous than voiceless ones. Finally, certain languages further distinguish between the sonority levels of liquids – rhotic vs. lateral. These finer-grained distinctions suggest a more detailed sonority scale, provided in (1b).

(1) The sonority levels of consonants
    a. The basic sonority scale (e.g., Clements, 1990):
        glides (s=4) >liquid (s=3) >nasals (s=2) >obstruents (s=1)
        (e.g., y,w)   (e.g., l,r)   (e.g., m,n)   (e.g., p,t,b,d,f,v,s,z)

b. A detailed sonority scale (e.g., Zec, 2007):

glides	>rhotics	>laterals	>nasals	>voiced fricatives
(s=8)	(s=7)	(s=6)	(s=5)	(s=4)>
(e.g., y,w)	(e.g., r)	(e.g., l)	(e.g., m,n)	(e.g., z,v)

voiceless fricatives	>voiced stops	>voiceless stops
(s=3)	(s=2)>	(s=1)
(e.g., f, s)	(e.g., b, d)	(e.g., p,t)

The sonority levels of segments play a critical role in shaping syllable structure. Recall from Chapter 6 that high-sonority segments (e.g., vowels) are preferred at the syllable's nucleus, whereas low-sonority segments (e.g., stops) are favored at the syllable's margins – codas and onsets. Our interest in this chapter specifically concerns the sonority restrictions on complex onsets (e.g., *bl* in *block*). To describe the structure of such onsets, let us inspect the sonority distance (Δs) between the onset's consonants. The examples in (2) list various onset types along with their sonority distance (Δs), calculated by subtracting the sonority level of first onset consonant (S_1) from the sonority level of the second (S_2; for simplicity, all calculations use the basic sonority scale in 1a). Onsets such as *bl* manifest a large rise in sonority ($\Delta s=2$); *bn* manifests a smaller rise ($\Delta s=1$); *bd* exhibits a sonority plateau ($\Delta s=0$), whereas *lba* has a sonority fall – a "negative" sonority distance ($\Delta s=-2$; see 2).

(2) The sonority distance in complex onsets (calculated according to the basic sonority scale in 1a)

	S_1	S_2	Δs
bl	1	3	2
bn	1	2	1
bd	1	1	0
lb	3	1	−2

Although each of these sonority distances is attested in some human language, languages vary considerably on the range of sonority distances that they allow. English, for example, typically requires sonority rises of at least two steps (e.g., *bl*) – while larger distances (e.g., *twin*, $\Delta=3$) are allowed, smaller distances (e.g., *pn*, $\Delta=1$) are not systematically tolerated.[1] Other languages, however, do permit those smaller distances. Ancient Greek allows obstruent-nasal onsets with a one-step rise in sonority (e.g., *pneuma* 'breath'; Steriade, 1982); Hebrew minimally tolerates sonority plateaus (e.g., *gdi* 'kid'), and Russian and Polish allow even sonority falls (e.g., Russian: *rzhan* 'zealous,' Halle, 1971; Polish: *rtęć* 'mercury,' Kenstowicz, 1994). Although languages clearly differ on the

---

[1] A notable exception to this generalization is presented by *s*-initial clusters (*sport, sky*) – similar exceptional behavior of *s*-initial onsets has been also observed in other languages (for explanations, see Wright, 2004).

range of sonority distances that they tolerate, all languages favor onsets with large sonority distances over smaller ones (Clements, 1990; Smolensky, 2006). For example, onsets such as *bl* are preferred to onsets such as *bn*, and *bn*, in turn, is preferred to *lb* (see 3). The following discussion illustrates these preferences in phonological alternations and typological evidence.

(3) The preference for large sonority distances
   ...Δs=2 ≻ Δs=1 ≻ Δs=0 ≻ Δs=−1 ≻ Δs=−2...
      bl   ≻   bn   ≻   bd   ≻   nb   ≻   lb

8.2.1 Evidence from phonological processes

Numerous phonological processes favor onsets with larger sonority distances. One example comes from the syllabification of word-internal consonant sequences (e.g., *alba*). In such cases, the grammar must choose between two possible syllable parses – either *a.lba* or *al.ba*. While the former, *a.lba*, has a highly marked onset of falling sonority, the latter, *al.ba*, avoids the marked onset at the cost of acquiring a coda (i.e., violating the NoCODA constraint). Since word-internal sequences require a choice to be made, they present us with a window into the grammar's sonority preferences. In fact, word-internal sequences might mirror such preferences more clearly than word-initial ones (e.g., *lba*). When a marked consonant sequence occurs at the beginning of a word (e.g., *lba*), markedness pressures can only be satisfied by adding or deleting a segment (e.g., *lba*➔*ləba* or *bla*➔ *ba*) – operations that would violate faithfulness restrictions – and the high toll exacted by the eradication of such sequences renders markedness pressures less likely to win, and hence, harder to detect. But when the same consonant sequence occurs word-medially (e.g., *alba*), the grammar now has "cheaper" repair weapons in its arsenal. And indeed, in such cases, sonority preferences are clearly evident: Languages overwhelmingly favor a parse that avoids onsets with small sonority distances (*al.ba*). Remarkably, such medial onsets are avoided even in languages that would otherwise tolerate them word-initially. Recall, for example, that Polish allows word-initial onsets of falling sonority. Nonetheless, sequences like *kormoran* are syllabified as *kor.mo.ran*, not *ko.rmo.ran* despite the fact that syllables like *mo* and *rmo* are both attested in this language (Kenstowicz, 1994).

The eradication of marked onsets in word-medial positions also interacts with processes of word formation. One highly cited example is presented by Donca Steriade's influential analysis of reduplication in Ancient Greek (1982; see 4). Ancient Greek generates perfect stems by reduplication, but the precise output depends on sonority distance. Stems beginning with onsets of rising sonority, such as *tlā* ('to endure'), form perfects with a prefix that copies the initial consonant of the stem, followed by the vowel *e* (e.g., *tlā*➔**te.tla**.men 'to endure'). This process preserves the original rising-sonority onset in the reduplicated output. In contrast, stems with smaller sonority distances (e.g., *sper* 'to sow') form the perfect by adding the prefix *e* to the stem, such that the illicit onset *sp* is eliminated (i.e., *es.par.mai*, instead of *se.spar.mai*). Like Polish, Ancient Greek tolerates onsets such as *sp* word-initially. But when reduplication renders such clusters word-medial and "cheaper" repairs become available, onsets with small sonority distances are eradicated, whereas better-formed onsets of rising sonority (specifically, onsets whose sonority cline minimally comprises a voiceless stop and a nasal consonant) are preserved. These observations suggest that phonological alternations favor onsets with large sonority distances.

(4) Perfect reduplication in Ancient Greek (from Steriade, 1982)
   a. Well-formed onsets:
      Stem *Perfect stem*
      tlā   te.tla.men   'to endure'
      krag ke.kra.ga   'to cry'
   b. Ill-formed onsets:
      Stem *Perfect stem*
      sper   es.par.mai   'to sow'
      kten   ek.to.na   'to kill'

8.2.2 Typological evidence

Further evidence for the preference for onsets with large sonority distances is presented by the distribution of onsets across languages. Although languages differ greatly on the range of onsets that they tolerate, languages that tolerate small sonority distances also tend to allow larger distances. Russian, for example, allows sonority falls (e.g., *rtut* 'mercury'), but it also admits sonority plateaus (e.g., *zveno* 'link') and rises (e.g., *kniga* 'book'). In contrast, languages that allow large sonority distances do not necessarily tolerate smaller distances.

An inspection of Greenberg's (1978) survey of ninety genetically diverse languages suggests reliable statistical regularities related to sonority distance (see 5). First, onsets with large sonority distances are more frequent than onsets with smaller distances: Of the 90 languages in the sample, 75 manifest large rises in sonority, 58 manifest a small rise, 44 have a sonority plateau, and only 12 tolerate a fall in sonority. Second, languages with smaller sonority distances tend to allow larger distances. Consider, for example, languages with sonority falls. Of the 12 languages with falling sonority, the grand majority (11/12) also allows a sonority plateau. Similarly, 41 of the 44 languages with sonority plateaus allow sonority rises, and most (57/58) languages with small rises also allow large rises.

The tendency of languages with small sonority distances to allow larger ones is not absolute. For example, Huichol (an Uto-Aztecan language spoken in Mexico) tolerates word-initial clusters of level and falling sonority (e.g., *ptiʔuzima'yata* 'he is working'; ***mtite'ruwa*** 'he who is reading') but exhibits no sonority rises (Greenberg, 1978; McIntosh, 1944). This language is among a handful of exceptions, indicated in the top right corner of the table. It is important to keep in mind, however, that the absence of a certain onset type (i.e., obstruent-liquid onsets) could also occur for reasons unrelated to sonority (e.g., the absence of liquids in the language). Whether such cases present true counterexamples to the sonority hierarchy is an important question that must be addressed on a case-by-case basis. Across languages, however, small sonority distances imply larger ones. The contingency of larger distances on smaller distances is a reliable statistical tendency. Moreover, this contingency holds true even when one controls for the preponderance of larger distances. Had the probability of languages with large rise in sonority been determined only by the frequency of large rises in the sample (75/90 = .83), then the probability that a language with small rises exhibits large rises should have been .83. But in the sample, the observed probability is much higher – 57/58 (.98) of the languages with small rises exhibit large rises, and the probability of this outcome given the overall probability of large rise (.83) is extremely low ($p<.0001$). This also holds for the contingency of small rises on plateaus (5b), and that of plateaus by falls (5c). These observations demonstrate that languages which allow onsets with small sonority distance tend to allow onsets with larger sonority distances, and this contingency is a genuine statistical fact that is inexplicable by the overall frequency of such onsets in the sample.

(5) The contingency of large sonority distances on smaller ones (data from Greenberg, 1978; reanalyzed in Berent et al., 2007a; for any given sonority distance, <+> indicates presence, <–>indicates absence)

a. Small sonority rises imply larger rises:

		Large rise	
		+	–
Small rise	+	57	1
	–	18	14

b. Sonority plateaus imply sonority rises:

		Rise	
		+	–
Plateau	+	41	3
	–	35	11

c. Sonority falls imply sonority plateaus:

		Plateau	
		+	−
Fall	+	11	1
	−	33	45

8.2.3   The grammatical markedness of onset clusters

The typological preference for large sonority distances converges with the evidence from productive phonological processes to suggest that the phonological grammar includes universal restrictions related to sonority distance. The precise nature of those restrictions has received distinct formulations (e.g., Clements, 1990; Kiparsky, 1979; Selkirk, 1984; Smolensky, 2006; Steriade, 1982; Zec, 2007; see also Hooper, 1976; Saussure, 1915/1959; Vennemann, 1972), but, in all accounts, markedness is inversely related to sonority distance. One can express this convergence by the hypothesis that the phonological grammar includes constraints that render onsets with sonority distance d less marked than onsets with a smaller sonority distance, $d − 1$ (see 6). Specifically, sonority falls (e.g., *lb*, $\Delta s = -2$) are more marked (i.e., dispreferred) compared to sonority plateaus (e.g., *bd*, $\Delta s = 0$), which in turn are more marked than small sonority rises (e.g., *bn*, $\Delta 1 = 0$); least marked on this scale are large sonority rises (e.g., *bl*, $\Delta s = 2$).

(6) The markedness of sonority distances in onsets
$\Delta s = D \succ \Delta s = D − 1$
e.g., $\Delta s = 2 \succ \Delta s = 1 \succ \Delta s = 0 \succ \Delta s = -1 \succ \Delta s = -2$

It is important to recognize that the hierarchy in (6) captures only the markedness of outputs (i.e., onsets) – it does not necessarily correspond to an actual grammatical constraint. While it is certainly possible that the grammar includes a constraint that specifically refers to sonority distance (e.g., the restrictions on the sonority of onsets and nuclei; see de Lacy, 2006; Prince & Smolensky, 1993/2004), the hierarchy in (6) could also emanate from the interaction of broader grammatical principles that do not explicitly concern sonority. For example, Paul Smolensky (2006) captures sonority restrictions using general constraints on the alignment of consonantal features with syllabic domains (see Box 8.1). In a nutshell, sequences like *lba* are avoided because the consonantal features of *b* are not aligned with the beginning of the syllable, and onsets like *bn* are disfavored relative to *bl* because the second consonant shares more consonantal features with the first. Despite having no explicit constraint on sonority (or sonority distance),

172 Phonological universals are core knowledge

this account predicts the preferences in (6). Our present discussion does not address the question of *why* onsets with small sonority distances are marked. Rather, our interest is the possibility that the relevant markedness restrictions are universal – that all grammars are equipped with principles concerning sonority (either explicitly or implicitly), and that such principles universally favor large sonority distances.

Box 8.1 Smolensky's (2006) account of the restrictions on complex onsets

Paul Smolensky shows how sonority restrictions can emanate from broader principles concerning the alignment of feature domains. Smolensky's proposal specifically accounts for two of the preferences concerning complex onsets:
(a) Unmarked onsets do not fall in sonority.
(b) Unmarked onsets have a steady rise in sonority.

These two preferences fall out primarily from two markedness constraints:COD ≡ ALIGN-L (C, σ): requires that the left edge of a feature domain [-φ] must coincide with the left edge of a syllable

F^0: For each feature $_\varphi$, an input segment that is [+$_\varphi$] (or [−$_\varphi$]) corresponds to an output segment that is the head of an output [+$_\varphi$] (or [−$_\varphi$]) domain.

A third constraint – F – mandates faithfulness between input and output features by demanding that each input and output have the same values for any given feature [$_\varphi$]

The tableau in (7) illustrates how COD ≡ ALIGN-L (C, σ) compels the language to avoid sonority falls. The tableau lists two inputs, a syllable with a liquid, obstruent, and a vowel (LOV) and one with an obstruent-liquid onset (OLV), represented in the top row of the left and right halves of the tableau. Each input segment is specified for four binary features that define its sonority – syllabic, vocoid, approximant, and sonorant; positive values define vowels, negative values are consonantal. The constraint COD ≡ ALIGN-L (C, σ) requires the consonantal (i.e., negative) feature values to be aligned with the left edge of the syllable. The outputs, listed in the leftmost column, indicate the parsing of each feature along the consonantal and vocalic domains. Within any given domain (e.g., consonantal), shared features are listed only once, as the domain heads. For example, negative syllabic feature of the two obstruents in OOV is specified once, as [–][+], rather than [–][–][+]).

Sonority restrictions: linguistic and typological evidence 173

(7) Sonority rises are preferred to falls

Output	Input	COD ≡ ALIGN–L (C, σ)	F	Input	COD ≡ ALIGN–L (C, σ)	F
syllabic vocoid approximant sonorant	– – + – – + + – + + – + /LOV/			– – + – – + – + + – + + /OLV/		
.LOL. [–] [+] [–] [+] [+] [–] [+] [+] [–] [+]		*! *				*!* **
.OOV. [–] [+] [–] [+] [–] [+] [–] [+]		☞	☺			*! *
.OLV. [–] [+] [–] [+] [–] [+] [–] [+]			*!* **		☞	

(8) Larger rises are preferred to smaller ones

Output	Input	[F⁰]³	F	[F⁰]²	[F⁰]¹	Input	[F⁰]³	F	[F⁰]²	[F⁰]¹
syllabic vocoid approximant sonorant	– – + – – + – – + – + + /ONV/					– – + – – + – + + – + + /OLV/				
.ONV. [–⌒] [+] [–⌒] [+] [–⌣] [+] [–] [+⌒]			*!		*		!*		*	
.OLV. [–⌒] [+] [–⌣] [+] [–] [+⌒] [–] [+⌣]	☞			* *	*	☞			* *	

An inspection of the OLV case (e.g., *bla*) shows that it incurs no violation of the COD ≡ ALIGN-L (C, σ) constraint: since *l* shares all of its consonantal features with *b*, no feature in *l* initiates a new consonantal domain. Accordingly, the optimal output for *bla* is the faithful one – *bla* – any other output would fatally violate the faithfulness constraint F. The situation is quite different for a liquid-obstruent input, such as *lba* (indicated in the second column). Here, the obstruent has two consonantal features that are not shared by the liquid (approximant and sonorant), so these features must initiate new consonantal domains to the right of the syllable's left edge, an output that incurs two violations of COD ≡ ALIGN-L (C, σ). In this case, an unfaithful obstruent-obstruent output (e.g., *bda*) would be superior to the faithful output (e.g., *lba*): Despite the violation of faithfulness, this output manifests no new consonantal feature in its second consonant, and consequently, it eliminates the violation of COD ≡ ALIGN-L (C, σ); (the remaining unfaithful candidate, OLV, is inferior, because its solution for the COD ≡ALIGN-L (C, σ) violation comes at a greater cost in terms of faithfulness).

The tableau in (8) illustrates the preference for a large rise in sonority. Here, we compare the representation of two inputs with a sonority rise – either a large rise, as in the obstruent-liquid onset (OLV, at the right half of the tableau), or a smaller one, in the obstruent nasal sequence (ONV, at the left half). As in the previous tableau, each segment is specified for the features syllabic, vocoid, approximant, and sonorant. The markedness constraint F^0 requires that each feature corresponds to the head of its domain, and domains are assumed to be left-headed, so negative consonantal features must be aligned with the left edge of the syllable. Accordingly, every negative feature value that the second consonant shares with the first would incur a single violation of F^0. Formally, F^0 is captured as a power hierarchy – a hierarchy of constraints that corresponds to the number of F^0 violations: A single shared value violates the constraint once at the level of F^{01}; two features incur a violation F^{02}, three violations incur a violation of F^{03}, and so forth. Languages can vary on their tolerance for such violations depending on the ranking of the faithfulness constraint F relative to the hierarchy. In this tableau, the faithfulness constraint F is ranked above F^{02}, so this language allows the second onset consonant to share two of its features with the first onset consonant – three shared features (or more) are not tolerated. In inputs such as *bl*, the second consonant shares only two features with the first, and consequently, the winning candidate is the faithful one. But in obstruent-nasal onsets such as *bna*, the nasal *n* shares with *b* three consonantal features – syllabic, vocoid, and approximant. Accordingly, the faithful output *bna* loses in favor of *bl* – an onset that achieves a tolerable violation of two shared consonantal features.

8.2.4 Sonority restrictions are candidates for core knowledge

Sonority restrictions are good candidates not only for grammatical universals, but also for principles of core knowledge, specifically. Recall from Chapter 3 that principles of core knowledge are idiosyncratic, universal, and functionally adaptive. The fact that sonority restrictions impose specific restrictions on the combination of phonological primitives distinguishes them from domain-general principles that broadly mandate structural simplicity or functional efficiency.

This is not to say that sonority restrictions are functionally unmotivated. One can certainly imagine how broad preferences on the alignment of feature domains (Smolensky, 2006) could be computationally advantageous, as well-aligned structures are potentially simpler. But unlike the preference for "single-feature interactions" (discussed in the previous chapter), where simplicity might well "run the show" on its own, the intricate restrictions on sonority appear less amenable to this explanation. To the extent that sonority preferences are grounded in simplicity, simplicity is more likely to exert its effect indirectly: Simpler grammatical principles might be favored, but simplicity may not obviate the need to encode sonority-related preferences in the grammar.

Similar arguments apply to the phonetic basis of sonority. Sonority restrictions are clearly well grounded in phonetic pressures, as they allow for a simultaneous production of consonants and vowels in a manner that optimizes rapid transmission while maintaining the intelligibility of the speech signal (Kawasaki-Fukumori, 1992; Mattingly, 1981; Ohala, 1990; Wright, 2004). The adaptive, functional value of sonority restrictions would explain why such restrictions could have been favored by natural selection. The clear adaptive value of sonority restrictions, however, does not mean that they can be reduced to phonetic pressures. Indeed, sonority distinctions are partly arbitrary, as the phonological notion of "more sonorous" varies across languages: While sonority restrictions in Ancient Greek are acutely sensitive to the voicing of obstruents (Steriade, 1982), English syllables appear oblivious to this factor.

Not only are sonority restrictions demonstrably distinct from phonetic pressures, they are sometimes even contradictory. Consider, for example, the following example from Somali, suggested by Paul de Lacy. In this language, voiceless stops become voiced in word-final position (e.g., /arak/➔a.rag, 'see'; de Lacy, 2006: 123). The voicing of coda consonants is remarkable because it is phonetically unmotivated: Voiced codas are harder to produce than voiceless ones. Sonority restrictions, however, offer a plausible grammatical explanation for this process. This time, however, the relevant restrictions target the sonority cline at the syllable's end. Like onsets, unmarked codas fall in sonority relative to the nucleus. Codas and onsets, however, differ on their preferred sonority slopes. While unmarked onsets exhibit a sharp decline from the nucleus (e.g., *pa* > *na*), codas favor a moderate one (e.g., *an* > *ap*; Clements, 1990). Because voiceless codas are less sonorous than voiced codas (see 1b), coda voicing yields a more moderate sonority decline from the nucleus, and consequently, it improves the well-formedness of the syllable. Coda voicing thus presents an example of a phonological process that obeys sonority even at a clear phonetic expense. Such processes demonstrate that sonority restrictions are autonomous from the functional demands of speech production. And indeed, in Chapter 6, we have shown that similar sonority restrictions apply even to sign languages.

The ubiquity of sonority restrictions across languages, their functional motivation, on the one hand, and their demonstrable autonomy from phonetic pressures, on the other, are consistent with their view as aspects of core phonological knowledge. If this hypothesis is correct, and universal grammatical restrictions are, in fact, active in all grammars, then one would expect to see their effect in the behavior of individual speakers. The next section explores this question using experimental methods.

8.3 Broad sonority restrictions are active in the grammars of individual speakers: experimental evidence

The typological and linguistic evidence reviewed in the previous sections suggests that languages systematically restrict the sonority distance in the onset – the larger the sonority distance, the less marked the onset (see 7). In line with this prediction, a large experimental literature shows that people are sensitive to the structure of onsets that are attested in their language. Sonority restrictions have been implicated in speech errors (e.g., Stemberger & Treiman, 1986), word games (e.g., Fowler et al., 1993; Treiman, 1984; Treiman & Danis, 1988; Treiman et al., 2002), reading tasks (e.g., Alonzo & Taft, 2002; Levitt et al., 1991), first- (Barlow, 2005; Pater, 2004) and second-language acquisition (e.g., Broselow & Finer, 1991; Broselow & Xu, 2004; Broselow et al., 1998), developmental phonological disorders (e.g., Barlow, 2001; Gierut, 1999), and aphasia (e.g., Buchwald et al., 2007; Romani & Calabrese, 1998b; Stenneken et al., 2005). These conclusions from native onsets, however, do not directly speak to the question of grammatical universals. A finding that English speakers, for example, favor *bla* to *lba* cannot tell us whether this preference is due to a grammatical constraint on sonority, or to some non-grammatical sources – the fact that *bla* is more familiar, or perhaps easier to perceive and produce. Even if the preference for *bla* is unequivocally linked to the grammar, a second question arises regarding the scope of this preference: Does the advantage of *bla* reflect a universal constraint that broadly favors onsets with large sonority distances (as in 9) or a narrow preference that only favors the range of sonority distances allowed in English?

(9) The markedness hierarchy of sonority distance (the box indicates sonority distances that are unattested in English)

...$\Delta s = 2 \succ$ $\boxed{\Delta s = 1 \quad \succ \quad \Delta s = 0 \quad \succ \quad \Delta s = -1 \succ \quad \Delta s = -2}$...

To adjudicate between these possibilities, we might turn to gauge people's preferences for onsets that are unattested in their language. Consider, for example, English speakers. English restricts the sonority distance of onset clusters to a minimum of two steps. It allows onsets such as *bl*, but it does not systematically tolerate smaller distances – smaller rises (e.g., *bn*, $\Delta = 1$), plateaus (e.g., *bd*, $\Delta = 0$), and falls (e.g., *lb*, $\Delta = -2$; the only exception to this generalization are *s*-initial clusters, see footnote 1, p. 167). Our question is whether English speakers are nonetheless sensitive to the markedness of these unattested onsets (the boxed section of 9) – do they prefer onsets that are universally unmarked to marked ones?

Several studies indeed report that as the markedness of the cluster increases, people are less likely to produce (Broselow & Finer, 1991; Davidson, 2000; 2006b) and perceive them (Moreton, 2002) correctly, and they are also less likely to judge the cluster as frequent across the world's languages (Pertz & Bever, 1975). But whether these difficulties are directly due to sonority distance is not entirely clear. Some authors have suggested that the difficulties in the production and perception of unattested clusters result not from sonority restrictions but from a host of articulatory and acoustic reasons (e.g., Davidson, 2006a; Fleishhacker, 2001; Ohala, 1990; Redford, 2008; Zuraw, 2007). Lisa Davidson and colleagues (2006), for example, have shown that English speakers are more likely to misarticulate the onset *vn* compared to *zm* and *zr* despite the fact that their sonority distance is identical ($\Delta s = 1$). So while it is clear that unattested onsets are not equally preferred, these findings leave open the question of whether these preferences are constrained by the universal sonority hierarchy.

In what follows, we address this question in two steps. We first review evidence demonstrating that speakers of various languages are sensitive to sonority restrictions on the structure of onsets that are unattested in their language. We next examine whether this finding can be captured by several non-grammatical explanations, including the statistical properties of linguistic experience and processing difficulties. Since non-grammatical explanations fail to capture the experimental findings, by elimination, we conclude that speakers are equipped with broad, perhaps universal, restrictions on sonority sequencing.

8.3.1 The effect of markedness on behavior: some predictions

Before we embark on the experimental journey, a brief review of our path is in order. Let us assume for the moment that all grammars include universal markedness constraints that yield the sonority preferences in (9). If this assumption is correct, then similar preferences ought to emerge in psychological experiments as well. But how are we to gauge such grammatical constraints from human behavior? How can we infer the markedness of two onsets from the responses they elicit?

In what follows, we will infer the markedness of phonological structures from the pattern of their identification. So let us first briefly consider how markedness and identification are linked. Past research has shown that people tend to misidentify onset clusters that are unattested in their language. For example, Mark Pitt (1998) observed that English speakers misidentify the input *tla* as *təla*. While these results do not specifically indicate why unattested onsets are misidentified, grammatical ill-formedness presents a possible explanation. Optimality Theory (Prince & Smolensky, 1993/2004) indeed predicts a systematic link between markedness and faithful identification. Unattested onsets, in this view, are structures that are unprotected by faithfulness constraints. It is the ranking of the markedness constraints against *tla* (here, I will just capture them schematically as M_{tla}) above the relevant faithfulness restrictions (F_{tla}) that bans such structures from existence. In fact, some versions of Optimality Theory (e.g., Anttila, 1997; Davidson et al., 2006) specifically predict that faithfulness to such inputs is proportional to their markedness – the more marked the onset, the less likely it is to be encoded faithfully (see Box 8.2). As a result, marked onsets are systematically recoded as less marked outputs. Since English typically repairs marked onsets by the epenthesis of a schwa – such repairs are frequent in loanword adaptation, for instance (e.g., *bnei brith*→*bənei brith*) – one would expect English speakers to apply similar repairs in experimental settings as well. Consequently, marked inputs such as *tla* should be misidentified as *təla*, as indeed they are (see 10).

(10) The effect of markedness on misidentification: an Optimality Theoretic account

Input=tla	M_{tla}	F_{tla}
tla	*	
☞təla		*

Note that in this account, misidentification is a grammatical reflex. People misidentify *tla* not because they cannot encode its phonetic form, or because

Box 8.2 Stochastic markedness effects

In this box, we consider how the markedness of various unattested onsets is linked to their repair. Let us begin by first considering the prediction of "classic" Optimality Theory (Prince & Smolensky, 1993/2004), illustrated in the tableau in (10)). The tableau shows that marked onsets (e.g., *tla*) are likely to undergo repair. While this framework explains the difference between attested onsets (which are allowed to be encoded faithfully) and unattested ones (which aren't), it offers no mechanism to distinguish between unattested onsets. In the case of English, for example, this framework predicts that unattested onsets (e.g., *bn, bd, lb*) will all be repaired, but the likelihood of repair is identical for all unattested onsets – *lba*, the worst of the worst on the sonority hierarchy, is just as likely to undergo repair as *bna*. Because each of these structures is situated below faithfulness (schematically, F), neither can be encoded faithfully (a probability of 0 correct identification in all cases).

Suppose, however, that the faithfulness constraint F can be promoted above its normal location – either because its position is subject to some stochastic variability, or because people could exert some cognitive control over the faithful encoding of the input. Either way, let us assume that the faithfulness constraint F can be promoted. For simplicity, let us further stipulate that the probability of promotion to each of the three unattested positions (2–4) along the markedness hierarchy is constant, p (see 11) (Anttila, 1997; Davidson et al., 2006). Although the probability of promotion by three steps (to the highest point ④) is theoretically equal to the probability of promotion by only one step (to point ②), every time F is promoted to a given position (e.g., to position ④), it allows not only for the faithful encoding of the most highly marked structure (e.g., sonority falls) but also of any other structure of lesser markedness. So every time F is promoted to step n, people should be able to correctly compute all structures of markedness n or lower, and consequently, they should, in principle, be able to identify it correctly (assuming identification is not further barred by extra-grammatical factors). For example (see 12), once *lb* is allowed, so is *bn*. By contrast, faithfulness to *bn* will not license *lb*, so the likelihood of correct encoding is higher for *bn* than for *lb*. This stochastic model would thus predict that, as the markedness of the onset increases, people should be less likely to encode it correctly.

(11) The possible positions of faithfulness relative to the fixed markedness hierarchy of sonority distance

④*falls ≻③* plateaus≻②* small rises≻ ┆ ①*large rise
┆ *F English state*

(12) The likelihood that an onset cluster (*bn* vs. *lb*) is encoded faithfully given two alternative placements of the faithfulness constraint *F* along the sonority hierarchy

F position	Ranking	bn	lb
②	F >>*bn	✓	X
④	F>>*lb	✓	✓

they are unfamiliar with it. Misidentification is neither a consequence of functional pressures nor does it result from unfamiliarity. Rather, misidentification occurs because *tla* is ill formed, and consequently, it is actively recoded by the grammar (e.g., *tla*→*təla*).

Such systematic markedness reflexes come handy in our hunt for universal markedness constraints. If markedness can trigger misidentification, then one can use misidentification as a gauge for markedness. All things being equal, higher misidentification indicates greater markedness. As always, the "all things" qualification is important because, as we will next see, misidentification can certainly occur for numerous reasons unrelated to markedness. Nonetheless, the phenomenon of misidentification presents an anchor to begin our investigation of universal grammatical restrictions on sonority. Specifically, if people represent the universal sonority hierarchy in (9), then as the markedness of the onset increases, people should be more likely to repair the onset by inserting a schwa, and consequently, identification accuracy should decrease. The experiments reported next test this prediction. Subsequent sections investigate whether markedness can, in fact, be the source of those findings.

8.3.2 Are English speakers sensitive to the sonority of onsets that are unattested in their language?

In a series of experiments, my colleagues – Donca Steriade, Vered Vaknin, then my student, Tracy Lennertz, and I examined whether speakers are sensitive to the sonority distances of onsets that are unattested in their language. In one study (Berent et al., 2007a), we presented English-speaking participants with three types of onset clusters (see 13). One set of onsets comprised mostly stop-nasal combinations (e.g., *bnif*), so such onsets had a small rise in sonority ($\Delta s = 1$) – smaller than the minimum rise required in

English ($\Delta s = 2$). Another set of onsets comprised two stops (e.g., *bdif*), so their sonority distance – a plateau – was more marked. The third, most marked set of monosyllables had sonorant-obstruent combinations, so these onsets had a fall in sonority. To examine the effect of onset structure specifically, we matched these three types of monosyllables for their rhyme. For each such monosyllable, we also constructed a matched disyllable, identical in all respects except that the two initial consonants were separated by a schwa (e.g., *bənif, bədif, ləbif*). Monosyllables and disyllables were mixed, and they were presented to participants aurally, spoken by a Russian talker (because Russian allows all these onset structures, the talker was able to produce all onsets naturally).

(13) A sample of the materials from Berent et al. (2007)

	Monosyllables	*Disyllables*
Small rise	*bnif*	*bənif*
Plateau	*bdif*	*bədif*
Fall	*lbif*	*ləbif*

If English speakers possess knowledge of the sonority hierarchy in (9), then as the markedness of the onset increases, people should be less likely to represent the monosyllables faithfully. Assuming, further, that ill-formed onsets are repaired by means of schwa epenthesis (e.g., *lbif*➔ *ləbif*), then people should specifically misidentify ill-formed monosyllables as disyllables. For example, people should misidentify the monosyllable *lbif* as the disyllable *ləbif*, they should be less likely to misidentify *bdif*, and least likely to misidentify *bnif*. And since *lbif* is recoded as *ləbif*, people might also incorrectly judge such marked monosyllables as identical to their disyllabic counterparts (e.g., *lbif = ləbif*). Crucially, the likelihood of misidentification should be monotonically linked to markedness: It should be strongest for sonority falls, followed by plateaus, and small rises.

8.3.3 People believe lbif *is disyllabic*

To test these predictions, my colleagues and I first presented participants with a syllable count task. In each trial, people heard one auditory stimulus – either a monosyllable or a disyllable – and they were asked to indicate whether it includes one syllable or two. Results showed that, as the markedness of the monosyllable increased, participants were less likely to classify the input as monosyllabic. Instead, marked onsets were misidentified as disyllables (see Figure 8.1).

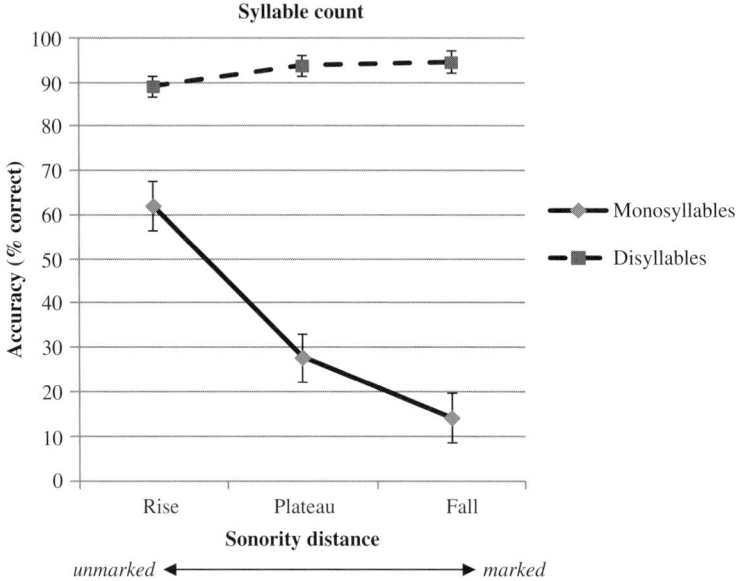

Figure 8.1 Response accuracy in the syllable count task (from Berent et al., 2007a). Error bars indicate 95% confidence intervals for the difference between the means.

Interestingly, the markedness of monosyllables also modulated responses to their disyllabic counterparts: Responses to *bənif* were less accurate than to *bədif*. The difficulty with *bənif* is clearly not due to its own ill-formedness, as *bənif* and *bədif* are both well formed. This difficulty with *bənif* is also not due to the phonetic properties of these materials. The schwa – the element that distinguishes disyllables and monosyllables – was not shorter in *bənif* than in *bədif*, and the difficulty persisted even after the duration of the schwa was statistically controlled. It thus appears that the misidentification of *bənif* results not from its own properties (phonetic or phonological) but rather from those of its monosyllabic counterpart. This possibility is indeed likely given that the task elicits a forced choice (one vs. two syllables). In this situation, responses to a disyllable might well be affected by the well-formedness of the monosyllabic alternative – the better formed the monosyllable, the more viable a contender it becomes, and therefore, the more difficult it is to select the correct disyllabic response. Ill-formed monosyllables like *lbif* are bad contenders, but better-formed ones like *bnif* might tempt people to seriously doubt whether the disyllable they had just heard (*bənif*) was actually a monosyllable, *bnif*. The possibility that *bnif* is a better contender than *bdif* provides converging evidence that people are sensitive to the well-formedness

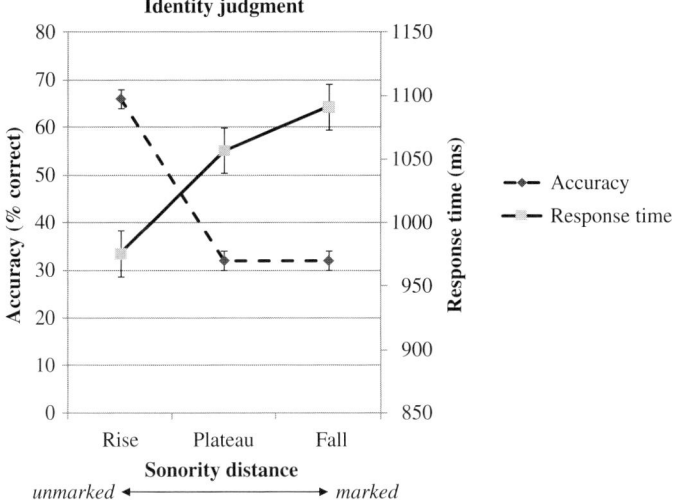

Figure 8.2 Response accuracy and response time to non-identity trials in the identity-judgment task (from Berent et al., 2007a). Error bars indicate 95% confidence intervals for the difference between the means.

of unattested onsets, and that ill-formed onsets are misidentified as their disyllabic counterparts.

8.3.4 Does lbif = ləbif?

The syllable count results demonstrate that marked monosyllables tend to be misidentified disyllables. But what kind of disyllables, precisely, are these? Is *lbif* specifically recoded as *ləbif* (rather than some other disyllable, e.g., *əlbif*)? And if *ləbif* is, in fact, the preferred mode of repair, how robust is this grammatical illusion? Will the *lbif = ləbif* delusion persist even when people hear these two items back to back and are explicitly asked to distinguish them?

To examine this question, we next subjected the same items to an identity task (AX). In each trial, participants were presented with two stimuli – either identical (two monosyllables: e.g., *lbif-lbif*, or disyllables, e.g., *ləbif-ləbif*), or non-identical (e.g., *lbif-ləbif; ləbif-lbif*), and they were simply asked to determine whether the two items were identical.

The responses to non-identical trials (e.g., does *lbif = lebif*?) are presented in Figure 8.2 (identical trials included two identical tokens, so in this case responses were not reliably affected by sonority distance). As the markedness of the monosyllable increased, people had greater difficulty discriminating the monosyllable from its disyllabic counterpart. In fact, on most trials (about 70 percent), people thought that sonority falls such as *lbif* were identical to their disyllabic counterparts. And on the rare occasions when people were able to correctly distinguish among non-identical items, they took longer to respond as the markedness of the monosyllable increased. Thus, not only are participants likely to misidentify marked monosyllables as disyllabic, but they even fail to discriminate between monosyllables and disyllables. The systematic misidentification of marked onsets is in line with the possibility that such onsets are grammatically marked.

8.3.5 Are small sonority distances systematically misidentified?

The previous section shows that the behavioral preferences of individual speakers converge with the putative universal restrictions on sonority. Before moving to determine the source of this convergence – whether it is, in fact, caused by universal grammar, or non-grammatical sources – let us first address some challenges to the empirical findings. The challenges considered here do not concern *why* marked onsets are misidentified. Rather, it is misidentification itself that is questioned. The concern is that people do not systematically misidentify onsets with small sonority distances. The difficulties with marked onsets reported above are all due to idiosyncratic aspects of the materials used in these specific experiments – either the choice of the disyllabic baseline, or the properties of the monosyllables. Once it has been established that misidentification is, in fact, a systematic phenomenon, subsequent sections will investigate its source.

8.3.5.1 Spurious effects of the disyllabic baseline

The experiments described in the previous sections gauge the markedness of various CCVC monosyllables by comparing how they are distinguished from CəCVC counterparts. Underlying this approach is the assumption that marked monosyllables are repaired epenthetically (e.g., *lbif→ləbif*), so the confusion of CCVC monosyllables with their CəCVC counterparts potentially reflects their markedness.

Epenthesis, however, is not the only possible form of repair. It is well known, for example, that marked onsets can also be repaired by prothesis (e.g., *lbif→əlbif*) – such repairs are frequently seen in loanword adaptation in numerous languages (e.g., Spanish: *sport→esport*). So it is certainly conceivable that participants in our experiments do not repair marked onsets by epenthesis alone (Peperkamp, 2007). However, this possibility, interesting as it may be, does not necessarily undermine our conclusion. If people recode *bnif* and *lbif* as *əbnif* and *əlbif*, then they will still experience difficulty with the identification of these inputs (since their mental representation, as *əbnif* and *əlbif*, is disyllabic, and it differs from the disyllabic baselines in identity judgment, *bənif* and *ləbif*). As long as the method of repair – epenthesis or prothesis – is consistent across onset types, the greater difficulty with marked onsets would still indicate that such onsets are more likely to undergo repair.

There is evidence, however, that the choice of repair might actually vary according to the markedness of the onset (Gouskova, 2001), and this possibility opens the door for a conspiracy theory that is quite damaging to our conclusions. Consider, for example, a scenario where onsets of falling sonority are repaired by epenthesis (e.g., *lbif*→*ləbif*), whereas onsets of rising sonority are repaired by prothesis (e.g., *bnif*→*əbnif*). Under this scenario, *lbif* is mentally represented as *ləbif*, and consequently, it is highly confusable with its counterpart, *ləbif* (see 14). The unmarked, *bnif*, in contrast, would be mentally represented as *əbnif* – quite distinct from the experimental baseline *bənif*, so this input would be *easy* to identify. Crucially, the superior identification of *bnif* (compared to *lbif*) is unrelated to markedness. In fact, *bnif* should still be easier to identify even if people were utterly insensitive to the markedness of its onset. If this conspiracy theory is true, then the difficulty with *lbif* might not be a feature of marked onsets, generally, but rather an artifact of our experimental design, which happens to compare *lbif* with the baseline *ləbif*.

(14) The representation and discrimination of *bnif* and *lbif* (a conspiracy theory)

CCVC input	Mental representation of the input	Discrimination from CeCVC
bnif	*əbnif*	Easy
lbif	*ləbif*	Hard

But several observations counter this possibility. First, although the linguistic evidence suggests that markedness modulates the choice of repair in loanword adaptation, the observed pattern is actually contrary to the conspiracy theory above: Marked onsets of falling and level sonority are *less* likely to elicit epenthetic repair than unmarked sonority rises, suggesting that, if anything, the results described above underestimate the actual rate of epenthetic repair for marked onsets (Gouskova, 2001). Moreover, the experimental evidence for misidentification replicates regardless of the choice of baseline. A follow-up AX experiment replicated the markedness effect even when all onsets were compared to prothetic counterparts (e.g, *bnif-əbnif*; *bdif-ebdif*; *lbif-əlbif*; Lennertz & Berent, 2011; for similar results in Spanish speakers, see Berent et al., 2011b). Finally, the disyllabic misidentification of marked syllables obtains in tasks that do not rely on the discrimination of CCVC from any particular disyllabic form – in syllable count, and even when people are simply asked to transcribe the monosyllables (Berent et al., 2009). These observations suggest that the misidentification of marked onsets is a genuine phenomenon that is inexplicable by the choice of the disyllabic baseline.

8.3.5.2 Spurious properties of the monosyllables

A second challenge attributes our findings to the choice of monosyllables. In the materials described so far, the worst-formed structures correspond to sonorant-obstruent combinations (e.g., *lbif*) whereas the best-formed onsets invariably comprised obstruent-sonorant sequences (e.g., *bnif*). It is thus possible that the advantage we observed is not general to all unmarked sonority distances but pertains only to these specific sequences (which happen to also resemble the onsets allowed in English).

But follow-up experiments challenge this explanation as well. These experiments examine the effect of markedness on two types of onsets that were both nasal-initial – either onsets of rising sonority (e.g., *mlif*) or sonority falls (e.g., *mdif*). These two types of onsets were compared in multiple tasks, using several sets of materials. Once again, however, onsets of falling sonority were more likely to be misidentified as disyllabic, and they tended to be (incorrectly) judged as identical to their disyllabic counterparts (Berent et al., 2009; Berent et al. 2010; Berent et al., 2011b; Berent et al., 2012a). The consistent preference of onsets of rising sonority – either obstruent or nasal initial – suggests that onsets with larger sonority distances are represented more accurately.

8.3.6 Why are marked onsets misidentified?

The results reviewed so far suggest that marked onsets are systematically misidentified, and this phenomenon is robust with respect to the specific structure of the materials (obstruent vs. nasal initial) and the choice of the disyllabic baseline. The convergence between the preferences of English speakers concerning onsets that are unattested in their language, on the one hand, and the cross-linguistic evidence, on the other, is in line with the hypothesis that sonority restrictions are universally active in all grammars. Convergence, however, is not sufficient to demonstrate causality. And indeed, misidentification can also occur for various non-grammatical reasons. Moreover, even if misidentification were caused by speakers' phonological knowledge, the relevant knowledge may not necessarily be universal grammar. In what follows, we consider these possibilities.

8.3.6.1 The role of phonetic factors

Why do speakers misidentify marked onsets? Our explanation of choice links such misidentifications to universal grammatical restrictions on sonority. We assert that the sonority hierarchy is universally represented in all grammars, but grammars differ on the tolerance of marked structures – some grammars (e.g., Russian) allow all these distances, whereas others (e.g., English) disallow portions of the hierarchy. Misidentification, then, is the result of universal grammatical knowledge. As noted above, however, auditory stimuli might be misidentified for various phonetic reasons.

Stimuli artifacts
The most trivial phonetic alternative attributes misidentification to spurious properties of our stimuli. Perhaps English listeners misidentify the monosyllabic materials used in these experiments because these particular stimuli were, in fact, disyllabic – they included acoustic cues that are universally interpreted as disyllabic, irrespective of linguistic knowledge.

This possibility is rather easy to rule out. Recall that the materials in Berent et al. (2007a) were recorded by a Russian talker. If these materials are fundamentally flawed, then any listener will classify these items as disyllabic. In contrast, if the materials are valid tokens of Russian monosyllables, then, unlike English participants, Russian participants should interpret those items as monosyllabic. A replication of the English experiments with Russian speakers supported the latter possibility (Berent et al., 2007a). Unlike English participants, Russian listeners identified the monosyllabic stimuli as such in over 90 percent of the trials, and they were also quite accurate in discriminating these items from their disyllabic counterparts in the AX task (accuracy was over 80 percent in all conditions). Similar results were also obtained with the nasal-initial stimuli (Berent et al., 2009).

The role of phonetic knowledge
The contrasting behaviors of the two groups of participants – English and Russian speakers – suggest that the misidentification of our materials by English participants is indeed due to their linguistic knowledge, rather than to stimuli artifacts. But what kind of knowledge is it?

Above, we suggested that the relevant knowledge is grammatical. In this view, it is the grammatical markedness of such clusters that actively triggers their recoding as less marked outputs, and consequently, their misidentification. According to an alternative scenario, misidentification occurs at an earlier processing stage. Indeed, before people can apply their grammatical phonological knowledge, they must first extract a viable phonetic representation of the input. To do so, they must accurately register the auditory form, extract the relevant phonetic cues, and map them onto phonetic and surface phonological forms. It is conceivable that the various types of onsets studied in these experiments are not equally easy for phonetic processing – marked onsets, such as *lbif*, could be harder to encode, as their acoustic properties might be compatible with the representation of *ləbif*. Because English listeners

188 Phonological universals are core knowledge

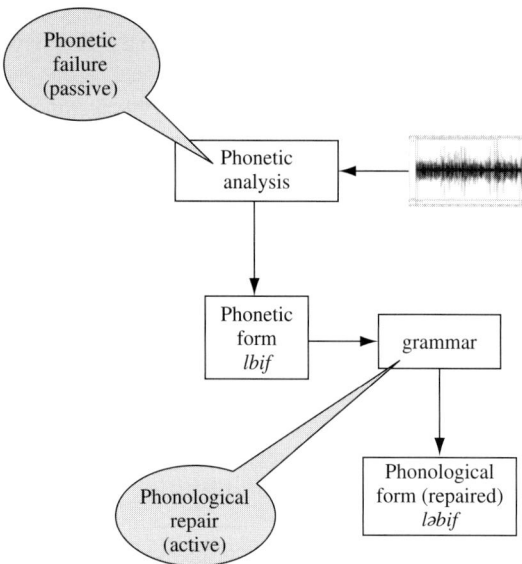

Figure 8.3 The phonetic vs. phonological accounts of misidentification

have no experience in the phonetic processing of such onsets, they might lack the phonetic knowledge necessary to parse the acoustic input *lbif* faithfully, and misidentification ensues.

Phonetic misidentification, however, differs from phonological recoding in two important respects. First, on the phonetic view, misidentification is guided by knowledge that is phonetic, rather than phonological. Second, these two views differ on how misidentification occurs. On the phonetic view, misidentification is a passive process – it reflects the failure to register the surface phonetic form of *lbif* from the acoustic input. But on the phonological alternative, identification is an active recoding. English speakers, in this view, can successfully extract the phonetic form *lbif* – misidentification occurs only at a subsequent grammatical stage that actively recodes *lbif* as *ləbif* in order to abide by grammatical constraints (see Figure 8.3).

The subtle distinction between these two tales of misidentification – the phonological and phonetic (see 15) – is absolutely crucial for my argument. Although it is certainly possible that marked structures might present challenges to phonetic encoding, it does not follow that misidentification solely occurs for phonetic reasons. Many people, however, both laymen and experienced researchers, assume without question that misidentification is a purely phonetic phenomenon. The tendency to equate misidentification and phonetic failure is so strong that they are often viewed as synonyms. Laymen often state that nonnative speakers misidentify their language *because* they cannot "hear" it "properly." Similarly, many prominent researchers assert that the difficulty in the identification of phonologically ill-formed structures results from a perceptual illusion. While the precise meaning of "perception" is rarely clarified, the intended interpretation, it appears, concerns phonetic, or perhaps even auditory processing. But passive phonetic failure is not the only possible source of misidentification. Ample research demonstrates that the identification of linguistic stimuli is affected by a host of grammatical factors. These considerations open the door for an alternative causal chain. In this view, people misidentify ill-formed phonological structures because their grammatical knowledge prevents them from faithfully encoding these structures. Phonological ill-formedness, then, is not the consequence of misidentification but rather its cause.

(15) Two tales of ill-formedness and misidentification
 a. Phonetic failure➔misidentification➔phonological ill-formedness
 b. Phonological ill-formedness➔misidentification

Why are ill-formed onsets misidentified: passive phonetic failure or active phonological recoding?
With a clear distinction between the phonological and phonetic sources of misidentification securely in place, we can now move to see how one can use experimental evidence in order to adjudicate between them. To anticipate the conclusions, ill-formed onsets are not invariably harder for phonetic processing. In fact, misidentification demonstrably occurs even when phonetic difficulties are entirely eliminated. These results suggest that the misidentification of marked onsets is not due to difficulties in extracting their phonetic forms.

Ill-formed onsets are not invariably harder to process The phonological and phonetic accounts disagree on one crucial issue – whether the phonetic form of ill-formed onsets is encoded accurately or deficiently. On the phonetic account, phonetic encoding is deficient, but on the phonological view, it is not necessarily so, as recoding occurs at a subsequent phonological stage. Since people typically base their behavior on the phonological, rather than phonetic, form, ill-formed onsets are often misidentified. Misidentification, however, is not inevitable. The phonological (but not phonetic) account predicts that the phonetic encoding of marked onsets is potentially free of distortion, so if participants could only be "convinced" to access the phonetic representation directly, then the difficulty with ill-formed onsets should vanish.

The possibility of directly consulting the phonetic form is actually not far-fetched. Past research has shown that people can attend to various non-distinctive phonetic properties, including aspiration and voice onset time

(Coetzee, 2011; Theodore & Miller, 2010), so it is conceivable that participants could access the phonetic form of marked clusters as well. The crucial question is what kind of phonetic representation is available for such onsets – whether it is faithful or distorted. If the phonetic form of marked onsets is accurate, then once people attend to the phonetic level, then they should be able to identify marked onsets accurately – as well as they identify their less marked onsets. The results of various experiments are indeed consistent with this conclusion (Berent et al., 2007a: Experiments 4–5; Berent et al., 2011b). Here, we consider one such example concerning the identification of nasal clusters (Berent et al., 2012a).

In these experiments, English speakers identified nasal-initial onsets with either sonority rise (e.g., *mlif*) or fall (e.g., *mdif*). To generate these monosyllables, Tracy Lennertz, Evan Balaban, and I first had a native English talker produce their disyllabic counterparts (e.g., *məlif* and *mədif*), and we next excised the schwa ə in several steady increments. This procedure, in turn, yielded a continuum ranging from a fully disyllabic form (*məlif* and *mədif*, at step 6) to the monosyllabic counterparts (*mlif* and *mdif*, at step 1). In accord with past research, we expected that as the duration of the schwa increases, people should be more likely to identify the input as disyllabic (Dupoux et al., 1999). Moreover, when provided with the monosyllabic endpoints (at step 1), sonority falls should be less likely than rises to yield monosyllabic responses. The critical question is whether this misidentification reflects the failure to register the phonetic form of such onsets or their active repair.

To adjudicate between these possibilities, we compared the identification of these onsets under two conditions, designed to promote attention to either the phonological or the phonetic form. In the phonological condition, people were asked to determine whether the input includes one beat or two – a proxy of the number of syllables. Because this judgment capitalizes on syllabification, participants are expected to consult the output of the phonological grammar. Regardless of whether the misidentification of sonority falls occurs at a phonetic or a phonological stage, both accounts predict that the phonological representation of *mdif* is typically disyllabic, and consequently, this condition should yield a higher rate of disyllabic responses to sonority falls compared to sonority rises. Our main question is whether misidentification can be eliminated when people are engaged in phonetic processing. If people do register the phonetic form of marked outputs accurately, then once accessed, sonority falls should no longer be misidentified. To encourage people to attend to the phonetic level, a second group of participants was instructed to monitor a particular phonetic attribute that is correlated with disyllabicity: whether or not the two initial consonants are separated by a vowel (i.e., vowel detection, e.g., "Is there a schwa between the *m* and *l* of *mlif*?"). Participants were administered the two tasks – vowel detection and beat count – in two counterbalanced orders, and they were informed of the incoming task switch ahead of time (to increase

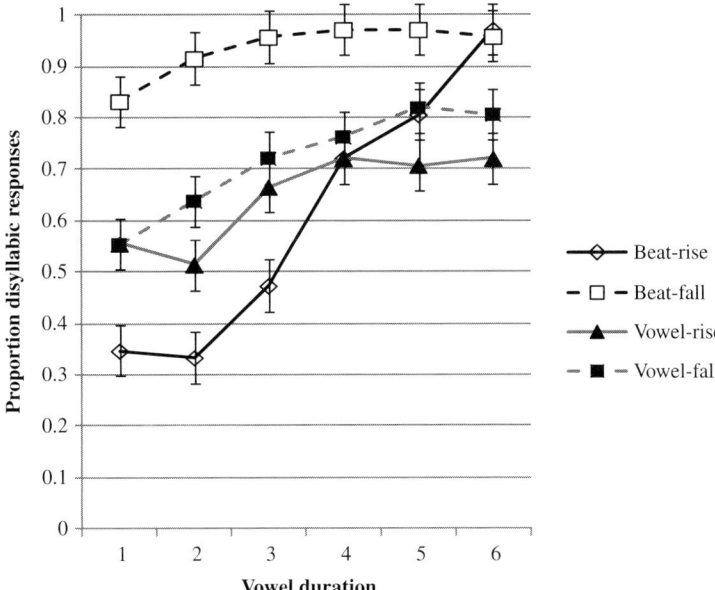

Figure 8.4 The effect of task demands on the misidentification of ill-formed onsets (from Berent et al., 2012a). Error bars reflect 95% confidence intervals constructed for the difference between the means of sonority rises and falls. Note: vowel = vowel detection; beat = beat counting; rise = sonority rise; fall = sonority fall.

their vigilance throughout the experiment). One group performed beat count followed by vowel detection; a second was given the reverse order. Of interest are results to the first block, in which performance is uncontaminated by carry-over effects.

An inspection of the findings (see Figure 8.4) suggests that the phonological and phonetic tasks yielded distinct patterns. When presented with pure monosyllables (step 1), participants in the phonological, beat-count task showed the normal misidentification of marked onsets: They misidentified sonority falls as disyllabic on over 80 percent of the trials, whereas sonority rises were mostly identified as monosyllabic (the rate of disyllabic identification was close to 30 percent). But when participants were instructed to perform the phonetic, vowel-detection task, sonority falls were no more likely to elicit disyllabic responses than sonority rises, that is, sonority falls were as likely to be misidentified as rises.

This latter finding is not merely due to the insensitivity of the vowel-detection task to the structure of the stimuli. Participants in this condition remained sensitive to the phonetic duration of the schwa – as the schwa duration increased, disyllabic responses were more likely. Moreover, there are independent reasons to believe that this condition promoted attention to phonetic detail. The evidence comes from the responses to fully disyllabic stimuli, at step 6. Unlike the previous steps, the disyllabic stimuli were produced naturally, without any splicing. Other research with these materials (Berent et al., 2010) has shown that people are highly sensitive to splicing. They interpret the bifurcation associated with splicing (at steps 1–5) as a cue for disyllabic responses, and consequently, the absence of splicing, at step 6, attenuates the identification of disyllabicity. To exploit this cue, however, participants must be tuned to phonetic attributes, so sensitivity to splicing is diagnostic of phonetic processing. Interestingly, the phonological and phonetic tasks differed on their sensitivity to this cue. While participants in the phonological, beat-detection task were oblivious to splicing (indeed, they were more likely to give disyllabic responses to the unspliced last step relative to the penultimate step in the rise continuum), participants in the vowel-detection task were less likely to yield disyllabic responses to the unspliced endpoint. This observation confirms that participants in the vowel-detection task effectively engaged in phonetic processing. Crucially, once the "phonetic mode" is engaged, marked onsets are no longer misidentified.

Together, these results suggest that the phonetic form of sonority falls is not invariably defective. Accordingly, their typical misidentification probably occurs because participants tend to consult their phonological representations – representations that are recoded by the grammar. Misidentification, then, is due to the active phonological recoding of such onsets, rather than to a passive inability to register their auditory or phonetic forms.

Ill-formed onsets are misidentified even in the absence of phonetic processing The results presented so far suggest that misidentification can be dissociated from the phonetic demands associated with acoustic processing – people can engage in acoustic processing without having marked onsets selectively misidentified. If the phonological account is correct, then it might be possible to also demonstrate the complementary side of the dissociation: Misidentification might occur even when acoustic processing is eliminated altogether. This prediction is indeed borne out by the results of a series of experiments with printed materials (Berent et al., 2009; Berent & Lennertz, 2010). In these experiments, participants were presented with a succession of two printed forms in alternating cases (e.g., *lbif-LEBIF*), and they were asked to determine whether the two forms were identical. Although the results with printed materials are less robust than with auditory stimuli, participants consistently take longer to distinguish marked onsets from their disyllabic counterparts (see Figure 8.5).

The persistent misidentification of printed marked onsets demonstrates that the difficulties in processing marked onsets are not confined to the auditory

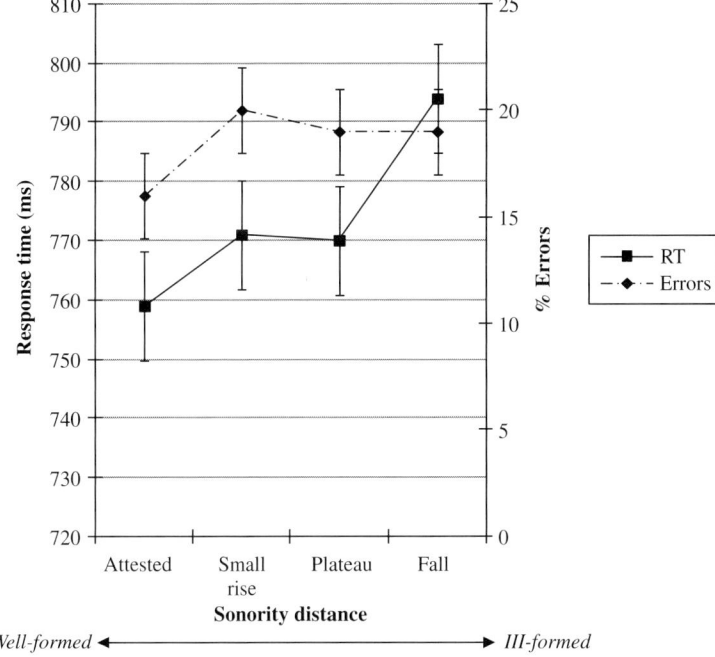

Figure 8.5 The effect of phonological ill-formedness on the identification of printed materials (from Berent & Lennertz, 2010, Experiment 1). Error bars indicate 95% confidence intervals for the difference between the means

modality. These results, together with the finding that marked onsets are not necessarily harder for phonetic encoding, suggest that their misidentification originates from an active process of recoding, triggered by people's linguistic knowledge.

8.3.6.2 The role of lexical analogy

Let us summarize our conclusions so far. In previous sections, I have shown that marked onsets are systematically misidentified, and such misidentifications cannot be blamed on either acoustic artifacts of the auditory stimuli or listeners' inability to extract their phonetic form. Misidentifications, then, do not passively "happen," but rather they are actively "promoted." Nonetheless, misidentifications are clearly modulated by linguistic experience. Russian speakers, whose language tolerates a wide range of onsets, can perfectly identify onsets that English speakers typically misidentify. Together, these results suggest that misidentifications reflect active recoding, promoted by some linguistic knowledge.

But what kind of knowledge is it? One possibility is that speakers are equipped with universal grammatical restrictions on sonority – restrictions that apply to equivalence classes of segments (e.g., obstruents) and generalize across the board, to any member of the class, familiar (e.g., *b*) or novel (e.g., *x*). Alternatively, participants could rely on knowledge of the English lexicon. A preference of *bn* over *lb*, for example, could be informed by the fact that many English onsets begin with an obstruent (e.g., *blow*), and some end with a nasal (e.g., *snow*), but none starts with a sonorant. The preference of *bn* over *lb* could thus reflect the statistical properties of the English lexicon, rather than universal grammatical knowledge. This possibility is also in line with computational results by Robert Daland and colleagues (2011), apparently showing that the English findings (Berent et al., 2007a) can be induced from the co-occurrence of features in the English lexicon.[2]

But there are several reasons to doubt that sonority preferences are solely the product of inductive learning. First, sonority preferences replicate even in languages whose onset inventory is far smaller than that of English. Consider Spanish, for instance. Although Spanish allows onset clusters with large sonority rises (e.g., *playa* 'beach,' *braso* 'arm'), it categorically bans any onset of smaller sonority distance. In this respect, Spanish differs from English, in which some *s*-obstruent exceptions are tolerated (e.g., *stop*), so Spanish speakers clearly have a more limited experience with onset clusters. Nonetheless, Spanish speakers exhibit the same markedness preferences documented in English (although their preferred form of repair is different; Berent et al., 2012d).

A yet stronger test for the role of lexical analogy is presented by languages that have no onset clusters altogether. Korean presents one such case. Korean disallows consonant clusters in the onset, so this language provides its speakers with no lexical experience that would favor onsets such as *bl*.[3] But despite lacking onset clusters altogether, Korean speakers are sensitive to the sonority profile of onset clusters: As the markedness of the onset along the sonority

---

[2] Whether these findings can, in fact, account for the original English data is not entirely clear. Daland and colleagues base their simulations on a set of experimental findings obtained using materials procedures that differ substantially from the ones in Berent et al. (2007a). Moreover, they present no statistical evidence that people in their experiments are sensitive to the structure of unmarked onsets, and their findings show no hint of grammatical repair. Accordingly, their computational results may not necessarily generalize to Berent et al.'s original findings.

[3] Although Korean does allow for word-initial consonant-glide combinations (e.g., /kwaŋ/, 'storage'), the glide in such sequences does not form part of the onset (Kim & Kim, 1991) – unlike true onset clusters, consonant-glide sequences are not subject to co-occurrence restrictions, and their mirror image, glide-consonant, is unattested in codas. Glides, however, do pattern with the following vowel with respect to phonological copying, assimilation, and co-occurrence restrictions (Kim & Kim, 1991; Yun, 2004), in the structure of the Hangul script and in speech errors (Kang, 2003), suggesting that the glide forms part of the nucleus, rather than the onset.

Experimental evidence for sonority restrictions   195

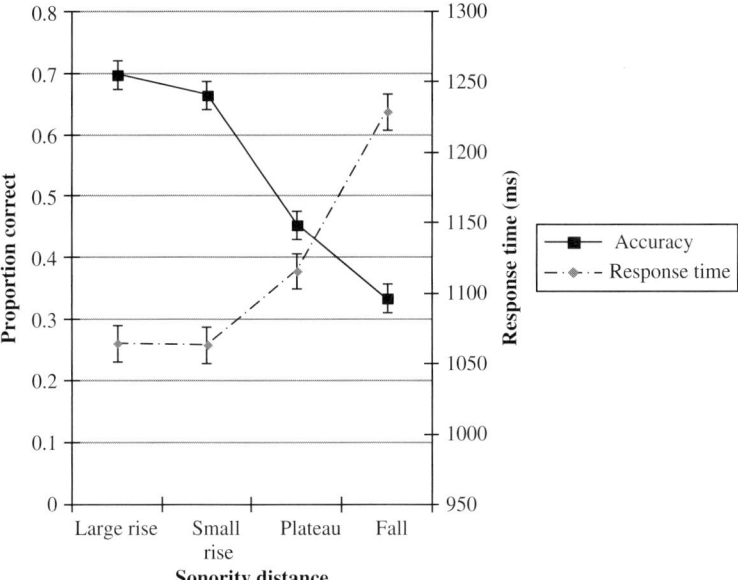

Figure 8.6 The sensitivity of Korean speakers to the sonority hierarchy in an identity judgment task (from Berent et al., 2008). Error bars indicate 95% confidence intervals for the difference between the means

hierarchy increases, Korean speakers are more likely to misidentify monosyllables (e.g., *lbif*) as disyllabic, and they tend to consider them as identical to their disyllabic counterparts (*lbif=ləbif*; see Figure 8.6). Moreover, auxiliary analyses suggest that the performance of Korean speakers is inexplicable by a host of extraneous factors – both phonological and phonetic. For example, the preference of Korean speakers is not due to their familiarity with English as a second language – not only was their sensitivity to the sonority hierarchy unaffected by their (self-reported) English proficiency, but their overall sensitivity in the experimental task (operationalized as d') was higher than native English participants. Similarly, performance was inexplicable by various phonetic and phonological properties of the Korean language (the phonetic release of initial stop-consonants, their voicing, the distribution of [l] and [r] allophones, the experience with Korean words beginning with consonant-glide sequences, and the occurrence of the CC sequence across Korean syllables; see supporting materials; Berent et al., 2008).

Subsequent experimental work by Jie Ren and colleagues (2010) has extended these findings to Mandarin – a language that likewise bans onset clusters, but additionally bans most codas, so its inventory of clusters of any types – either onset clusters or ones occurring across syllables – is highly limited. Ren and colleagues showed that Mandarin speakers nonetheless favor onsets of rising sonority (e.g., *bl*) over sonority falls (e.g., *lb*). Follow-up investigations by Xu Zhao and me (Zhao & Berent, 2011) have established that Mandarin speakers are in fact sensitive to the entire hierarchy of onset clusters – sonority rises are preferred to plateaus, and these, in turn, are preferred to onsets of falling sonority (e.g., *bn*≻*bd*≻*lb*). Moreover, computational simulations by Bruce Hayes (forthcoming) suggest that such preferences are unlearnable by an inductive-learning model in the absence of a substantive bias to attend to sonority-relevant features. Together, these results demonstrate that sensitivity to the sonority hierarchy of onsets does not result from lexical analogy alone.

8.4 Summary and conclusions

This chapter examined whether speakers possess core phonological knowledge that imposes universal grammatical constraints on phonological forms. To this end, we conducted an in-depth analysis of a single case study – the restrictions on sonority sequencing in complex onsets. A review of the linguistic evidence suggested that sonority restrictions present a plausible candidate for a putative grammatical universal. To further determine whether sonority restrictions are, in fact, active universally, in the grammars of all speakers, we next turned to experimental evidence. We saw that marked onsets tend to be systematically misidentified – the more marked the onset, the more likely it is to be misidentified. These results establish that onsets that are disfavored across languages are also dispreferred by individual speakers. Crucially, speakers extend these preferences even to onsets that they have never heard before.

Not only do speakers' sonority preferences converge with those predicted on grounds of markedness, but they also diverge from several non-grammatical explanations. We showed that responses to marked monosyllables (e.g., *lbif*) are independent of the choice of the disyllabic baselines (e.g., *ləbif* vs. *əlbif*). Similarly, the typical misidentification of marked onsets is inexplicable by their phonetic properties – either spurious auditory artifacts of a specific set of materials, or the phonetic characteristics of marked onsets, generally. Misidentification, then, results not from a passive failure to encode the phonetic form of marked onsets but rather from a process of grammatical repair – a process that actively recodes the surface phonetic form to abide by linguistic knowledge.

Additional analyses established that the relevant knowledge does not concern the statistical properties of the lexicon. The preference for unmarked onsets indeed remains significant after their similarity to attested items is statistically controlled. In fact, sonority preferences have been documented even when speakers' language lacks any onset clusters altogether, as in the case of Korean and Mandarin Chinese. By elimination, then, we concluded

Summary and conclusions

that the knowledge governing the preferences for unattested onsets must be grammatical in nature. And since this grammatical knowledge is not based on inductive learning, it is likely due to principles that are broadly shared across speakers.

These conclusions are all consistent with the possibility that sonority restrictions form part of core phonological knowledge. Core knowledge systems manifest unique universal design that is innate (i.e., they are not induced from experience), adaptive, and largely invariant across individuals despite variations in experience. Sonority-sequencing principles meet each of these characteristics. Sonority-sequencing principles are demonstrably in place despite the absence of opportunities for inductive learning. Accordingly, such principles are likely to be broadly present in many grammars, perhaps even universally. Like other principles of core knowledge, sonority-sequencing principles are also functionally adaptive, as large sonority distances optimize the articulation and perception of spoken onsets. Nonetheless, sonority principles cannot be reduced to functional pressures. Sonority constraints do not merely express a generic preference for sequences that are easy to produce/perceive, nor do they vaguely disfavor structural complexity. Rather, they impose precise constraints on the sequencing of phonological primitives. And in some instances, sonority restrictions can be shown to apply even when they conflict with phonetic pressures. These observations are in line with the view of sonority restrictions as idiosyncratically phonological principles, autonomous from the phonetic component.

These conclusions, however, also leave numerous unanswered questions. The possibility that some aspects of core knowledge might not depend on inductive learning does not necessarily mean that such principles are independent of experience. It is indeed perfectly possible that in order for sonority-sequencing principles to manifest themselves, they might require some critical phonetic triggers. Whether such triggers are indeed required and what they might be remain entirely speculative at this point. It is conceivable, for example, that markedness constraints on onset clusters might require experience with consonant clusters of some type (e.g., clusters occurring across syllables, e.g., *abda*). But the role of such triggers should not be confused with learning. I do not propose that people infer the well-formedness of *bla* from the fact that *bl* onsets (or some other structurally analogous onsets, e.g., *pn*) are frequent in their language. Rather, it is the experience with the phonetic properties of some consonant combinations that might trigger the markedness hierarchy of onsets. Unlike inductive learning, phonetically based triggers could be informative even if none of those consonant sequences occurred in onset positions, and even if their frequency was all balanced (e.g., the frequency of *lb*=*bl*). Since the experimental evidence for the restrictions on onset clusters are based on languages that all allow some form of consonant clusters, it remains unknown whether similar preferences might be available to speakers of languages with no clusters at all (purely CV languages).

More generally, these observations underscore the possibility that principles of core knowledge may not be experience-independent. Grammatical restrictions such as the constraints on sonority represent the final state of a mind-brain system that is shaped by multiple sources – ones that are either internal to the grammar or external. This scenario allows for the possibility that, rather than having the grammar fully formed at birth, core phonological knowledge might be configured only gradually, throughout development. How the phonological grammar is implemented in the brain, and how it is configured in ontogeny and phylogeny, is the topic of the last part of this book.

Part IV

Ontogeny, phylogeny, phonological hardware, and technology

9 Out of the mouths of babes

> Previous chapters have examined the hypothesis of core phonology by inspecting the phonological grammars of adult speakers. Core phonology, however, potentially encompasses not only mature phonological systems but also the initial state of the grammar. It is the early onset of core phonological knowledge that might form the scaffold for mature phonological systems by providing a universal set of primitives and combinatorial principles. In this chapter, we examine whether universal primitives and principles are in fact active in early phonological systems. We demonstrate that at birth, infants possess algebraic computational machinery commensurable with the algebraic powers of adult grammars. The phonological preferences of infants and young children are likewise consistent with several of the primitives and markedness constraints seen in adult phonological systems; some of those preferences extend to structures that are unattested in the child's language, and a handful is documented in early infancy. While the available findings are insufficient to fully evaluate the core knowledge hypothesis, they are nonetheless consistent with this possibility.

The evidence presented in the previous chapters suggests that, by the time humans reach adulthood, they possess an algebraic phonological grammar, equipped with representations and constraints that are putatively universal. How do human adults converge on this shared phonological system?

The answer, according to the core knowledge hypothesis, is that people are innately equipped with core phonological knowledge. Core knowledge determines the initial state of the grammar ("the initial state") and, together with experience and nonlinguistic constraints, it guides mature phonological systems to converge on a similar final state ("the final state of the grammar"). If this hypothesis were correct, then one would expect early phonological systems to exhibit shared primitives and constraints. In its strongest form, the hypothesis of core phonological knowledge would predict that the phonological grammar is fully formed at birth; a weaker version might assume a later onset; this weaker version places the genesis of the phonological grammar somewhere after birth, but before the acquisition of language-particular phonotactics, near the child's first birthday. Regardless of its precise onset, both views assert that core phonological knowledge should be present early in life. In what follows, we test this hypothesis.

We begin by examining the computational properties of infants' phonotactic knowledge. If infants possess core phonological knowledge, then they must also be endowed with an algebraic computational system that is commensurable with that of the adult – a system that, *inter alia*, operates on variables and supports across-the-board generalizations. Next, we move to examine what primitives and combinatorial principles guide such generalizations. We first investigate whether infants encode their phonological experience using the same primitives found in adult phonologies – elements such as features, consonant and vowel segments, and syllables. Finally, we gauge the constraints on the combinations of those primitives by comparing infants' phonotactic preferences to the ones evident in adult systems across languages.

9.1 Computational machinery

Productivity is the defining feature of phonological knowledge: Adult learners extend their phonological knowledge across the board, in a manner that allows them to produce and comprehend a large number of patterns they have never heard before. For example, adult speakers encode the occurrence of identical consonants (e.g., *state*) – they distinguish such sequences from ones with non-identical segments (e.g., *skate*), and generalize the identity relation to novel instances (e.g., *snane*; e.g., for speakers of English: Coetzee, 2008; German: Domahs et al., 2009; Hebrew: Berent et al., 2002). In earlier chapters, I argued that these generalizations require algebraic computational machinery that encodes equivalence classes and represents their relations by means of operations over variables. If core phonology is active in early development, then young infants should be likewise endowed with such computational mechanisms. The existing literature suggests this is indeed the case.

A classic study by Gary Marcus and colleagues (1999) has shown that 7-month-old infants can learn and generalize algebraic rules. In this study, infants were presented with sequences that include two identical syllables, and the location of these syllables was manipulated (see 1). One group of infants was presented with ABB sequences – sequences with identical syllables at their end (*ledidi, wedede*); a second group was presented with ABA sequences (*ledile, dewede*). Recall (from Chapter 5) that identity is a formal relationship that binds the occurrence of any two members of a class (e.g., the B class) by a variable. If infants encode this algebraic relation, then they should be able

to extend it across the board to any novel sequence, including ones that share no common segments or features with familiarization items.
(1) Rule-learning by 7-month-old infants: the design of Marcus et al.'s (1999) experiment

	Familiarization examples	Generalization (test)
ABB group	ledidi, wedede	bapoba vs. bapopo
ABA group	ledile, dewede	

To test this prediction, following familiarization, infants were next presented with two novel test sequences with either an ABA or ABB structure (*bapoba* vs. *bapopo*). Note that these sequences share no segments with training items. Likewise, test items were matched for their feature-similarity to the ABB and ABA familiarization sequences. In the absence of statistical cues, the distinction between the two items can reflect only the encoding of their structure. Remarkably, 7-month-old infants were sensitive to this relationship: They attended reliably longer to novel items that were inconsistent with the structure of familiarization examples compared to consistent ones. For example, infants familiarized with ABB sequences attended longer to novel ABA items compared to novel ABB items.

Subsequent research has shown that the capacity to encode syllable-identity is present practically at birth. A series of experiments by Judit Gervain and colleagues (2008) compared the brain responses of neonates (mean age of three days) to ABB sequences and ABC controls using an infrared spectroscopy, a noninvasive technique that gauges changes in the concentration of oxygenated and deoxygenated hemoglobin (oxyHb and deoxyHb, respectively) in the blood from the scattering and absorption of near-infrared light reflected from the infants' skull. Results showed that ABB sequences elicited a stronger metabolic response, dominated by the left hemisphere, and this response increased monotonically with exposure to such sequences. Subsequent research has shown that neonates also distinguish AAB from ABB sequences (Gervain et al., 2012), suggesting that they encode not only the presence of identical elements but also their specific word locations (the left vs. right edge). This finding suggests that two fundamental features of mature phonological grammars – the capacity to encode relations and bind them to positions – are present at birth. Remarkably, infants' capacity to learn and generalize reduplication rules does not apply indiscriminately to any auditory input. While 7-month-olds and neonates were highly sensitive to the reduplication of syllables, they were unable to learn restrictions on the identity of musical notes (Gervain et al., forthcoming; Marcus et al., 2007). Thus, humans are equipped with algebraic machinery for rule-learning from birth, and this machinery is preferentially tuned to linguistic inputs.

9.2 Gauging core phonology: some ground rules

The computational power to attain broad generalization is undoubtedly a *sine qua non* for phonological competence. Nonetheless, merely having the capacity to generalize is not sufficient. Indeed, phonotactic patterns are systematically constrained. Mature phonotactic systems include a finite, potentially universal set of representational primitives, and the combinations of those primitives are restricted by a putatively universal set of grammatical constraints. Together, these representational primitives and constraints form the core phonological system. Our question here is whether this core knowledge is in fact active in the initial state of the phonological grammar.

Before we move to review the evidence, it is important to clarify our goals and define some ground rules. First – a few disclaimers. This chapter strictly focuses on the earliest known aspects of core phonological knowledge. While phonological development is undoubtedly shaped by numerous nonlinguistic factors, including statistical learning and motor articulatory constraints (MacNeilage, 2008), this review considers these factors only inasmuch as they present alternative explanations for core phonological principles. To gauge the properties of the initial state of the phonological system, the discussion will be further narrowed to the earliest documented stages of phonological development, ignoring, for the most part, evidence from older children (for review, see Demuth, 2011). Finally, as in the rest of the book, this chapter concerns only phonotactics, and for the sake of clarity, it describes the phenomena of interest by focusing on a few representative studies – a full review of the literature falls beyond the scope of this chapter.

Gauging the characteristics of the initial phonological state is indeed not a simple matter. Despite much research effort on language acquisition, the evidence concerning core phonological principles is rather scarce, and the distinction between such principles and nonlinguistic pressures is difficult to establish. To begin making some progress, we will therefore proceed in two steps (see 2). At an initial approximation, we will first consider only whether infants' behavior is *consistent* with the knowledge identified in the mature, adult grammar as the putative system of core phonology. Whether infants do, in fact, possess core phonological knowledge is a different, much harder question.

(2) Gauging core phonology: a two-stage approach
    a. Are the phonological preferences of infants *consistent* with mature phonological grammars?
    b. Do the phonological preferences of infants reflect core phonological knowledge? Specifically, is this knowledge:

- Universal:
  - General across languages
  - Early – attested prior to the acquisition of language-particular phonotactics
- Algebraic: encoding equivalence classes of discrete phonological elements and their relations
- Robust across perception and production

To address this second question, one must determine the universality of the principles guiding infants' behavior and their domain-specificity. While there is no foolproof method to decide on these issues, some rules of thumb might be useful. To demonstrate that the knowledge implicated in infants' behavior is universally present in the initial state of the grammar, one must rule out the possibility that this knowledge directly mirrors the statistical properties of linguistic experience. Ideally, one would like to document the relevant knowledge across learners of different languages, but absent such evidence, one could also show that this knowledge is evident at an age in which knowledge of language-particular phonotactics (e.g., principles that contrast English and French phonology) is still undetectable (e.g., within the first six months of life). Ruling out the contribution of extra-linguistic factors is likewise difficult, but at the very minimum one must demonstrate that the relevant knowledge is algebraic in nature (e.g., it concerns phonological equivalence classes of discrete elements, rather than analog, continuous phonetic entities), and it applies to both perception and production.

With these guidelines in mind, we can begin the excursion into the phonological mind of infants. Section 9.3 gauges the role of the mature phonological primitives in the initial state; section 9.4 examines whether the constraints on their combinations include some of the markedness reflexes attested in mature phonological systems.

9.3 Phonological primitives

All phonological systems encode phonotactic restrictions by appealing to the feature composition of segments; many restrictions further differentiate consonants from vowels and limit their co-occurrence in the syllable. Phonological features, consonants and vowels, and syllables are thus primitives of the adult phonotactic knowledge. In what follows, we examine whether those primitives play a role in the phonological system that is active in early infancy.

9.3.1 Features

Phonological features are the smallest primitives in the phonological system. Practically every phonological process is conditioned by the feature composition of phonological elements. For example, the final segments in *cat* and *dog* contrast on their voicing feature – *t* is voiceless, whereas *d* is voiced, and this contrast is critical for explaining their distinct phonological behaviors (e.g., the devoicing of the plural suffix in *cats* and its voicing in *dogs*). Our interest here is in whether phonological features are represented in the initial state of the phonological system.

The answer, as it turns out, is surprisingly difficult to come by. Although it is well known that young infants can distinguish between segments that differ by a single feature (e.g., the voicing contrast in *ba* vs. *pa*; Eimas et al., 1971), and they can detect contrasts that even their parents are unable to discern (e.g., Werker & Tees, 1984), this ability, impressive as it is, does not necessarily demonstrate the representation of phonological features. Indeed, phonological features have some clear acoustic and phonetic correlates. The phonological feature of voicing, for example, correlates with variations on voice onset time – the lag between the release of a consonant closure and the onset of the vibration of the vocal cords associated with voicing (Lisker & Abramson, 1964), and this acoustic correlate, in turn, can be further linked to several phonetic features related to phonological voicing (i.e., the phonetic categories of voiced, voiceless unaspirated, voiceless aspirated; Keating, 1984). Accordingly, infants could distinguish *ba* from *pa* either by relying on phonological features (e.g., voicing) – features that are discrete and algebraic – or by detecting their acoustic or phonetic correlates – properties that are analog and continuous. Indeed, infants who can distinguish *ba* from *pa* nonetheless fail to distinguish among words that contrast on voicing (e.g., *beak* vs. *peek*; Pater et al., 2004). To determine whether phonological features are represented in the initial state, it is thus necessary to establish not only whether the infant detects a given contrast but also whether the contrast is phonological or phonetic.

One way to dissociate between these possibilities is to show that infants' ability to differentiate between two segments specifically depends on shared phonological features. Several studies have indeed found that segments that share a feature are more likely to be grouped together as members of a single class (e.g., Hillenbrand, 1983; Saffran, 2003a). Subsequent research by Amanda Seidl and colleagues suggests that infants perceive two segments as similar *only* when they share a phonological feature in their language (Seidl & Cristia, 2008; Seidl et al., 2009).

One such demonstration exploits the fact that distinct languages differ on their phonological features. Consider, specifically, the contrast between oral and nasal vowels in English and French (see 3). English and French speakers produce both nasal (e.g., [æ̃]) and oral (e.g., [æ]) vowels. In the case of English vowels, however, the contrast between nasal and oral vowels is entirely predictable from their context: English vowels are always nasalized before a nasal consonant (e.g., *ban* [bæ̃n]), and they are oral elsewhere (e.g., *bad* [bæd]).

Since no pair of English words differs solely on their nasality, in English this feature is considered as phonetic, rather than phonemic, hence, the nasal and oral versions of a vowel (e.g., cf. [æ̃] and [æ]) are allophones. In French, however, the very same nasality contrast is sufficient to contrast between different words (*bas* [bæ] 'low' cf. *banc* [bæ̃] 'bench'), so unlike English, in French, nasality is a phonological feature, and the nasal and oral vowels (e.g., [æ̃] and [æ]) are distinct phonemes.

(3) The role of nasality in English vs. French
   a. English: nasal vowels appear only before nasal consonants:
      ban [bæ̃n] vs. *bad* [bæd]
   b. French: two words can contrast only on vowel nasality:
      bas [bæ], 'low'
      banc [bæ̃], 'bench'

The different roles of nasality in English and French can help gauge infants' sensitivity to phonological features. If infants encode words in terms of their phonological features, then French infants should be sensitive to contrasts in nasality that English infants ignore. The experiment by Seidl and colleagues specifically examined whether French- and English-learning infants differ on their ability to learn phonological regularities concerning vowel nasality. In these experiments (Seidl et al., 2009), 11-month-old infants were exposed to CVC sequences that systematically varied the manner of articulation in the coda consonant (stop vs. fricative) depending on the nasality of the vowel. For one group of participants, nasal vowels were followed by a stop consonant (e.g., [kæ̃p]) whereas oral vowels were followed by a fricative (e.g., [dæz]); a second group was presented with the reverse coupling (see 4).

(4) Testing the role of the nasal features in English and French: an illustration of the design of Seidl et al.'s (2009) experiments

	Familiarization			Test
	Vowel	*Coda*	*Example*	
Group 1	Oral	Stop	kæp, vɔp	pɛk pɛ̃k
	Nasal	Fricative	dæ̃z, vɔ̃p	
Group 2	Nasal	Stop	kæ̃p, vɔ̃p	pɛv pɛ̃v
	Oral	Fricative	dæz, vɔp	

To determine whether infants have acquired this regularity, participants were next presented with novel test items: either ones consistent with the regularity presented in the familiarization, or inconsistent items. All test items featured vowels and codas that did not appear in the familiarization. For infants in the oral vowel-stop/nasal vowel-fricative condition, consistent items comprised combinations of oral vowels with stop consonants (e.g., pɛk) or nasal vowels with fricative consonants (e.g., p̃ɛv), whereas inconsistent items had either nasal vowels followed by stops (p̃ɛk) or oral vowels followed by fricatives (e.g., pɛv); for the second group, consistency was reversed. If infants encode segments only by means of their acoustic features, then English and French infants should not differ in their ability to encode the relevant pattern. In contrast, if infants encode nasality as a phonological feature, then this feature should be available to learners of French (where the contrast is phonemic) but not English (where the nasal contrast is allophonic). Accordingly, French infants should be better able to track the regularity than English participants. This is precisely the result observed by Seidl and colleagues with 11-month-old infants.

The failure of infants from English homes to learn regularities concerning the nasal feature is unlikely to be because infants are oblivious to allophonic contrasts – other research has shown that infants (Maye et al., 2002; McMurray & Aslin, 2005) and adults (e.g., Allen & Miller, 2004; Theodore & Miller, 2010) are highly sensitive to allophonic distinctions. It is also unlikely that the differences between the English and French groups are solely due to differences in their familiarity with nasal vowels. Statistical analysis suggests that English- and French-learning infants are exposed to nasal vowels to a similar extent (Seidl et al., 2009). It thus appears that the selective sensitivity of French (but not English-learning) infants to the nasality feature is specifically due to their distinct roles in their phonological systems. To further support this conclusion, Seidl et al. have repeated the experiment with infants of 4 months – an age at which the distinction between allophonic and phonemic contrasts is still not fully established. Unlike their older counterparts, 4-month-old infants from English-speaking families were able to encode the relevant contrast, presumably due to their reliance on phonetic or acoustic representations.

These results suggest that, by the end of the first year of life, infants possess the ability to encode phonological features and ignore non-distinctive phonetic correlates of those features. But because the status of a feature as distinctive requires the acquisition of a particular phonological system (e.g., French), these findings cannot attest to the properties of the initial state of the grammar. Whether the very capacity to encode phonological features is present at earlier development remains to be seen.

9.3.2 Consonants and vowels

Consonants and vowels are different "phonological animals." Earlier in Chapter 4 we have seen numerous manifestations of the distinction between consonants and vowels in the mature phonological system. Here, we will show that a similar contrast is present in children's phonology. A very early distinction between consonants and vowels is evident in the ability of newborn infants to extract the rhythmic patterns of different languages. In a typical experiment, an infant is first exposed to speech in one language (e.g., French). Next, the infant is presented with either some more speech from the same language or from a different language (e.g., Russian). Results consistently show that infants as young as four days of age can distinguish between these two languages. Subsequent studies have shown such discrimination even *in utero* – in fetuses at 33–51 weeks gestational age (Kisilevsky et al., 2009).

How can a newborn infant tell French from Russian? Jacques Mehler and colleagues were able to rule out several extraneous explanations for this phenomenon. In particular, infants do not simply rely on a change in talkers, as similar results are obtained when the two languages are presented by a single bilingual talker (Mehler et al., 1988), and when the language-switch is compared against a control condition that alters the talker while maintaining the same language (Nazzi et al., 1998; Ramus et al., 2000; for similar conclusions in fetuses, see Kisilevsky et al., 2009). It is also unlikely that infants distinguish Russian from French only by detecting differences in the inventories of the sounds that occur in the two languages. First, infants can discriminate between these two languages when the speech is filtered in a manner that eliminates many segmental cues (Mehler et al., 1988). Moreover, subsequent research has shown that adults can distinguish between two languages even when speech samples are edited so that the two languages contrast only on the succession of consonants and vowels (Ramus & Mehler, 1999). Like adults, newborn infants are able to discriminate between languages that differ in their rhythmical properties (e.g., English vs. Japanese). In fact, infants can distinguish between languages of distinct rhythmical groups (e.g., English vs. Japanese) even when neither language matches their maternal tongue (French), but they fail when the rhythmical characteristics of the two languages are similar (e.g., English vs. Dutch: Nazzi et al., 1998; Ramus et al., 2000). Accordingly, it appears that newborns distinguish between languages by tracking their rhythm, and since linguistic rhythm is defined by the sequencing of consonant and vowel categories, these results imply the representation of these two categories. While these findings do not determine how, precisely, these categories are represented – whether these are abstract discrete phonemic categories, or analog phonetic classes that register the precise duration of consonants and vowels – they do suggest that some precursor for consonant/vowel distinction is present practically at birth.

Not only do infants distinguish consonants from vowels, but they also assign them distinct roles. Recall that, in the adult grammar, consonants carry the burden of lexical distinctions (vowels, in contrast, play a greater role in conveying prosodic and syntactic information; Nespor et al., 2003). A series of experiments by Thierry Nazzi and colleagues suggests that the division of labor between consonants and vowels is already in place in the phonological system of 20- to 30-month-old infants (e.g., Nazzi, 2005; Nazzi et al., 2009). In these experiments, infants first learned to associate three novel objects (A, B, C) with three different novel names (e.g., /duk/, /guk/, /dɔk/). The names were selected such that one member of the triplet, the target (e.g., /duk/), differed from the remaining items by one feature, either a consonantal feature (e.g., /duk/ vs. /guk/) or a vocalic one (e.g., /duk/ vs. /dɔk/). If consonantal information plays a greater role in lexical access, then the alternative that shares the target's consonants (and differs by a single vocalic feature, i.e., /duk/ vs. /dɔk/) should be considered more similar to the target than the one that differs by a single consonantal feature (e.g., /duk/ vs. /guk/).

To test this hypothesis, the experimenter next presented infants with the target object (e.g., an object called /duk/) and the two other objects (named /guk/ and /dɔk/), and participants were asked to pick up the object "that goes with this one." As expected, infants were more likely to pick up the object whose name shared the target's consonants (i.e., /dɔk/) over the same-vowel alternative (e.g., /guk/). Subsequent research showed that the tendency to lump the same-consonants names is not due to an inability to differentiate among those names (Nazzi et al., 2009). Likewise, the greater salience of same-consonant names was independent of the position of the consonants and vowels in the word (Nazzi et al., 2009), the syllabic role of the consonants (onsets vs. codas; Nazzi & Bertoncini, 2009), their place of articulation (labials vs. coronals; Nazzi & Bertoncini, 2009), manner of articulation (stop vs. fricatives; Nazzi & New, 2007), and the number of consonants relative to vowels in the language (Nazzi et al., 2009). These results thus suggest that infants are biased to weight consonantal information more heavily in lexical contrasts.

The very early sensitivity to the consonant/vowel contrast in marking the rhythmical properties of languages and their role in defining lexical distinctions, later in development, suggests that infants encode consonants and vowels as distinct phonological primitives.

9.3.3 Syllables

A third building block of phonological systems – both early and mature – is the syllable. The syllable is at the center of the phonological edifice, as it marks the intersection between sub-syllabic units – features and segments – and larger prosodic constituents. At each of these levels, there are phonological constraints that appeal to the syllable. Specifically, at the sub-syllabic level, syllables form the domain of phonotactic restrictions on the co-occurrence of segmental features. At the supra-syllabic level, syllables are called by restrictions on prosodic structure. These observations suggest that the syllable is a phonological primitive. In what follows, we document such constraints in the early phonological systems of infants and young children.

9.3.3.1 Syllables are the domain of early phonotactic knowledge

Syllables are at the center of numerous phonotactic restrictions. English, for example, allows sonorant-obstruent sequences like *lb* to occur, but it strictly restricts their location. While such sequences are perfectly well formed across syllables (e.g *el.bow*), they are ill formed at a syllable's onset (e.g., *lbow*). To encode such phonotactic knowledge, infants must possess the capacity to represent syllables.

Several sets of findings suggest that, by the end of the first year of life, infants extract knowledge of the particular phonotactics of their native language. For example, 9-month-old infants from English- and Dutch-speaking families can discriminate syllables that are attested in their languages from unattested ones (Friederici & Wessels, 1993; Jusczyk et al., 1993). Specifically, Dutch infants listen longer to legal Dutch syllables that feature the sequence *br* in the onset (e.g., /bref/) compared to illegal syllables that manifest the same sequence in the coda (e.g., /febr/; Friederici & Wessels, 1993). Similarly, Dutch infants tend to disfavor illicit Dutch syllables over licit Dutch syllables that happen to be ill formed in English (e.g., *vlakte*), whereas English-speaking infants favored licit English syllables over Dutch patterns (Jusczyk et al., 1993).

Not only do infants possess knowledge of syllable phonotactics but they further put it to use in the segmentation of speech (Mattys et al., 1999; Mattys & Jusczyk, 2001). We all know too well how difficult it is to spot our white car amidst a parking lot full of other white vehicles – a red-car lot would have rendered the search far less agonizing. In the same vein, spotting a syllable amidst a continuous speech stream depends not only on the syllable's own properties but also on its context. Syllables flanked by consonants that form ill-formed syllables are more likely to "pop out." For example, the CVC syllable *gaffe* is detected more readily when it is flanked by consonants that would form illicit onset and coda clusters (e.g., bea**n**.gaffe.**h**old, including the illicit onset *ng* and the illicit coda *fh*) compared to a context in which no such cues are available to mark the boundaries of the CVC target. Remarkably, this sensitivity to phonotactics obtains for 9-month-old infants. And when older infants (14-month-olds) are presented with sequences whose phonotactic structure is illicit in their language (e.g., *ebzo*, illicit in Japanese), they tend to "repair" it in perception (e.g., as *ebuzo*; Mazuka et al., 2012). While these findings do not determine how, precisely, such knowledge is encoded – whether children store the occurrence of specific syllable instances (e.g., *block*, *boy*) or whether they encode syllables by abstract equivalence classes – these results are nonetheless consistent with the possibility that, by the age of 9 months, infants extract some knowledge of syllable structure.

9.3.3.2 Syllables define prosodic structure

In addition to their role in defining phonotactic restrictions, syllables are also the building blocks of prosodic structure. Reports from numerous languages suggest that early words are limited to a maximum of two CV syllables (i.e., a binary foot, e.g., *baby*) – longer words are truncated to fit the disyllabic template (e.g., *potato*➔ [te:do], Pater, 1997a; see also Demuth, 1995; Gerken, 1994; Ota, 2006; Prieto, 2006). Truncation, however, is not an arbitrary process. Rather, the output of truncation is subject to numerous phonotactic and prosodic restrictions (e.g., Fikkert & Levelt, 2008; Gnanadesikan, 2004; Pater, 1997a, 1997b; Pater & Werle, 2003).

Consider, for example, children's strong preference for trochees – binary feet with initial stress (e.g., *báby*). Monitoring the first words produced by a Hebrew-learning boy close to his first birthday, Galit Adam and Outi Bat-El (2009) documented a strong bias toward trochees: This child was less likely to attempt producing iambic targets (e.g., *todá* 'thanks') than trochaic ones (e.g., *sáfta* 'grandmother'), and, once attempted, iambic targets were more likely to be truncated into a monosyllable (e.g., *todá*➔*da*) than trochaic ones. The bias toward trochaic targets is remarkable given that trochees are actually less frequent than iambs in adult Hebrew speech. Adam and Bat-El attribute this bias to a universal preference for the trochee. In support of this possibility, they point out that this preference has been documented in several other languages in which trochees are outnumbered by iambs (e.g., Catalan: Prieto, 2006; French: Allen, 1983). Feet, however, typically comprise precisely two syllables (Hayes, 1980). Accordingly, the representation of such domains could indicate the representation of syllables. I deliberately use the modal "could" because this finding also has an alternative explanation. In this view, the early preference for a trochee reflects a preference for an unparsed C′VCV template that does not concern syllables at all (Fikkert & Levelt, 2008). Similarly, like the evidence from phonotactics (reported in section 9.3.3.1 above), these findings are ambiguous as to whether children encode syllables as an equivalence class.

The findings of Amalia Gnanadesikan, however, strongly suggest that children specifically represent the syllable. Analyzing the outputs of her 2-year-old English-learning daughter, Gnanadesikan (2004) shows that the first syllable of the target word is routinely replaced by the dummy syllable *fi* (e.g., *umbrella*➔ [fi-bɛyˈ]; see 5). Remarkably, this substitution applies to a wide range of targets that differ on their segmental contents, weight, and shape (e.g., CV, CVC, CCV, VC). The fact that these syllable instances are all treated alike, and that differences between them are ignored with respect to this generalization, suggests that they form an equivalence class. The linguistic behavior of this child implies the representation of an algebraic constituent equivalent to the adult syllable.

(5) First-syllable substitution by the dummy syllable *fi* (data from Gnanadesikan, 2004)
Umbrella [fi-bɛyˆ]
Mosquito [fi-giDo]
Christina [fi-dinˆ]
Rewind [fi-wayn]

In summary, syllable-like units play an early role in development. Like adult syllables, the units seen in the child's early production form equivalence classes and serve as the domain of various phonological restrictions. One notable limitation of these findings, however, is that they are all observed at ages in which the effect of language-particular knowledge is already in place. Although this fact does not necessarily mean that the syllable is induced from linguistic experience – indeed, the prosodic preferences of children are at times inconsistent with the statistical structure of adult speech – the documentation of such knowledge earlier in development would be highly desirable. In the following section, we review evidence suggesting that, within the first six months of life, infants possess preferences on syllable structure. Those findings, in conjunction with the results described here, suggest that some precursor of the syllable could be present already at the initial state.

9.4 Universal combinatorial principles: some markedness reflexes

The evidence reviewed in the previous sections suggests that the initial state of the grammar might include several phonological primitives – features, consonants and vowels, and syllables. We now turn to investigate what grammatical principles might govern the combinations of those primitives. Specifically, we examine whether the initial state also includes knowledge of universal markedness restrictions. We first review some evidence suggesting that markedness restrictions are highly ranked in the initial state of the grammar; we next turn to inspect some specific manifestations of markedness.

9.4.1 Markedness constraints are highly ranked in early grammars

Grammatical markedness constraints are believed to play a double duty in the course of human development. In the adult grammars, markedness restrictions limit the range of attested phonological systems – the linguistic and experimental evidence reviewed in previous chapters is consistent with this possibility. But in addition, markedness universals may also guide the acquisition of phonological systems by the child. To shape the acquisition of phonological systems, markedness restrictions must be active in the initial state of the grammar. In fact, there is reason to believe that the effect of markedness in early grammars might be even more pronounced than in the final, mature state. The argument, outlined by Paul Smolensky (1996; see also Davidson et al., 2004) is a logical one. Smolensky shows that many languages are unlikely to be learned unless learners are a priori biased against marked structures. We first review this argument and then describe some experimental findings that are consistent with its predictions.

9.4.1.1 A learnability argument

Adult grammars vary on their tolerance for marked structures. Vietnamese, for instance, allows onsets with the velar-nasal stop ŋ (e.g., *ŋgu* 'sleep') whereas in English, such segments are possible only in codas (e.g., *sing* /sɪŋ/). To acquire the adult grammar, not only must children therefore know that ŋ-onsets are marked (e.g., relative to n), but they must also determine whether their own language happens to tolerate such onsets. Figuring out those facts is not trivial. Because Optimality Theoretic grammars only constrain the outputs of the grammar (a principle known as the Richness of the Base), it is formally impossible to ban ŋ-onsets by simply excluding such words from the lexicon (i.e. by banning inputs with ŋ-onsets form occurring). Accordingly, the English ban on ŋ onsets can only be expressed by constraint ranking. Specifically, English grammar must rank the markedness constraint against ŋ-onsets, $M_ŋ$ above the constraints that enforce faithfulness to such inputs, $F_ŋ$, whereas Vietnamese grammar exhibits the opposite ranking (see 6).

(6) Onset restrictions in English vs. Vietnamese
    a. Vietnamese grammar (ŋ-onsets are allowed): $F_ŋ >> M_ŋ$
    b. English grammar (ŋ-onsets are disallowed): $M_ŋ >> F_ŋ$

Our question here is whether the English ranking can be learned from the evidence available to the language learner. To learn the English ranking, children must have positive evidence demonstrating that ŋ-onsets are illicit. Such evidence could come from processes where ŋ is expected in the onset, but it fails to occur in that position. This would be the case when a word ending with an ŋ coda is re-syllabified, such that ŋ is now expected as an onset (aŋ+a→a.ŋa). If learners noted that expected ŋ-onsets are systematically changed to *n* (aŋ+a→a.na), then they might conclude that such onsets are banned and consequently rank the ban on ŋ-onsets above the relevant faithfulness preference ($M_ŋ >> F_ŋ$). English, however, presents no such cases, so the $M_ŋ >> F_ŋ$ ranking cannot be learned from experience. More generally, inventories in which marked structures (e.g., ŋ) do not alternate with their unmarked counterparts (e.g., n) do not allow the child to infer that marked structures are disfavored by their target grammar. Since such putatively unlearnable inventories are actually quite frequent across languages, Smolensky (1996) concludes that markedness constraints must outrank faithfulness constraints already in the initial state of the phonological grammar. The findings from a landmark study by Peter Jusczyk, Paul Smolensky, and Theresa Allocco (2002) are consistent with this prediction.

9.4.1.2 Experimental evidence

To determine the ranking of markedness and faithfulness constraints in the initial state of the grammar, Peter Jusczyk and colleagues examined the sensitivity of young infants (age 4.5 months) to markedness and faithfulness constraints. The specific case study concerned nasal place assimilation. Across languages, nasal coronals frequently assimilate to their labial neighbors. For example, an input consisting of *in* and *po* is likely to assimilate to *impo*. Such alternations (i.e., cases where inputs is modified) suggest that the relevant markedness constraints (schematically, M_{np}) outrank faithfulness pressures (F_{np} ; see 7; for an alternative explanation, see de Lacy, 2006).

(7) Nasal place assimilation

in+po	M_{np}	F_{np}
inpo	*	
☞impo		*

While nasal place assimilation is universally favored, English does not provide its learners with evidence for this preference: Some English words exhibit assimilation across morphemes (e.g., *in+polite* →*impolite*), but others do not (e.g., *input*). Moreover, given the young age of participants in these experiments (4.5 months, an age in which knowledge of language-particular phonotactics is typically undetectable), it is unlikely that they have acquired this knowledge from their linguistic experience – either their exposure to adult language, or from their attempts at producing it, attempts which occasionally give rise to such erroneous outputs (e.g., neck→ŋɛk; Smith, 2009). Our question here is whether such young infants nonetheless prefer this alternation. To the extent that young infants systematically favor nasal place assimilation, in the absence of evidence such behavior might indicate that M_{np} is highly ranked already in the initial state of the grammar.

To test this possibility, Jusczyk and colleagues presented infants with an alternation between inputs consisting of two syllables (e.g., *in+po* vs. *im+po*) and an output. In the critical condition (see 8I), the outputs were always identical to the input (i.e., faithful), but they differed on their markedness: In the harmonic condition, the output consonants shared the same place of articulation (e.g., a labial consonant, *impo*) whereas in the disharmonic condition, the two consonants disagreed on place (e.g., the coronal-labial sequence in *inpo*). Results showed that infants attended reliably longer to the harmonic condition, indicating that, once faithfulness is held constant, infants favor unmarked harmonic outputs – ones that share the same place of articulation – to marked, disharmonic ones.

(8) Nasal harmony materials and results from 4.5-month-olds (from Jusczyk et al., 2002 and Davidson et al. 2006)

This table compares three alternations that differ with respect to the violation of markedness (M) and/or faithfulness (F) constraints – alternations that respect markedness constraints (harmonic) are indicated in the second column, whereas ones violating markedness (i.e., disharmonic alternation) are listed in the third column. Each such alternation is illustrated by an example, and its harmony relative to those constraints is indicated (* indicates constraint violation; ✓ marks outputs that do not violate the relevant constraint). The looking times associated with each such condition are indicated as well (in seconds).

Comparison	Harmonic alternation	Disharmonic alternation
I. Markedness only	im+po→impo M✓F✓ M=15.23	in+po→inpo M*F✓ M=12.37
II. Faithfulness only	im+po→impo M✓F✓ M=15.36	im+po→uŋkə M✓F* M=12.31
III. Markedness vs. faithfulness	in+po→impo M✓F* M=16.75	in+po→inpo M*F✓ M=14.01

Many adult languages, however, favor such harmonic outputs even when they alter the input – such processes violate faithfulness in order to satisfy markedness. Of interest, then, is whether such preference to avoid markedness at the cost of faithfulness violation is already present in the initial state. Juszcyk and colleagues examined this question in two stages: first they determined whether infants obey faithfulness; next they investigated whether faithfulness will be violated to satisfy markedness.

To examine infants' sensitivity to faithfulness, Juszcyk and colleagues presented infants with two types of sequences (see II in 8). In both cases, the outputs exhibited place harmony (satisfying markedness), but in the harmonic sequence, these outputs were identical to the input (satisfying faithfulness) whereas in the other, they did not. Results showed that infants attended longer to the faithful sequences. Given that infants of this age are sensitive to faithfulness, one can now move to examine whether infants will sacrifice their preference for faithful outputs in order to satisfy markedness constraints that favor place harmony. To this end, Jusczyk and colleagues compared two outputs: One of these inputs achieves place harmony at the cost of violating faithfulness; the other satisfies faithfulness to the input at the cost of violating markedness (see

III in 8 above). Results suggested that, despite the violation of faithfulness, infants favored unmarked outputs to marked ones.

While the results from this single pioneering study await replication and extensions to additional markedness manifestations, the findings suggest that infants as young as 4.5 months of age possess rudimentary markedness preferences that favor place assimilation, and that these early preferences outrank faithfulness preferences in early grammars. In the following section, we further examine the role of two additional classes of markedness restrictions in the initial state: CV-structure constraints and the sonority hierarchy.

9.4.2 Syllable markedness

The typological research reviewed in Chapter 6 suggests that mature grammars systematically favor certain syllable types over others: Syllables with onsets are preferred to ones without them (CV≻V), simple onsets (e.g., CV) are preferred to complex ones (CCVC), and open CV syllables are preferred to syllables that include a coda (e.g., CVC). Finally, all syllables require a nucleus – a high-sonority segment, preferably a vowel. The available findings suggest that similar preferences are active in early phonological systems.

One source of evidence is presented by the order in which these various syllable structures are first produced. If the early phonological grammar is equipped with constraints that disfavor marked syllable types, then, other things being equal, markedness should be inversely related to the order of acquisition. The findings from normally developing children are generally consistent with this prediction: CV syllables are typically produced before CVC syllables, which, in turn, are produced before syllables with complex onsets and codas (Bernhardt & Stemberger, 1998; 2007; Fikkert, 2007; Jakobson, 1941; Levelt, Schiller & Levelt, 1999; Smith, 2009). These developmental trends, however, are subject to counterexamples (e.g., *strong* [strɔn] produced before *cat* [kæʔ]; Smith, 2009) and considerable individual differences (Bernhardt & Stemberger, 2007). Moreover, these preferences could reflect functional articulatory pressures (MacNeilage, 1998; 2008) and experience (e.g., Levelt et al., 1999; Ota, 2006), rather than grammatical markedness constraints. Other observations, however, dissociate markedness preferences from these extra-grammatical factors.

To distinguish markedness preferences from articulatory pressures, one would like to determine whether the preferences seen in production are also evident in perception – a convergence would imply that such preferences are represented as abstract, amodal knowledge. The onset of such knowledge could further attest to its origins: If the preferences seen in perception are shaped by articulatory practice, then they should only emerge after the onset of babbling in the second half of the first year of life (MacNeilage, 2008). If, however, syllable-structure preferences are evident in perception within the first months, then an articulatory basis is unlikely. Such findings would likewise challenge the possibility that markedness preferences are induced by tracking the frequency of various syllable frames in the adult language, as knowledge of the specific phonetic categories of a language (Werker & Tees, 1984) and its phonotactics (e.g., Friedrich & Friederici, 2005; Jusczyk et al., 1993) is typically observed only toward the end of the first year. Remarkably, some aspects of syllable structure are detectable already in the first two months of life. In a pioneering study, Josiane Bertoncini and Jacques Mehler (1981) gauged the phonological preferences of infants by monitoring their sucking responses to spoken syllables. Infants (age 2 months) were first presented with a repeatedly spoken syllable (e.g., *tap, tap, tap* . . .) for a period of about five minutes. After the infant has habituated to the speech stream (as determined by the decline in the infant's sucking rate), he or she was next presented with either the same syllable (e.g., *tap*) or a variant of the initial syllable (generated by permuting its consonants, e.g., *pat*). If the infant has securely represented the initial syllable, then he or she should detect the change, and the novelty should elevate his or her sucking rate. Of interest is whether infants' ability to detect the change depends on the well-formedness of the syllable (see 9). Results suggested that infants were indeed responsive to the change given well-formed CVC syllables (e.g., *tap*➔*pat*). Similarly, infants were able to detect a change in complex VCCCV strings (e.g., utʃpu➔upʃtu), as long as these strings could be parsed into well-formed syllables – syllables that include a vocalic nucleus (utʃ.pu). But when presented with strings that lack a vowel (e.g., tʃp), infants failed to detect the change (e.g., tʃp➔pʃt) despite the fact that these ill-formed strings featured the very same consonant sequences as the encodable, VCCCV sequences (e.g., utʃpu➔u pʃtu). These results suggest that infants encode only sequences that can be parsed into unmarked syllables including a vocalic nucleus – marked syllables that lack a vowel cannot be reliably encoded.

(9) The effect of syllable structure on the detection of alternations: the design of Bertoncini and Mehler's (1981) study
   a. Well-formed syllables:
      • Simple syllables: *tap*➔*pat*
      • Complex syllables: utʃ.pu➔upʃ.tu
   b. Ill-formed syllables: tʃp➔pʃt

A ban on marked syllables can also explain some of the syllable-structure preferences seen in later development. The universal preference for an onset could account for infants' difficulty in segmenting vowel-initial words from continuous speech (e.g., Mattys & Jusczyk, 2001; Nazzi et al., 2005; for some phonetic explanations, see Seidl & Johnson, 2008). Similarly, the ban on codas might explain why 14-month-old Dutch infants show greater sensitivity to coda addition (e.g., *pa*➔*pat*) compared to coda deletion (e.g., *pat*➔*pa*; Levelt, 2009). Although, taken at face value, each of these results is open to alternative explanations, and the critical finding from early infancy awaits replication and extension, the evidence as a whole opens up the possibility that at least some syllable-markedness constraints might be active in early development.

9.4.3   The sonority hierarchy

Another markedness preference that is present in the child's early grammar is related to the sonority hierarchy, reproduced in (10). Recall that across languages, onsets typically favor segments of low sonority, whereas high-sonority segments are preferred in codas (Clements, 1990). Similar sonority preferences are evident in children within their first years of life.
(10) The sonority hierarchy of consonants:
    Glides (e.g., *y, w*)>liquids (e.g., *l, r*)>nasals (e.g., *n, m*)
    >fricatives (e.g, *f,v*)>stops (e.g., *p,b*)
One set of findings concerns the simplification of adults' multisyllabic words in language production. As discussed earlier (in section 9.3.3), children in their second year of life limit the prosodic structure of words to a maximum of a disyllabic trochee (e.g., *baby*): Disyllabic iambs are reduced into monosyllables (*garage*➔ [ga:dʒ] cf. *garbage* [ga:bɛdʒ]), and longer, trisyllabic inputs are truncated into disyllabic ones (e.g., *potato* [te:do]; examples from Pater, 1997b) . To perform these simplifications, children must choose which segments should "survive" in the simplified output. Remarkably, these choices reflect systematic preferences, constrained by sonority. Other things being equal, children preserve the onset and rhyme of the final syllable (see 11). And indeed, when the final onset is unmarked (either an obstruent or a nasal), children's productions target the final syllable (e.g., *abacus*➔[æ:ʃɪʃ]; *animal*➔[æmʊ]). But when the final syllable has a high-sonority liquid, and it is preceded by a syllable with a less marked onset (an obstruent or a nasal), children now opt for the penultimate, unmarked onset and combine it with the final rhyme (e.g., *broccoli*➔[baki]; *buffalo*➔ [bʌfo]; Pater, 1997b).
(11) Sonority preferences constrain the simplification of multi-syllabic words: low-sonority onsets (obstruents or nasals) are required
    a. By default, children's productions target the final syllable (including both its onset and rhyme):
        abacus➔[æ:ʃɪʃ]
        animal➔[æmʊ]
    b. When the onset of the final syllable is a liquid, and the penultimate onset is lower in sonority, the penultimate onset wins:
        broccoli ➔[baki] [*bali]
        buffalo➔ [bʌfo] [*bʌlo]

Similar sonority-driven preferences are evident in cluster reduction. Children in their first and second years of life often reduce complex onsets to simple ones (e.g., *clean*➔[kin], *please*➔[piz]; *friend*➔[fɛn]; Pater & Barlow, 2003). Once again, the choice of the surviving segment is constrained by its sonority: Children typically maintain the least sonorous member of the cluster (Barlow, 2005; Gnanadesikan, 2004; Ohala, 1999; Pater & Barlow, 2003; Wyllie-Smith et al., 2006), and in so doing, they maximize the sonority rise between the onset and the nucleus. The preference for onsets with large sonority rise has been documented in numerous studies with children acquiring English (Ohala, 1999; Pater & Barlow, 2003; Wyllie-Smith et al., 2006), Spanish (Barlow, 2005), Dutch, Norwegian, Hebrew (Yavas et al., 2008), and Polish (Lukaszewicz, 2007). Nonetheless, sonority preferences can be overridden by other factors, including both grammatical and extra-linguistic pressures. Grammatical pressures require complex onsets to share the same place of articulation (Kirk, 2008); ban fricatives (Pater & Barlow, 2003); and constrain the reduction of a segment by its role in the syllable (Barlow, 2001; Gierut, 1999; Goad & Rose, 2004). Reduction might also be shaped by the ease of articulation (Kirk & Demuth, 2005) and the acoustic properties of the input (Demuth & McCullough, 2009). Because the ranking of sonority constraints relative to sonority-unrelated pressures can differ across children and change with development, the role of sonority restrictions is difficult to unequivocally support from naturalistic observations.

Experimental manipulations, however, allow one to control for such factors. A pioneering experiment by Diane Ohala examined the production of complex onsets and codas whose structure was carefully selected. Results suggest that cluster simplification obeys the sonority-sequencing principle (Ohala, 1999; see 12). Specifically, when complex onsets are simplified, fricatives are much more likely to survive when they are followed by a (more-sonorous) nasal (e.g., *snuf*➔*suf*) compared to a less-sonorous obstruent (e.g., *skub*➔*kub*). In contrast, when fricatives appear at the coda, paired with either stops or sonorants, the pattern reverses, as the more-sonorous element is now preferred to the less-sonorous one (e.g., *fisk*➔*fis*; *valk*➔*val*). The sensitivity of reduction to the syllabic context – onset vs. coda – is significant, because it rules out the possibility that simplification is governed only by the articulatory demands associated with certain segments (Blevins, 2004). Instead, such findings are readily explained by sonority preferences. Across languages, onsets and codas exhibit opposite sonority profiles – onsets favor a large rise in sonority, whereas codas manifest a moderate decline in sonority (Clements, 1990). These restrictions on sonority sequencing, coupled with the fact that fricatives are more sonorous than stops (see 10), explain children's preference of fricatives over stops in codas, but not onsets.

(12) Sonority preferences constrain reduction in elicitation experiments (from Ohala, 1999)
   a. Onset reduction preserves the lower-sonority segment:
      snuf➔*suf*
      skub➔*kub*
   b. Coda reduction preserves higher-sonority segment:
      fisk➔*fis*
      valk➔*val*

While such findings are consistent with the proposal that children encode markedness restrictions that are related to sonority, these results are limited inasmuch as they are obtained only in production. This limitation opens up the possibility that the asymmetric preference for certain consonant sequences might be due to contextual articulatory factors, rather than syllable structure per se. Furthermore, the confinement of such findings to onsets that are attested in the child's language do not allow one to dissociate the effect of markedness pressures from familiarity. To address these limitations, it is important to examine whether children extend markedness constraints to syllable structures that are unattested in their language, and whether the preferences seen in production also apply in perception. The pioneering studies of Ohala (1999) and Pertz & Bever (1975) have begun addressing these questions, but their results were unclear, as children in these experiments failed to generalize their knowledge to many untested onsets.

Subsequent research, however, observed that children aged 3–5 years systematically favor onsets with large sonority distances over smaller ones even when all clusters were unattested in their language, and even when they were not required to articulate those stimuli (Berent et al., 2011a). This experiment employed a modified version of the identity judgment task, using the same materials employed previously with adults (Berent et al., 2007a). In each trial, the child was presented with two characters (a baby chimp doll and a baby gorilla doll) engaged in an imitation game: One character uttered a word and the second attempted to imitate it. These words were monosyllables with an unattested onset cluster whose sonority distance was manipulated – either a sonority rise, plateau, or fall (e.g., *bwif, bdif, lbif*), along with their disyllabic counterparts (e.g., *bəwif, bədif, ləbif*). Successful imitations shadowed the first item precisely (e.g., *lbif-lbif*) using two identical tokens, so these imitations were readily identified by children. Our main interest, however, concerns unsuccessful imitations – trials in which monosyllabic targets were paired with their disyllabic counterparts (e.g., *lbif-ləbif; ləbif-lbif*). Of interest is whether the child's ability to detect such changes depended on the sonority profile of the onset.

(13) An illustration of the imitation task with monosyllabic targets (Berent et al., 2011a). Note: participants were also presented with "successful imitation" trials consisting of disyllables, and "unsuccessful imitations" further paired these items in the reverse order (those additional trials are not shown here).

		Unmarked	Marked
Successful imitation	Rise-fall	*bwif-bwif*	*lbif-lbif*
	Rise-plateau	*bwif-bwif*	*bdif-bdif*
	Plateau-fall	*bdif-bdif*	*lbif-lbif*
Unsuccessful imitation	Rise-fall	*bwif-bəwif*	*lbif-ləbif*
	Rise-plateau	*bwif-bəwif*	*bdif-bədif*
	Plateau-fall	*bdif-bədif*	*lbif-ləbif*

Previous research with adults suggests that onsets with small sonority distances are repaired by appending a schwa (e.g., *lbif*→*ləbif*) – the smaller the sonority distance, the more likely the repair (Berent et al., 2007a). Consequently, adults tend to misidentify marked onsets with small sonority distances as their disyllabic counterparts (e.g., *lbif*=*ləbif*). If similar markedness pressures and repair mechanisms are operative in the child's grammar, then as the markedness of the onset increases, children's accuracy should decrease. To examine this prediction, the experiment compared marked onsets to less marked counterparts. There were three such comparisons (see 13): One condition compared sonority rises and falls, another compared rises and plateaus, and the final condition compared sonority plateaus and falls.

Results showed that performance was modulated by the markedness of the onset (see Figure 9.1): In each of the three conditions, children were better able to detect the epenthetic change with unmarked onsets (e.g., distinguish *bwif* from *bəwif*) compared to marked ones (e.g., distinguish *lbif* from *ləbif*). Auxiliary analyses suggested that the difficulty with marked onsets was not due to their statistical properties, nor was it captured by several phonetic characteristics of those items. While the findings from 3–5-year-olds cannot attest to the initial state of the grammar, they do suggest that young children possess broad sonority preferences that they extend to structures that are unattested in their language. The convergence between the sensitivity to sonority profile in perception and production opens up the possibility that sonority-related preferences might be active early in life.

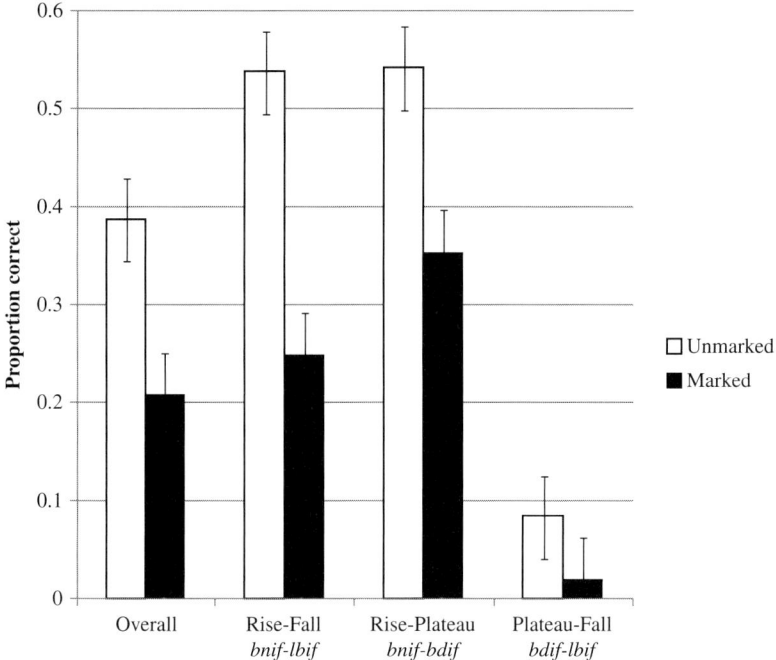

Figure 9.1 The effect of markedness on response accuracy to unattested onsets in the "unsuccessful imitation" condition (Berent et al., 2011a). Error bar indicates confidence intervals constructed for the difference between means.

9.5 Conclusions

This chapter outlined many intriguing parallels between the phonological knowledge of adult speakers and the early phonological instincts of young infants. We have seen that the algebraic computational machinery of core phonology is active practically at birth. There are also some hints that phonological primitives and markedness restrictions might be operative in early development.

With respect to primitives, infants' behavior is consistent with the possibility that the initial state of the grammar encodes features, consonants and vowels, and syllables. The encoding of phonological features is supported by infants' ability to contrast allophonic and phonemic variations. Features further distinguish consonants from vowels, and this distinction, in turn, allows newborns (and even preborn fetuses) to contrast languages from different rhythmical groups and supports older infants' ability to rely on consonants for lexical access. Consonants and vowels form syllables, which define the domain of phonotactic preferences and constrain the structure of words produced by young children. Some of the main findings and the age at which they are observed are summarized in (14).

(14) Evidence for phonological primitives in early development
   a. Features:
      - Older infants (11-month-olds) distinguish between allophonic and phonemic contrasts – they learn phonotactic regularities only when they concern features that are phonologically contrastive in their language (Seidl et al., 2009).
   b. Consonants and vowels:
      - Newborn infants (and even preborn fetuses) distinguish between language groups, defined by the duration of consonants and vowels (Kisilevsky et al., 2009; Mehler et al., 1988; Nazzi et al., 1998; Ramus et al., 2000).
      - Older infants (20–30 months) selectively attend to consonants (but not vowels) in accessing their lexicon (Nazzi, 2005; Nazzi & New, 2007; Nazzi & Bertoncini, 2009; Nazzi et al., 2009).
   c. Syllables:
      - Syllables are the domain of language-particular phonotactic regularities (~9-month-olds: Friederici & Wessels, 1993; Jusczyk et al., 1993; Jusczyk et al., 1994; Mattys et al., 1999; Mattys & Jusczyk, 2001).
      - Syllables are the domain of universal phonotactic regularities (2 months: Bertoncini & Mehler, 1981).
      - Syllables define prosodic restrictions on early words (1 year: e.g., Adam & Bat-El, 2009).
      - Syllables can be selectively targeted for substitution in word production (2 years: e.g., Gnanadesikan, 2004).

Infants' and young children's preferences are also consistent with some of the markedness reflexes seen in mature grammars (see 15). We reviewed principled learnability arguments suggesting that markedness constraints must be highly ranked in the initial state of the grammar, and we showed that this prediction is borne out by the behavior of 4-month-old infants. Like adults, 2-month-old infants further favor unmarked syllables that include a vocalic nucleus; young children produce unmarked syllables before they master more marked syllables, and their production and perception preferences are consistent with sonority restrictions.

(15) Evidence for markedness-reflexes in early development
   a. Early grammars rank markedness constraints above faithfulness pressures (Jusczyk et al., 2002).
   b. Syllable markedness:
      - Children often produce unmarked syllables before they produce marked ones (e.g., Bernhardt & Stemberger, 1998).

Conclusions 225

- Infants favor unmarked syllables in perceptual experiments (2 months: Bertoncini & Mehler, 1981).
c. Sonority restrictions:
- The production of early words respects sonority restrictions (1–3 years: e.g., Gnanadesikan, 2004; Pater, 1997b).
- Children identify onsets with large sonority distances more accurately than onsets with smaller distances even when these onsets are all unattested in their language (3–5 years: Berent et al., 2011a).

While those findings are all *consistent* with the possibility that universal phonological primitives and markedness constraints are active in the initial state, whether infants *are*, in fact, equipped with such core phonological knowledge is harder to tell. Most of the evidence discussed in this chapter concerns knowledge that is attested in the child's language and is documented at ages in which linguistic experience plays a demonstrable role. Moreover, the grand majority of those results are confined to language production experiments. Such findings do not allow one to determine whether the child's preferences reflect production constraints or linguistic knowledge, nor can they decide on whether the relevant knowledge concerns principles that are putatively universal or ones induced from the child's linguistic experience.

A few studies, however, have suggested that very young infants (under 6 months of age) are sensitive to a handful of phonological primitives (e.g., Mehler et al., 1988; Nazzi et al., 1998; Ramus et al., 2000), as well as markedness constraints on syllable structure (Bertoncini & Mehler, 1981) and place assimilation (Jusczyk et al., 2002). Because such preferences are seen in perception, and observed in very young infants at ages where the effect of linguistic experience is still undetectable, they are unlikely to be due to either linguistic experience or articulatory constraints. Older children have likewise been shown to favor unmarked structures to marked ones despite the fact that those marked variants are either less frequent than their marked counterparts (Adam & Bat-El, 2009) or utterly unattested in the child's language (Berent et al., 2011a; Ohala, 1999; Pertz & Bever, 1975). Such evidence hints at the possibility that several primitives and markedness reflexes are active in early development. The number of such demonstrations, however, is very small, and most findings do not dissociate grammatical and functional explanations for the results. So while these results are consistent with the possibility that core phonological knowledge is active in the initial state, the evidence necessary to fully evaluate this hypothesis is still incomplete.

10 The phonological mind evolves

In previous chapters, we have considered the possibility that the human mind possesses a system specialized for phonological patterning. How special is this capacity? Is it unique to humans, or shared with other animals? To address this question, we examine three defining features of human phonological systems: (a) their reliance on algebraic computational machinery, (b) their assembly by the conjunction of learning and universal, substantive constraints, and (c) their tendency to optimize analog phonetic pressures using algebraic means. We next proceed to investigate whether those three capacities are available to nonhuman animals. Anticipating the conclusions, neither algebraic machinery nor the capacity to shape communication patterns by both learning and innate knowledge are uniquely human, as each of these separate capacities is widely attested in the animal kingdom. But surprisingly, few species combine them in their natural communication systems, and no comparable case is attested in nonhuman primates. Precisely because these two ingredients (algebraic machinery and substantive constraints) are widely available to nonhumans, their unusual conjunction in human phonology is likely due to some modification to the human genome and brain that regulates the spontaneous, systematic capacity of humans to engage in phonological patterning.

10.1 The human phonological instinct from a comparative perspective

The discussion in this book has so far concerned itself with the specialization of the phonological mind – whether humans possess a specialized mechanism, equipped with innate universal constraints that specifically target the structure of phonological patterns. The question of specialization is important because it touches on the age-old debate concerning the origins of human knowledge – whether our knowledge and beliefs are induced from experience, or shaped a priori by our biology. The results presented so far open up the possibility that the phonological mind might indeed be specialized in this manner. We have seen that disparate phonological systems manifest a uniform design that distinguishes them from nonlinguistic patterns, speakers of different languages exhibit knowledge of these principles even when they are unattested in their language, and there is some evidence that this design is present already in early development. All these results suggest that the phonological mind might be a specialized system of core knowledge. But once the possibility of specialization arises, a new question immediately comes to mind: Is this design special? Is the makeup of the phonological mind similar to other systems of animal communication, or is human phonology unique in some way?

While the questions of specialization (does a system X have a special design, distinct from other systems?) and uniqueness (is the design of system X shared across species?) are logically distinct (e.g., a specialized system could well be shared across species) – these two questions are nonetheless related, as they are both linked to our primeval preoccupation with the constraints on human knowledge and their origins. And, if the design of the phonological mind did manifest continuity with animal communication, then one would further wonder about its source – does it reflect homology evolving through descent with modification from a common ancestor, or analogous systems that have developed independently of each other? A final, highly controversial question is how the human phonological mind evolved, and the role of natural selection in this process. These three questions (see 1) are at the center of this chapter. Because the continuity of the phonological system is a question that is logically prior to any discussions of its evolution, and because the state of knowledge on this topic far exceeds our understanding of language evolution, the discussion in this chapter mostly focuses on the continuity of the phonological mind – its evolution will be briefly considered in the final sections. My review of the very large animal literature will be strategically planned, rather than exhaustive. Likewise, in line with previous chapters, the discussion will typically be limited to phonotactics (for discussions of continuity in various aspects of rhythmic and tonal organization, see Jackendoff & Lerdahl, 2006; Patel, 2008; Pinker & Jackendoff, 2005).

(1) The evolution of the human phonological mind: some questions
   a. *Continuity*. Is the human capacity for phonological patterning shared with nonhuman animals?
     (i) Shared computational machinery
     (ii) Shared substantive constraints
   b. *Homology*: Do the putative nonhuman precursors of the human phonological system present a homology or analogy?
   c. *Selection*: How did the human phonological mind evolve: was it the target of natural selection?

10.2 Is phonological patterning special?

How special is the human capacity for phonology? At first blush, the possibility that phonology is a uniquely human instinct appears to be countered by numerous challenges. Various animal species spontaneously engage in aural communication that exhibits hierarchical patterning, akin to human phonological systems. Just as human utterances comprise phonemes, grouped into syllables, which, in turn, give rise to large prosodic units – metrical feet and prosodic words – so do birds weave their songs from a fixed set of notes, these notes give rise to syllables, and syllables, in turn, form motifs (see Figure 10.1). In both birds and men, aural communicative patterns are shaped by multiple constraints – both learned and innate. Finally, just as human phonological systems differ on their specific phonotactics (e.g., Russian allows syllables like *lba*, but English doesn't) but share universal constraints (e.g., all languages disfavor syllables like *lba*) – so do Swamp Sparrows, for instance, constrain their syllable structure by multiple principles, some being shared across many populations of bird species, whereas others vary across populations (Balaban, 1988a; Lachlan et al., 2010; Marler & Pickert, 1984).

Animals are also sensitive to many aspects of human speech. Various talented animals (e.g., chimpanzees, gorillas, parrots, and dogs; Gardner & Gardner, 1969; Kaminski et al., 2004; Patterson, 1978; Pepperberg, 2002) can produce and comprehend human words, both spoken and signed. Like humans, many animal species, including chinchillas (Kuhl & Miller, 1975), macaques (Kuhl & Padden, 1983), and birds (Dooling et al., 1995) perceive consonants categorically; animals can distinguish typical vowel exemplars (e.g., a typical instance of the /i/ category) from atypical ones (Kluender et al., 1998), produce formant structure in their own species-specific calls (Riede & Zuberbühler, 2003), and attend to formant-modulation in identifying both their species-specific calls

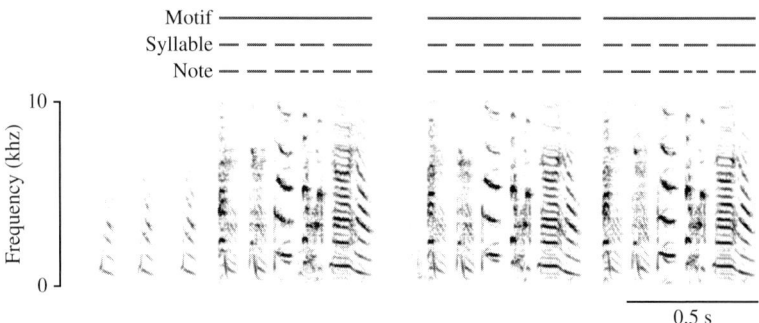

Figure 10.1 The hierarchical structure of the Zebra Finch song (from Berwick et al., 2011)

(Fitch & Fritz, 2006) and human speech (Lotto et al., 1997). Animals' sensitivity to the structure of speech also allows them to track its statistical properties and discern certain aspects of its rhythmical structure (Yip, 2006). Specifically, statistical learning of spoken syllables has been demonstrated in both rats (Toro & Trobalón, 2005) and cotton-top tamarin monkeys (Hauser et al., 2001; Newport et al., 2004), and cotton-top tamarins can further exploit the rhythmical properties of speech in order to distinguish between languages (e.g., Dutch and Japanese; Ramus et al., 2000).

In view of such observations, one might reasonably doubt whether there is anything special about the human capacity for phonological patterning. Such a sentiment might explain the widely held view that phonology likely lies outside the subset of the faculty of language that is domain- and species-specific (Fitch & Hauser, 2004; Fitch et al., 2005).

But there is reason to believe this conclusion might be premature. Merely showing that an animal can produce and identify certain phonological patterns does not necessarily demonstrate that the knowledge and representations that are guiding its performance are comparable to humans' (Pinker & Jackendoff, 2005; Trout, 2003). Indeed, an animal could successfully imitate limited aspects of human phonology while relying on weaker computational mechanisms that do not support the full scope of phonological knowledge and generalizations. Alternatively, an animal that possesses powerful computational mechanisms might fail to deploy them in the production of human sound patterns for reasons unrelated to phonology per se (e.g., because they lack the human articulatory tract or the ability to fully control their own articulatory mechanisms). Our real question here is not whether animals can mimic some aspects of human speech, but rather whether their capacity to pattern meaningless elements in their own system of communication (either spontaneous or learned) is comparable to humans'. This answer is not always clear from the available evidence.

Addressing this latter question requires a closer look not only at the mechanisms mediating animal communication but also at the ones governing phonological computations in humans. Existing cross-species comparisons have grossly underestimated the phonological capacities of humans. Such comparisons typically focus on either phonetic processing (chiefly, categorical perception) or some very rudimentary aspects of phonological processing, such as generic rule-learning or statistical-learning. Missing from this discussion is an in-depth analysis of the mechanisms supporting human phonological systems – not only those responsible for language variation but also the ones promoting cross-linguistic universals. Such inquiry can only be conducted in the context of an explicit account of the human phonological mind. The following section outlines such an account – we next proceed to review the animal literature in light of this proposal.

10.2.1 Species-specificity in phonology: some candidates

Throughout this book, I have argued that the human capacity for phonological patterning has three main characteristics (see 2). First, phonology is an algebraic system that allows humans to generalize their knowledge across the board, even to instances that they have never heard before. This capacity for unbounded productivity has been specifically attributed to several computational characteristics of the grammar – the capacity to encode equivalence classes (Chapter 4), to combine those classes to form hierarchical structures, and to encode their relations using variables (Chapter 5). Of interest, then, is whether nonhuman species possess the computational machinery that supports unbounded productivity, and whether this capacity can be put to use in forming meaningless patterns of communication – either patterns occurring naturally, in the animal's own communication, or ones acquired from human phonological systems.

(2) What aspects of phonology might be uniquely human? Some candidates
  a. Algebraic machinery that supports unbounded productivity:
    (i) Equivalence classes and their hierarchical organization
    (ii) Rules: operations over variables that stand for entire equivalence classes
  b. The capacity to shape patterning by both learning and universal, substantive constraints:
    (i) Universal representational primitives
    (ii) Universal combinatorial constraints
  c. Algebraic optimization of analog phonetic pressures

Human phonological systems, however, are identifiable not only by their general computational characteristics but also by the presence of broad, perhaps universal, substantive constraints on phonological patterns. Such constraints limit the range of phonological primitives that are attested in human languages and restrict their combinations. While distinct languages appear to share the same constraints, the ranking of those constraints varies as a result of the human capacity to learn phonological patterns. Our question here, then, is whether meaningless patterns in nonhuman communication are jointly shaped by learned and substantive constraints.

Although nonhuman communication systems might well abide by substantive species-specific constraints, it is unlikely that the specific substantive constraints on human phonology are broadly shared with nonhumans. Indeed, human phonologies are shaped by the phonetic properties of their channel – either speech (for most human language) or manual articulation (for sign languages) – and consequently, cross-species variations in articulatory mechanisms and auditory perception are bound to give rise to variations in patterns. Put bluntly, a species with no lips will not constrain the labial place of articulation. Such obvious variations in external functional pressures, however, obscure deeper potential similarities in design of the patterning system itself. All phonological systems are grounded in their phonetic channel. While previous chapters have invested much effort in dissociating grammatical phonological constraints from phonetic pressures, the ability to link them should not be taken for granted. The functional grounding of phonology bridges two distinct formats of representations – one is analog and continuous; another is digital and algebraic. Phonetic pressures apply to representations that are continuous and analog (e.g., voice onset time – the continuous acoustic dimension that marks the discrete phonological feature of voicing). But once these pressures become "phonologized," they are restated as algebraic restrictions representations that are digital and discrete.

The ability to restate analog phonetic pressure in algebraic terms is remarkable. To use an analogy, algebraic optimization could be likened to the building of a ball from lego blocks – the blocks are discrete, and the laws of their combinations are stated algebraically, but each such block is designed to optimize the construction (e.g., it is sufficiently strong and small to allow the approximation of a large rounded object), and the constraints on their combination are designed to approximate the continuous surface of a ball. This is not a trivial feat. Even if nonhuman species possess some of the algebraic machinery necessary to attain productivity, and their sound patterns are subject to substantive phonetic constraints, it remains to be seen whether these two capacities can be spontaneously combined to give rise to algebraic restrictions that are phonetically grounded.

The following sections examine whether these three traits of the human phonological mind are shared with our nonhuman relatives, both close and distant. Before we embark on the journey, an important caveat must be acknowledged. To allow for a broad cross-species comparison, the following discussion resorts to a rather broad definition of "phonology" – for the present purposes, we consider phonology as a system that patterns meaningless elements, regardless of whether this meaningless pattern has any extension (i.e., whether it is linked to specific concepts). In so doing, we purposely disregard a crucial aspect that sets human phonology apart from any known system of animal communication. All human languages exhibit duality of patterning – a patterning of both meaningful elements (i.e., words patterned into sentences) and meaningless elements (i.e., phonemes patterned to form words; Hockett, 1960). No known system of animal communication manifests this trait – while some systems have patterns at the level of meaning, and others pattern meaningless elements, no nonhuman system exhibits patterning at both levels. Accordingly, when the role of phonology is situated within the language system as a whole, no system of animal communication is truly comparable to human phonology. But because our interest here is in the capacity to pattern meaningless elements, specifically, for the present purposes, I will disregard this fundamental difference. With this major caveat in mind, we can now move to consider whether the above three aspects of human phonology are present in the communication patterns of nonhumans.

10.2.2 Computational machinery: what is shared?

At the center of the human capacity to generalize phonological patterns across the board is the ability to combine equivalence classes, to extend those equivalence classes to novel tokens, and to operate over such classes. For example, the requirement that the prosodic unit of a foot must minimally include two syllables (e.g., McCarthy & Prince, 1993) entails the representation of the syllable as a class of equivalent elements whose members (e.g., *dig*, *black*, *arc*) are all treated alike, irrespective of their shape or familiarity. Phonological equivalence classes are further grouped hierarchically. For example, onsets and rhymes form syllables – and the process can iterate, combining syllables into metrical feet, and feet into prosodic words. Moreover, people can apply various operations over such classes and encode their relations using variables (e.g., identity relations, *baba*). The question we consider next is whether these three computational tools – the formation of equivalence classes, their grouping into hierarchical structures, and the capacity to encode their relations in rules – can be deployed by nonhuman animals in the patterning of meaningless elements.

10.2.2.1 Equivalence classes

Many animals encode equivalence classes – both classes that are naturally occurring and ones that are learned. Numerous studies have demonstrated the use of equivalence classes in lab settings (for a review, see Schusterman et al., 2003). For example, rhesus monkeys spontaneously contrast singular and plural sets: They distinguish one apple from several apples while ignoring the distinction between two sets of apples (Barner et al., 2008). Such cases suggest that these animals can encode all members of a set (singular vs. plural) alike and disregard their differences for the purpose of the relevant generalization.

Animals also form classes in their natural communication systems. Many primate species spontaneously produce a variety of acoustically distinct calls in specific contexts (e.g., for specific predators or foods; for review see Hauser, 1996). Vervet monkeys, for instance, produce acoustically distinct calls for leopards, pythons, and martial eagles (Seyfarth et al., 1980). But while the "eagle" call is typically emitted in the presence of martial eagles, it can also extend to other bird species – both raptor (e.g., hawk eagle, snake eagle) and nonraptor birds (e.g., vulture, stark), and younger vervet monkeys further generalize those calls to geese and pigeons. This behavior could suggest that these stimuli form a single semantic category (e.g., "birds") whose members are all associated with a single call that elicits a single response (i.e., "look up," "run into dense bushes"). Note, however, that members of the class are not strictly equivalent – martial eagles, for instance, are more likely to elicit the "eagle" call than hawk eagles (Seyfarth et al., 1980). More importantly, the relevant class concerns the call's meaning (the class of bird-like entities), rather than its form (i.e., the "phonological" structure of the call). But unlike the vervet call, human languages (e.g., English) form equivalence not only of words' meanings (e.g., all even numbers, 2,4,6 …) but also of their forms (e.g., "any syllable," including the instances "two," "four," "six," but also "boy," "girl," etc.). While much evidence suggests that animals represent some concepts as equivalence classes (Hauser, 1996; Schusterman et al., 2003), it is unclear whether they spontaneously deploy this capacity in the phonological organization of their own calls.

Given the paucity of phonological equivalence classes in naturally produced calls, it is not surprising that they rarely combine to form hierarchies akin to the phonological hierarchy of human languages (e.g., phonemes → syllables → feet → phonological words). A strong contender for such a hierarchical organization is presented by birdsong. Many birds form their songs by chaining notes into syllables, which, in turn, give rise to larger motifs (see Figure 10.1). But while birds and men both encode hierarchies, they do not necessarily both embed equivalence classes. In the case of humans, there is evidence that syllables and their constituent phonemes each form equivalence classes, as phonological systems include constraints that apply to all members of the "syllable" and "onset" classes alike. But it is uncertain whether birds encode their syllables and notes in this fashion. While several bird species perceive the syllable as a unit (Cynx, 1990; Franz & Goller, 2002; Suge & Okanoya, 2010), such evidence does not necessarily require equivalence classes. And indeed, syllable-size elements can be encoded in multiple ways (see Chapter 4). Our intuition that *pen* forms a part of *pencil* could reflect either a hierarchical relation among equivalence classes (i.e., "a single syllable X forms part of a disyllable X+Y") or the chunking of individual units ("pen" and "cil" each forms a chunk of "pencil," but these chunks are not represented alike at any cognitive level).

The contrast between humans and birds is even clearer at the sub-syllabic levels. In the case of humans, sub-syllabic constituents form equivalence classes – an onset, for example, comprises a large number of instances that are clearly distinguished from each other (e.g., /b/ and /d/ in *big* vs. *dig*). In contrast, birds (specifically, Swamp Sparrows) perceive notes (constituents in their own syllables) categorically (Nelson & Marler, 1989), so the sub-syllabic constituents of bird-syllables appear to form phonetic categories (akin to the category of /b/, including [b1], [b2], etc.), rather than phonological ones (e.g., "phoneme," such as /b/, /d/, /f/, etc.). Summarizing, then, humans represent multiple layers of algebraic encoding – sub-syllabic units are equivalence classes, and those, in turn, combine to form larger equivalence classes of syllables. Birds, in contrast, encode their syllables as chunks, but the evidence reviewed so far does not demonstrate that they represent equivalence classes, let alone ones that are hierarchically structured (we will consider other evidence for equivalence classes along with our discussion of rules, later in this chapter).

While such isolated calls cannot determine whether animals spontaneously encode equivalence classes of meaningless elements in natural communication, many animals can learn artificial rules that combine meaningless equivalence classes in lab settings, and some animals can even deploy them spontaneously. These demonstrations of rule-learning, reviewed below, offer further evidence that various animal species (including birds) do possess the computational machinery necessary to form equivalence classes.

10.2.2.2 Rules

Various species of animals can represent and learn rules, and this capacity is evident in both rules employed spontaneously (either in communication or the encoding of conceptual knowledge) and rules learned in laboratory settings. While the animal literature has used the term "rule" in numerous ways, in line with Chapters 4–5, here, we will limit the discussion of "rules" to operations over variables that stand for entire equivalence classes. Such operations could constrain either the sequencing of equivalence classes (regularities that may or may not appeal to variables), or their relations (e.g., identity) – where the encoding of variables is demonstrably necessary (see Chapter 5).

Rules used spontaneously by animals

Rules used to represent knowledge. Of the various types of rules used spontaneously by animals, most striking is the complex ability of various species to use rules in reckoning their navigation (for review, Gallistel, 1990). Indigo Buntings, for example, compute their migratory path based on the constellation of stars and on the earth's magnetic field (Emlen, 1975; 1976). Because of the earth's motion, however, such calculations must compensate for the apparent motion of different stars, whose rate and direction varies depending on their distance from the North Star. The ability of the birds to compute such calculations and apply them to novel constellations they have not encountered before suggests that they possess complex computational machinery that uses abstract placeholders (i.e., variables, Emlen, 1975).

Honeybees likewise rely on complex computations in calculating the location of food sources relative to the solar ephemeris, which they communicate to their peers by means of their waggle dance. But like the constellation of the stars in the night sky, the position of the sun can acquire a large number of values, some of which have not been observed by the bee. Strikingly, bees can compute the position of the sun based on two pieces of information – the sun's past location (observed by the bee), and the passage of time (since the last observation of the sun's location), which, in turn, is estimated by the bee's circadian clock (Dyer & Dickinson, 1994). Such computations allow the bee to predict the sun's azimuth in the morning sky despite never actually experiencing the sun in that particular location (bees in that experiment were allowed to see the sun only in the late afternoon). Moreover, the bee reckons this information and communicates it in her dance despite the fact that, at the time of signaling, the overcast sky did not allow either the signaling bee or the perceiver bee to actually see the sun's location.

While the behaviors of the Indigo Bunting and honeybees imply knowledge of complex rules, those rules are not directly used in communication. Indigo Buntings use rules to calculate their navigational path, rather than to communicate it (e.g., in their song). Likewise, honeybees might well rely on rules to compute the location of food, but they communicate this knowledge by means of a dance that is clearly analog and continuous (e.g., the distance from the food source correlates with duration of the bee's waggling; Dyer & Seeley, 1991). Accordingly, both cases suggest the capacity to use algebraic rules for the mental computation of information, but not for its communication.

Rules used in vocal communication. Other animals exhibit complex vocal patterns in their natural communication, but whether those patterns are generated by rules is not entirely clear. Various reports indicate that primates restrict the sequencing of their naturally occurring calls and distinguish such sequences from their constituents (e.g., complex calls and their parts are used in different contexts). Wedge-capped capuchin monkeys, for example, exhibit five classes of naturally occurring calls which are combined in restricted ways (Robinson, 1984): Chirp-squaw combinations are attested, whereas the reverse sequencing is not. Similar restrictions on naturally occurring call sequences have been observed in putty-nose monkeys (Arnold & Zuberbuhler, 2006), Campbell monkeys (Ouattara et al., 2009), and chimpanzees – both wild (Crockford & Boesch, 2005) and captive (Marshall et al., 1999). While these cases certainly suggest that nonhuman primates constrain the sequencing of specific elements in their natural communication, they do not make it clear whether such restrictions apply to equivalence classes of perceptibly distinct elements, akin to the classes of "all consonants" (e.g., /b, d, g/) or phonetic categories whose tokens are indistinguishable (e.g., tokens of the /b/ phoneme). Similar questions arise with respect to the complex (possibly meaningless) patterns in the learned song of humpback whales. Human and computational analyses suggest that these vocalizations exhibit a hierarchical structure that includes long-distance dependencies among non-adjacent elements (Miksis-Olds et al., 2008; Payne & McVay, 1971; Suzuki et al., 2006), but it is presently unknown whether whales represent those dependencies by linking equivalence classes.

236   The phonological mind evolves

While the evidence for rules in the natural communication of mammals, and most significantly, primates, is scant, clearer support for natural rule-learning is presented by birdsong. As noted earlier, it is uncertain whether the song's constituents – entities such as "syllable," and "motif" – are encoded as equivalence classes (e.g., "any syllable" or specific instances of those elements: Gentner, 2008; Gentner & Hulse, 2000), and consequently, any restrictions governing the organization of those constituents could potentially reflect either reliance on rules (operations on variables, standing for equivalence classes) or restrictions that strictly concern the co-occurrence of specific instances. Birds, however, might also rely on other constituents in encoding their song, and those offer stronger evidence that distinct notes are represented as equivalent.

The evidence comes from the restrictions on the syntactic structure of Swamp Sparrow songs. Like the contrast between syllables in English (where *bla*, but not *lba* is attested) and Russian (where both *bla* and *lba* are allowed), so do Swamp Sparrows from different communities exhibit distinct dialects, marked by different "phonological" patterns. While birds from New York State favor syllables that begin with note I and end in note VI (i.e., I_VI), birds from Minnesota manifest the opposite order – VI_I. Crucially, this preference is seen irrespective of the contents of the intermediate elements (marked by "_"). Pioneering experiments by the biologist Evan Balaban have shown that New York birds prefer the I_VI order even when the intermediate note is replaced by material from the song of their Minnesotan conspecifics (see Figure 10.2).

These results are extremely significant because they imply that birds spontaneously represent their own song by means of an abstract I_VI rule, where "_" could stand for any note – an equivalence class. Moreover, this rule-based account is superior to several alternative explanations that attribute the findings to either holistic gestalts or narrow statistical learning. Specifically, because the intermediate note in these stimuli was excised from songs recorded in nature ten years

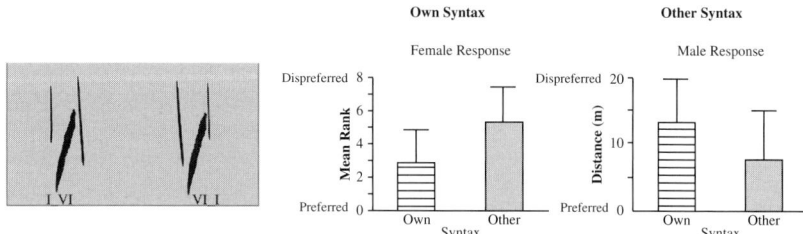

Figure 10.2 Learned variations in song patterns of Swamp Sparrows (from Balaban, 1988a). Syllable structure of New York vs. Minnesota birds (left), and the preference of New York birds for their own syntax (right). Note: preferences are plotted as ranked order, so lower numbers indicate preference.

earlier and delivered in the unfamiliar Minnesotan syntax and "accent," the New York birds' preference for the I_VI sequence could not have been based on familiarity with an acoustic template of the song as a whole. Similarly, the birds' preference is unlikely to be due to statistical learning. On that statistical account, the birds did not register the intermediate note by an equivalence class, but instead, they only encoded the position of notes I and VI relative to syllable edges (marked by #, e.g., #I vs. #VI). This account, however, would predict a similar preference for four-note syllables (e.g., I, IV, III, VI), but this prediction is incorrect – additional experimentation demonstrates that birds show no such preference (Balaban, 1988b). Taken as a whole, then, the most likely explanation for these results is that Swamp Sparrows represent their own song by means of a I_VI rule, where _ is a variable that stands for an equivalence class.

In the case of human phonology, however, equivalence classes not only ignore the distinction between class members but also generalize to novel class instances, including phonemes and features that are unattested in speakers' language. Accordingly, one wonders whether the Swamp Sparrows' classes are likewise open-ended, that is, whether these birds can freely extend the I_VI to any note, familiar or novel. Balaban's own findings do not fully determine the scope of those classes, but other findings suggest that they are quite broad (Marler & Peters, 1988). The evidence comes from experiments with song sparrows. Song sparrows are genetically related to the Swamp Sparrows, but their song structures differ: while Swamp Sparrows manifest a single phrase, comprising a single, multi-note syllable, repeated in a regular tempo, song sparrows' song typically includes multiple phrases. Phrase alternation is thus a defining feature of song sparrows' song. Remarkably, song sparrows generalize this preference to synthetic songs whose phrases comprise novel syllables (taken from recordings of their Swamp-Sparrow relatives). The generalization to foreign syllables was not simply due to the failure to distinguish conspecific from heterospecific syllables, as song sparrows clearly favored their conspecific syllables to heterospecific ones when the songs included either one or three phrases. But when the song consisted of two phrases, heterospecific and conspecific syllables were equally acceptable (indexed by the tendency of males to imitate those synthetic recordings), suggesting that the birds encoded an abstract notion of a phrase alternation that generalizes across the board, even to foreign syllables.

Together, these findings suggest that birds represent their song by means of rules that combine equivalence classes, and they can generalize those classes broadly, even to note-instances that they have never heard before. To my knowledge, these results present the strongest demonstration of rules in natural communication. Remarkably, no parallel findings have been reported with any primates.

Rules learned in laboratory settings

While there is currently no clear evidence that primates naturally encode phonological rules (i.e., algebraic restrictions on the patterning of meaningless equivalence classes in their natural communication), they appear capable of learning rules in laboratory settings. Rhesus monkeys trained on the "greater than" rule (e.g., 5>1) on the numbers 1–9 can generalize the rule to novel instances (10–30), irrespective of superficial differences in density, surface area, and perimeter (Cantlon & Brannon, 2006; for converging evidence from single-neuron recordings, see Bongard & Nieder, 2010). For our present purposes, however, more significant is the ability of animals to learn rules concerning meaningless elements, akin to the ones used in human phonological systems. Existing research has indeed documented such abilities in various species, and with various types of rules, including first-order restrictions on the sequencing of equivalence classes, and second-order restrictions on their relationships.

Learning sequential rules. One demonstration of the learning of sequential restrictions is presented by ordinal rules. In one experiment (Chen et al., 1997), rhesus monkeys were first trained on the ordinal relation of four visual stimuli (stimuli 1–4) presented at four different positions (A–D, see 3). Next, the monkeys were tested for their ability to relearn two derived lists of items. Both derived lists maintained the same items used in training, but disrupted their pair association, such that item pairs that were adjacent in training (e.g., A1–B1; A2–B2) were never adjacent in the derived lists (e.g., A2–B1). If, however, rhesus monkeys encode abstract ordinal position (e.g., item 2 appeared in the first position, A), then they should be better able to learn novel lists that preserve the items' original positions (e.g., item 2 appearing in position A; item 1 appearing in position B) even if their combination (A2B1) never appeared in training. The results support this prediction. The ability of the monkeys to learn and generalize such relationships is particularly striking given that 4-month-old human infants fail to learn them in a similar task (Lewkowicz & Berent, 2009). Further evidence for sequential learning is presented by the ability of cotton-top tamarin monkeys to learn and generalize a rule that concatenates a base word with a suffix, akin to inflection (e.g., *dog* +*s*→*dogs*) and distinguish the suffixation rule from prefixation (e.g., *s*+*dog*; Endress et al., 2009). The source of such generalization is not entirely clear – either the monkeys could have learned an equivalence class corresponding to "any word" or, alternatively, they could have formed a simple association between two specific elements – the suffix (s) and the right edge of the acoustic stimuli (marked as #, e.g., *dog#s, cat#s*, etc.). Nonetheless, these findings are certainly in line with the possibility that monkeys encode the sequential ordering of equivalence classes – a capacity that is central to many phonotactic restrictions (e.g., the contrast between *blog* and *lbog*). Accordingly, the capacity to learn sequential rules might not be unique to humans.

(3) Ordinal relations learned by rhesus monkeys (Chen et al., 1997)
   Training lists:
   List 1: A1➔B1➔C1➔D1   bird➔flower➔frog➔shells
   List 2: A2➔B2➔C2➔D2   tree➔weasel➔dragonfly➔water
   List 3: A3➔B3➔C3➔D3   elk➔rocks➔leaves➔person
   List 4: A4➔B4➔C4➔D4   mountain➔fish➔monkey➔tomato
   Testing sequences:
   Position maintained: A2➔B1➔C4➔D3   tree➔flower➔monkey➔person
   Position changed: B3➔A1➔D4➔C2   rocks➔bird➔tomato➔dragonfly

Learning relations. Not only can animals learn rules concerning the sequencing of abstract classes, but they can also encode second-order restrictions on their relations. Research in this area has considered two types of relations – identity (e.g., X➔XX) and recursion (X➔AXB). The capacity to encode and generalize such relations is significant because, as shown in Chapter 5, it requires the representation of variables. Specifically, to encode identity, XX, learners must be able to bind instances of the two X categories by a variable – if the first occurrence is instantiated by "ba," for example, so should the second (e.g., *baba*). Similar mechanisms are required for the encoding of rules such as A_nB_n (e.g., *baga*, *babagaga*, *bababagagaga*). While such rules have been used to gauge the representation of recursion, learners could, in fact, represent this regularity in several ways (Berwick et al., 2011). They could either represent it recursively (X➔AXB, where the procedure for constructing the category X invokes itself) or simply track the number of As and Bs (Berwick et al., 2011). Either way, however, learners must use variables to ensure that all members of the category (A and B) are instantiated by the same number of tokens (e.g., AABB; A=2; B=2), and, in some cases (e.g., *babagaga*), ensure that all As (and Bs) are instantiated by the same member. The ability to learn such rules has been demonstrated across numerous species.

Honeybees, for instance, have been shown capable of learning identity relations in a Y-maze (Giurfa et al., 2001). During training, bees were presented with a sample stimulus (either A or B) and learned to choose the maze-arm marked by the same stimulus (A stimulus➔A arm; B stimulus➔B arm). Remarkably, the bees generalized the identity function to novel stimuli, C, D, even when the training and test stimuli spanned different modalities (e.g., colors and odors). The capacity to learn second-order relations among variables has been also observed in several vertebrate species. The learning of identity-rules has been reported in free-ranging rhesus monkeys (Hauser & Glynn, 2009) and rats (Murphy et al., 2008). Likewise, the recursive A_nB_n rule has been examined in starlings (Gentner et al., 2006) and Zebra Finches (van Heijningen et al., 2009). Some of these demonstrations, however, do not make it clear whether the animal relies on broad algebraic operations over variables, or more limited heuristics that track the co-occurrence of specific training instances (see also Corballis, 2009; van Heijningen et al., 2009). For example, in the case of free-ranging rhesus monkeys, one class of training items consisted of aggressive calls, whereas test items were novel instances of the same call (Hauser & Glynn, 2009; a similar approach was adopted with starlings; Gentner et al., 2006). Although training and test items were acoustically different, it is unclear whether the animal represented them as such. Indeed, humans (Eimas & Seidenberg, 1997) and birds (Nelson & Marler, 1989) are known to engage in categorical perception, a process that largely eliminates the distinction between acoustically distinct tokens (e.g., between two instances of /b/, e.g., [b1], [b2]). To the extent the perceptual system is indifferent to the distinction between training items (e.g., between $[B_1]$ and $[B_2]$), then the extension of training on $A_1B_1B_2$ items to $A_2B_2B_2$ test items would reflect categorical perception (of B_1 and B_2) rather than true generalization.

Other demonstrations, however, address these worries by instituting more sophisticated controls. Van Heijningen and colleagues (2009) examined the ability of Zebra Finches to distinguish instances of the recursive A_nB_n rule from $(AB)_n$ foils. To ensure that generalization is not based on the failure to distinguish training and test items, these researchers sampled the training and test items from two distinct sets of the birds' own vocal elements (flats, slides, highs and trills). For example, training items might have consisted of sequences of "flats" and "slides," whereas testing items would comprise "highs" and "trills." Despite these tighter controls, at least one of the Zebra Finch subjects was able to generalize to distinguish novel A_nB_n items from $(AB)_n$ foils (e.g., to differentiate AABB test sequences from ABAB foils). Although a closer inspection of this bird's performance suggested that it relied on the occurrence of identical elements (e.g., the occurrence of BB in AABB, but not in ABAB), rather than the recursion rule per se, the ability of this bird to detect identity suggests that it is able to encode auditory sequences by means of abstract relations among variables.

10.2.2.3 Summary: algebraic machinery in animal communication
Summarizing the discussion so far, many animal species can learn rules that constrain the sequencing of equivalence classes in laboratory settings. Such rules concern both (meaningful) concepts (e.g., numerosity) as well as meaningless sequences, and they include both restrictions on the sequences of abstract classes and their identity relations. The capacity to spontaneously encode concepts by rules is also implicated by the complex navigational systems of birds and bees. Such results make it clear that the computational

machinery necessary to encode human phonotactic generalizations – the capacity to encode equivalence classes and represent their relations – is available to nonhumans. But precisely because this powerful algebraic machinery appears to be in place, it is remarkable how rarely it is used spontaneously in the patterning of *meaningless* elements. The only clear evidence for "phonological" rules in natural communication is presented by birdsong. While it is possible that future research might substantiate this capacity in other species (e.g., whales), those remarkable exceptions would still prove the rule: Many nonhuman species possess the capacity to encode equivalence classes and learn rules that operate on such classes, but few deploy this capacity for learning patterns of meaningless elements in their natural communication, and none of these species includes primates (Fitch, 2010). Precisely because the computational machinery that is at the core of human phonotactics is available to many species, its infrequent deployment in learning the species' own communication and its apparent absence in nonhuman primates are remarkable.

10.2.3 *The joint contribution of substantive constraints and learning: humans and birds*

Having the right computational machinery is undoubtedly necessary to support the phonological talents of humans, but it is apparently insufficient. The evidence reviewed in previous chapters suggests that, despite their diversity – the product of learning from experience – phonological systems are shaped by universal substantive constraints that limit the range of representational primitives in phonological systems as well as their combinations. While, not surprisingly, perhaps, there is no evidence that the particular substantive constraints on human grammars (e.g., NoCODA) are active in other species, it is nonetheless possible that their natural vocal patterning is shaped by substantive constraints of their own. The question we next address is whether the vocal patterns of nonhuman animals exhibit the capacity for constrained variation.

In the case of nonhumans, the evidence for innate restrictions on vocal communication is quite pervasive – it is actually vocal learning that is harder to come by (Hauser & Konishi, 1999; Hauser et al., 2002). Although there are reports of natural vocal learning in several mammals, including whales (Payne & McVay, 1971; Suzuki et al., 2006), elephants (Poole et al., 2005), and chimpanzees (Crockford et al., 2004; Marshall et al., 1999), the spontaneous learning of algebraic restrictions on complex multisyllabic vocalization is arguably absent in primates (Fitch, 2010). Unlike our close evolutionary relatives, however, passerine birds present ample evidence for vocal communication that is learned, yet highly constrained, in a species-specific manner.

So birdsong offers fertile grounds to probe for learned algebraic constraints that are "phonetically" grounded.

Many species of passerine birds acquire their vocal communication by learning from a tutor, and the songs of different bird communities can exhibit marked variations, ranging from the particular notes that are selected by each community to their number and their ordering in the syllable (Marler & Pickert, 1984). Recall, for example, that the songs of Swamp Sparrows from New York and Minnesota contrast on their syntactic organization – New York birds favor the I_VI order, whereas in Minnesota, it is the VI_I order that is typical. Similarly, the most frequent notes in the song of New York birds are I and VI, whereas birds from Minnesota favor notes I, II, and V; in the New York sample, three-note syllables are most frequent, whereas in Minnesota, four-note sequences are preferred. But despite these systematic learned variations, birdsong is nonetheless subject to universal structural restrictions that are likely innate. As in the case of human systems, the evidence comes from two main sources: typological studies and poverty-of-the stimulus arguments.

Taking a typological perspective, Lachalan and colleagues (2010) used the techniques of cluster validation statistics to gauge for species-specific universals in the songs of dozens of individual birds, sampled from geographically distinct communities of three different species – Chaffinches, Zebra Finches, and Swamp Sparrows. Although the data exhibited considerable variation, the notes and syllables of each single species nonetheless clustered into a small number of broad categories of universal species-specific primitives. Additional analyses revealed species-specific clustering in the statistical co-occurrence of those units as well. Together, these statistical analyses suggest that, despite their geographic diversity, members of each single species exhibit universal, species-specific constraints on the selection of phonological units and their sequencing (for similar conclusions, see also Nelson and Pickert, 1984).

The over-representation of certain structures across bird species mirrors the asymmetric distribution of phonological structures across human languages, asymmetries that might suggest the presence of universal structural restrictions (see Chapter 6). The strongest evidence for grammatical phonological universals, however, comes from cases in which speakers favor the unmarked variant over the marked one in the absence of experience with either (discussed in Chapters 7–8). Such observations suggest that the experience available to language learners is insufficient to explain the structure of the grammar that is ultimately acquired, an argument that falls within the broad category of "poverty-of-the-stimulus argument" (see Chapter 3; Chomsky, 1980; for experimental evidence, see Crain & Nakayama, 1987; Crain et al., 2005; Gordon, 1985; Lidz et al., 2003). In the limiting case, a grammar demonstrably emerges *de novo* in the absence of any linguistic input altogether. Although linguistic regenesis is rare, several cases show that, when children are grouped in a community, deprived of any linguistic input from adults, those children develop a language of their own, and their invention conforms to the same structural constraints evident in adult languages.

Earlier (in Chapter 2), we reviewed several reports of the spontaneous emergence of structured phonological systems among deaf signers. Recall, for example, the case of the Al-Sayyid Bedouin sign language (ABSL, Sandler et al., 2011), a nascent sign system documented in the Al-Sayyid Bedouin village in the south of Israel. In its inception, this young sign system exhibited various aspects of syntactic and morphological structure (e.g., it requires sentences to exhibit a Subject-Object-Verb order), but it lacked any evidence for phonological organization. Signs were typically iconic and holistic, and their structure violated many phonological restrictions documented in many mature sign languages. For example, while sign languages universally require all signs to exhibit movement (a constraint attributed to sonority restrictions; Sandler, 1993), adult signers of ABSL typically produced static signs. Remarkably, however, children born to this community spontaneously abided by such putatively universal phonological constraints. Unlike the holistic static signs of the older generation, the signs of the younger generation manifested movement and comprised phonetic features (e.g., handshape). And just as in spoken languages, adjacent segments often undergo assimilation when they disagree on their feature values (e.g., *in+possible→impossible*), so does the disagreement in handshape trigger assimilation in the ABSL.

The examples reported by Wendy Sandler and colleagues do not document how, precisely, phonological patterning has emerged in the ABSL community – whether it has been propagated by children or adults. But other reports of language birth suggest that linguistic patterning is the invention of children. This, possibly, is most clearly documented in the case of Nicaraguan Sign Language – a language invented by several generations of deaf children in Managua (Senghas & Coppola, 2001; Senghas et al., 2004). These children were grouped together in school for the purpose of elementary education and vocational training in Spanish, but rather than acquiring spoken language, they gradually developed a sign language of their own. The signs produced by the initial cohort of students exhibited little evidence of grammatical organization, but with time, subsequent cohorts have altered those signs to abide by various morphosyntactic restrictions that distinguish them from the gestures produced by their hearing counterparts and unite them with other natural languages. For example, when asked to depict a cat that, "having swallowed a bowling ball, proceeds rapidly down a steep street in a wobbling rolling manner," hearing people typically use a holistic spiral gesture, whereas signers combine two discrete movements for manner of motion and path of motion, respectively – a distinction present in many spoken languages. Moreover, compared to the initial cohort, later generations of signers are far more likely to rely on this grammatical combinatorial organization than on holistic gestures. Such cases suggest that grammatical structure, including phonological patterning, is an inherent property of language that is shaped by universal constraints, and these restrictions emerge spontaneously in the languages invented by children.

Similar cases of the spontaneous regenesis of patterning have been documented in birdsong. It is well known that birds will maintain certain aspects of their species-specific songs even when reared in isolation (e.g., Marler, 1997; Marler & Sherman, 1985). Similarly, when isolate birds are exposed to impoverished input that lacks syntactic organization, they spontaneously converge on their species-specific syntax. For example, white-crowned sparrows exposed to synthetic songs that comprised only a single phrase (e.g., repeated whistles, repeated buzzes, etc.) spontaneously produced structured songs that invariably begin with a whistle followed by buzzes and trills – an organization that mirrors their species-specific syntax (Soha & Marler, 2001). The reemergence of universal structural characteristics in the absence of any input suggests that at least some aspects of substantive constraints manifest themselves irrespective of experience. Nonetheless, the songs of isolate birds are clearly abnormal in many respects, a situation analogous to the impoverished grammatical structure of the first generation of children in nascent sign languages. The cases of Nicaraguan sign systems and the Al-Sayyid Bedouin sign language, however, suggest that across generations, these impoverished systems drift toward structures favored universally in existing systems. As it turns out, birds follow precisely the same trajectory.

The evidence comes from a study that examined the song of young Zebra Finches tutored by a mature male who was reared in isolation (Fehér et al., 2009). As expected, this isolate bird was hardly a great teacher, as its song exhibited various abnormalities that distinguish it from the song of typical Zebra Finches. But interestingly, the new generation of pupils outperformed their teacher. Rather than perpetuating these abnormalities by copying them verbatim, the pupils altered various aspects of the tutor's song. And as these pupils matured, they were "promoted" to tutor a subsequent generation of unrelated pupils, and those pupils were, once again, paired with subsequent generations. Remarkably, across generations, the song gradually drifted toward the normal song pattern.

The spontaneous regenesis of birdsong, along with the typology of naturally occurring song systems, demonstrates that, like human phonology, birdsong is shaped by innate substantive constraints that limit the range of representational primitives and their combinations. Nonetheless, the expression of these putatively innate constraints requires a social setting and some minimal experience. Reared in isolation, neither humans nor birds exhibit normal sound patterns. This, in turn, suggests that the substantive structural constraints on "phonological" patterns (either human phonology, or species-specific sound patterns, in nonhumans) are not a ready-made blueprint. Rather, the "phonological" genotype acts as a recipe for the assembly of a grammatical phenotype (human or nonhuman) in a manner that is informed by some minimal triggering conditions. Once these triggers are in place, however, substantive restrictions would canalize the patterning of the input in a highly restricted manner that spontaneously converges with the patterns characteristic of typical adult systems.

10.2.4 Algebraic optimization of analog phonetic pressures

The discussion so far has made it clear that the sound patterns of humans and birds are shaped by species-specific substantive constraints. Our next question concerns the link between these constraints and the "phonetic" channel – the properties of the production and perceptual systems that mediate communication. Recall that, in the human case, phonological systems are constrained by broad, possibly universal restrictions that are algebraic, but those algebraic restrictions are firmly rooted in phonetic pressures. The phonological constraint against complex onsets, for example, is grounded in several phonetic facts that favor simple onsets – simple onsets can be articulated more rapidly and perceived more readily than complex ones (Mattingly, 1981; Ohala, 1990). But once these phonetically grounded constraints enter the phonological system, they apply in an algebraic manner. And indeed, such constraints extend across modalities, they apply across the board to large classes of elements – familiar or new – that are all considered equivalent, and these constraints take effect even when their consequences in a given situation happen to be phonetically undesirable (e.g., see Chapters 2, 6). These observations suggest that humans have the capacity to use algebraic means in the service of phonetic pressures that are analog – the capacity for "algebraic optimization." Our question here is whether that capacity for algebraic optimization might be shared with other species.

The available evidence on this question is extremely scarce. Here, we will specifically consider the findings from oscine birds, where these questions have been most amply explored. There are several suggestions that song structure is shaped by several motor constraints. Roderick Suthers and Sue Anne Zollinger, for example, outline several links between a bird's articulatory apparatus and the structure of its song (Suthers & Zollinger, 2004). They note that small birds, such as Canaries, can produce rapid, uninterrupted trills by using a pattern of short "mini-breaths" between syllables. But for larger birds, such as the Mockingbird, those brief mini-breaths are insufficient, so they must interrupt their singing periodically in order to replenish their air supply. The difference between the continuous song structure of the canaries and its non-continuous imitation by the Mockingbird is thus due to respiratory differences between these species. Similarly, the ability of birds to produce various species-specific patterns, such as two-voice syllables (in brown thrashers), is attributed to their ability to independently modulate the two sides of their vocal organ, the syrinx.

Like human phonology, then, birdsong is subject to substantive restrictions that are grounded in motor pressures. But whether those pressures can, in fact, be "phonologized" onto algebraic constraints, akin to "onset," for example, remains unknown. This is not because birds simply fail to represent algebraic rules. Earlier, we have analyzed in detail the capacity of Swamp Sparrows to learn rules (either I_VI, or VI_I, for New York vs. Minnesotan birds, respectively) that target an equivalence class (__) for any note. The question at hand, then, is not whether birds encode algebraic restrictions, but rather whether such restrictions are phonetically grounded.

To address this question, let us consider again the case of Swamp Sparrows. One might speculate that these birds favor notes I and VI at the syllable's edges because these notes manifest rapid acoustic modulations compared to the intermediate notes (especially notes III and IV). Accordingly, the confinement of notes I and VI to the edges of the syllable increases its perceptibility – a restriction that resembles the human ONSET constraints in both substance and form. Concerning substance, both species appear to favor edges with rapid acoustic modulation – either consonants, especially stops (for humans), or the notes I and VI (for birds). And, in both species, the restrictions on edges are linked to an algebraic constraint on syllable structure. But there are nonetheless some important differences between the human and avian constraints. In the human case, the onset constraint answers to the algebraic and phonetic masters *simultaneously*, as the restriction on onsets is *both* algebraic (applying to an equivalence class) and phonetically grounded. In contrast, for birds, the algebraic and phonetic restrictions are enforced separately. The restriction on edges is phonetically grounded, but not algebraic (as each edge is represented by a single note, not an equivalence class), whereas the restriction on the intermediate note is algebraic (as _ forms an equivalence class), but phonetically arbitrary. Indeed, some members of that class (notes II and VI) exhibit weak acoustic modulation, whereas others (notes II and V) are comparable to the edges. What is missing here, then, is the capacity for "phonologization" – to meet phonetic pressures by restrictions that are algebraic.

Even if humans and birds might share not only the capacity for algebraic patterning of their natural communication but also its grounding in phonetic pressures, the ability to negotiate phonetic pressures and algebraic patterning is exceedingly rare, and entirely unattested in primates.

10.2.5 Conclusion: what is special about human phonology?

Summarizing the discussion so far, the human phonological mind has three defining features: (a) it runs on an algebraic machinery that allows for the representation of discrete combinatorial structure, hierarchical organization, and operations over variables; (b) it is jointly shaped by both learning and substantive universal restrictions that are grounded in phonetic pressures; and (c) it optimizes those analog phonetic pressures using algebraic means.

We know close to nothing about the role of (c) in nonhuman communication systems, but (a) and (b) are both attested, although their prevalence in humans and nonhumans differs greatly. While innate, species-specific calls are quite widespread in animal communication, learning is typically modest. Birdsong is the only well-documented case of a natural communication system that is jointly shaped by both learned and innate constraints, including constraints that are algebraic. In most other species, however, algebraic mechanisms are not clearly implicated in natural communication even when they are clearly displayed in laboratory settings. It is precisely because the ability to encode algebraic operations is widely present that its rare deployment in natural communication is significant. And even if it turned out that birds exhibit all three properties (a)–(c), the absence of these characteristics in our phylogenetically closest relatives, the great apes, would indicate that the emergence of these capacities in human evolution was independent from nonhuman species. At the very best, then, human and nonhuman "phonologies" are independent traits (analogies or homoplasies), rather than ones that descended from a common ancestor (i.e., a homology; Ridley, 2008).

10.3 The evolution of the phonological mind

The previous discussion has outlined some fundamental discontinuities between the phonological capacities of humans and the natural communication systems of nonhuman species. In light of these conclusions, two questions immediately come to mind. First, why does the human phonological system have this peculiar design? Second, how did this design emerge in the human lineage? We will consider these two questions in turn.

10.3.1 Is algebraic optimization adaptive?

To understand why human phonologies manifest the design that they do, we might inspect its functionality. Assuming that functionally adaptive designs are more likely to emerge in the course of evolution, one might wonder what functional advantages are conferred by the capacity to use algebraic means to optimize the functional pressures on communication. Considering first the role of functional grounding, here it seems quite obvious that a system that fits its channel is advantageous. Many authors have pointed out that unmarked phonological structures are easier to produce and perceive – they minimize articulatory cost while maximizing perceptual distinctiveness (Lindblom, 1998), they abide by the jaw cycle constraint (MacNeilage, 1998; 2008; MacNeilage & Davis, 2000), and they optimize the parallel transmission of consonants and vowels and their perception by maximally modulating the acoustic signal (Kawasaki-Fukumori, 1992; Mattingly, 1981; Ohala, 1990; Wright, 2004).

While the optimization of phonetic pressures would clearly benefit the production and perception of vocalization, the value of discrete algebraic representations is not immediately obvious. If vocal communication is to abide by the analog continuous pressures of the articulatory and auditory channels, then why shouldn't it pattern signals that are likewise analog and continuous? Why, instead, do human phonological patterns consist of elements that are digital and discrete? Not only does an algebraic mechanism of patterning appear superfluous, but it might also seem inefficient. And indeed, algebraic design and phonetic optimization can conflict in their demands. Recall, for example, that the adherence of Egyptian Arabic to its segment inventory (which includes voiced /b/, but not voiceless labial stops /p/) gives rise to geminates (e.g., *yikubb* 'he spills') that are harder to produce than their unattested counterparts (e.g., *yikupp; Hayes, 1999). So if an algebraic design exerts a functional cost, then what benefit would possibly give rise to the fixation of such a system in humans?

The answer, according to Martin Nowak and David Krakauer (1999), is that this design supports productivity – it allows talkers to express an unlimited number of words that will be optimally perceived by listeners. Their conclusions are based on a computer simulation of a language game. In the game, two participants, a talker and a hearer, attempt to maximize their ability to communicate with each other. Results show a clear tradeoff between the number of distinct words and their perceptibility: As the number of words increases, the risk of mutual confusion increases. If the language were to use only holistic signals, then the chances of confusion would be high, and consequently, this language would allow for only a handful of distinct words. Once the language allows for discrete sounds that are patterned together, however, the number of possible words increases substantially. The reliance on algebraic combinations of discrete sounds thus allows speakers to coin an unlimited number of words.

Given that phonetic grounding and algebraic patterning each confers distinct advantages, their conjunction, in the form of an algebraic system whose primitives and combinatorial principles are phonetically grounded, allows its users to abide by phonetic pressures while supporting unbounded productivity.

10.3.2 How did algebraic optimization emerge in human evolution?

The comparative animal literature makes it clear that the two ingredients of algebraic optimization – the capacity to encode algebraic operations and to abide by substantive phonetic pressures – are each present in many species. It also appears that the combination of these two ingredients has an adaptive advantage of its own. In view of such considerations, one next wonders how this combination might have emerged in the human lineage, and whether this combination was the target of natural selection. While a definitive answer to this question is very difficult to ascertain, we can nonetheless evaluate the plausibility of positive selective pressures by comparing this hypothesis against some alternative explanations.

One such alternative questions the feasibility of natural selection of grammatical universal constraints. Nick Chater and colleagues (2009) have argued that language exists in a state of constant flux – it is perpetually changing in the course of its transmission across generations. Being a "moving target," language will not benefit from innate constraints imposed by natural selection, as a structural feature favored at one point in time may no longer be beneficial once language structure has changed. Chater and colleagues, however, acknowledge that these results hold only if universal grammatical constraints are functionally arbitrary. It is precisely because universal principles are assumed to be arbitrary that their utility might diminish once the language has changed. But, at least with respect to phonology, there is ample evidence that grammatical constraints are not functionally arbitrary. And further modeling work by the same authors (Christiansen, Reali & Chater, 2011) suggests that, unlike arbitrary constraints, functionally motivated features of language could potentially become innate through natural selection. Whether the fixation of universal grammar in humans was due to natural selection remains to be seen (for competing accounts, see Fodor & Piattelli-Palmarini, 2010; Gould & Lewontin, 1979; Pinker & Bloom, 1994), but the results of Chater and colleagues certainly do not undermine this possibility.

Another challenge to selection-based accounts is presented by alternative explanations that attribute the design of the phonological mind to self-organization. This possibility is supported by computer simulations that capture substantive typological universals by relying on principles of self-organization, driven by phonetic pressures alone. These models mimic cross-linguistic preferences concerning the size of vowel inventories and their distribution in phonological space (Oudeyer, 2006), and they even give rise to preferences for unmarked syllable frames and sonority restrictions (Oudeyer, 2001; Redford et al., 2001). Such results are typically presented as evidence that universal markedness constraints are not represented in the grammar. If phonological systems are not shaped by grammatical universals, then questions regarding their evolution are obviously moot.

But these simulations do not effectively demonstrate that a universal phonological grammar is superfluous. Instead, all that is shown is that phonetic constraints might account for typological regularities across languages. Previous chapters, however, make it clear that the case for grammatical universals is not limited to typological regularities. Much evidence – both linguistic and experimental – demonstrates that phonological universals are active in the grammars of individual speakers, including adults, and potentially infants, and they systematically favor unmarked structures over their marked counterparts even when speakers have had no experience in the production or perception of such structures, and even when the input lacks phonetic properties altogether (e.g., for printed words). Accordingly, markedness preferences are algebraic constraints that cannot be subsumed by phonetic pressures, and they do not depend on experience with the processing of those particular structures. There is currently no evidence that such phonological constraints can emerge spontaneously.

In fact, the comparative animal literature directly speaks against this possibility. The findings reviewed in this chapter clearly show that the ingredients of algebraic optimization – algebraic operations and phonetic pressures – are quite pervasive in the animal kingdom. It is specifically the combination of these two ingredients – the capacity for algebraic optimization – that is extremely rare in animal communication. If self-organization were responsible for the emergence of the human capacity for algebraic phonological patterning, then one would expect this capacity to emerge in the communication systems of several other species, especially species that exhibit each of its two ingredients separately. But while the ingredients are quite common, their conjunction is rarely attested in learned vocal communication. The rare occurrence of this universal design feature of human phonology is most likely due to a modification of the human genome that allowed for the capacity for phonological patterning. Precisely how our genome was altered – whether this was due to random variation (neutral evolution), natural selection for some nonlinguistic function, or natural selection for a linguistic function, specifically – is unknown (for some competing explanations, see Fodor & Piattelli-Palmarini, 2010; Gould & Lewontin, 1979; Pinker & Bloom, 1994). We also don't know what other modifications might have been necessary to put the capacity for algebraic optimization to use in the service of communication. Full-blown communication would undoubtedly require many additional skills, including the capacity to implement these abstract phonological patterns in a particular phonetic channel (spoken or manual) and the pragmatic skills to engage in the social exchange of messages – skills that might have well required additional genetic modifications beyond those necessary to support the phonological grammar alone. Our interest here, however, specifically concerns the origins of the phonological grammar itself. At the heart of human phonological grammars is the ability to use algebraic means to optimize phonetic pressures. The rarity of this trait, even when each of its ingredients is independently attested, suggests some evolutionary genetic change to the human lineage as its cause.

11 The phonological brain

> Previous chapters have suggested that the human mind is equipped with core phonological knowledge – a system specialized for the computation of phonological structure. This chapter examines what brain mechanisms mediate phonological computation and evaluate their presumed genetic underpinnings. While the findings suggest that a neural phonological network certainly exists, they cannot determine whether this network is specialized for phonology. An answer to this question hinges on how specialization is defined, and, more generally, how cognitive explanations are linked to neuroanatomical models. Existing neuroanatomical models presently lack an explicit account of that link. I thus conclude that specialization, in general, and the hypothesis of core phonology, specifically, can be presently evaluated primarily at the functional, cognitive level. Neural data can be profitably correlated with functional findings, but they can rarely falsify functional hypotheses concerning specialization.

11.1 Individuating cognitive functions: functional specialization vs. hardware segregation

At the center of this book is the question of specialization: Are human minds equipped with a system specialized for phonological patterning? The previous chapters present several observations that are consistent with this possibility. We have seen that distinct phonological systems share design principles that distinguish them from nonlinguistic systems, that knowledge of grammatical universals is evident even when they concern structures unattested in one's language, and that the capacity for phonological patterning emerges spontaneously, in the absence of a model. Not only are phonological constraints universal and possibly innate, but they are also demonstrably distinct from nonlinguistic pressures, most notably, the phonetic pressures governing the processing of aural stimuli and their production. Functional specialization, however, should be further mirrored at the neural level. If the mind has a specialized computational system dedicated to phonological patterning, then one would expect this special "software" to require a specialized brain "hardware" that mediates phonological computation. The brain networks that support phonological computation could potentially present another test for the specialization of the phonological mind.

While the expectation that functional specialization should have some correspondence in the organization of the brain is uncontroversial, the precise nature of this correspondence is far less clear. Following Gary Marcus and Hugh Rabagliati (2006), we will distinguish between two views of specialization. A strong position requires a one-to-one isomorphism between cognitive function and biological hardware (see Figure 11.1a). In this view, functionally specialized systems should run hardware circuits that are entirely distinct and non-overlapping. Thus, if some specialized system S_1 exists at the cognitive level, then it should be possible to individuate this system at the level of the brain. In the case of the phonological system, there should be a brain network whose components are exclusively dedicated to the computation of phonology. Moreover, the assembly of this system (in development) and its online operation (in the final, adult state) should be controlled by genes whose entire *raison d'être* is the regulation of language functions in the brain. Specialization at the functional cognitive level should thus be transparently discernible from the organization of the brain and its genetic regulation. More generally, any two systems, S1 and S2, are said to be specialized at the functional level only if they can be segregated from each other at the level of the hardware.

In a second, weaker, hypothesis, functional systems individuated at the functional level are not necessarily segregated at the level of "hardware" (see Figure 11.1b). While this view still requires that functional systems can be each

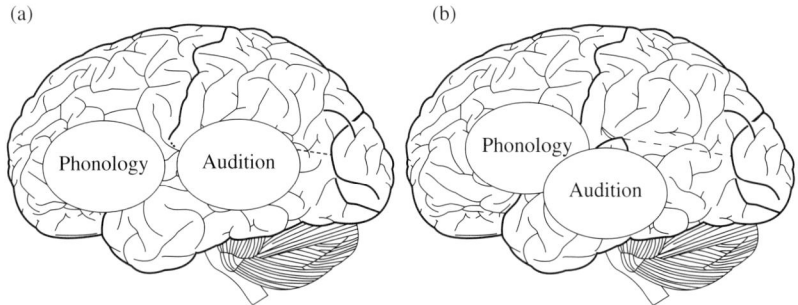

Figure 11.1 Cartoon illustrations of the relationship between two cognitive functions – phonology and audition – and their hardware implementation: either full segregation of the relevant brain substrates (a) or partial overlap (b). Any similarity between localizations in this cartoon and actual brains is purely accidental

linked to distinct brain networks, the relevant networks need not comprise nonoverlapping pieces of hardware. A phonological system, for example, might share some (or all) of its components with related nonlinguistic substrates (e.g., audition, motor control), and its operation might be regulated by genes that are expressed at multiple sites, and whose impact is not confined to linguistic brain functions. But despite having no piece of hardware (brain substrates or genes) that is exclusively dedicated to linguistic computations, human brains may be nonetheless innately predisposed to the computation of phonological structure. All healthy human beings would manifest phonological competence that is narrowly constrained by universal principles, this system might preferentially engage an invariant brain network across different individuals, and the functioning of this network could be linked to specific genetic changes unique to humans, such that newly generated random mutations that disrupt these changes will systematically disrupt the language function.

Although this weaker account of neural organization is sometimes considered inconsistent with functional specialization, I believe this view is mistaken. In fact, it's the strong view's insistence on complete neural segregation and its extreme characterization of innateness that are incompatible with modern genetics (see also Marcus, 2006). But for now, I will defer discussion of these claims until the relevant evidence is laid out. I will consider the evidence in two steps (see 1). In the first step, we will review the literature concerning the phonological network and its genetic control. We will begin the discussion by identifying the phonological network of spoken language. We will examine what brain areas mediate phonological computation in healthy individuals and review the consequences of their disruptions in language disorders. After the principal "biological actors" are introduced, we will next move to gauge their specialization for phonology. One test for specialization is the invariance of the system across modalities. To the extent that the system mediating the processing of spoken phonology is dedicated to phonology, rather than to audition or speech per se, then one would expect the key "hubs" of phonological computation to be active across modalities, for both signed and spoken languages. Such similarities, however, could also emerge for reasons unrelated to phonology. Indeed, sign and spoken languages might share computational routines that are domain-general, such as categorization, chunking, hierarchical organization, and sequencing. To illuminate the nature of the overlap, one would like to further evaluate the role of those "phonological" regions in processes that are clearly non-phonological. Music presents a handy baseline. If the neural phonological network is truly segregated, then the mechanisms dedicated to phonological computation should be segregated from those mediating computation in the musical domain. Another test for the specialization of the phonological system concerns its genetic regulation. We will review some hereditary phonological

disorders and examine whether phonological deficits segregate from non-phonological impairments.
(1) Are human brains specialized for phonological computations?
   a. Do human brains include a phonological network?
      (i) Is there a brain network that mediates phonological processing in normal individuals?
      (ii) Are disruptions to that network associated with phonological disorders (congenital or acquired)?
   b. Is the phonological network strictly specialized?
      (i) *Robustness across modalities*: Does the phonological network of spoken language mediate phonological processing in sign language?
      (ii) *Specificity*:
         • Are the components of the phonological network implicated in the processing of music?
         • Do congenital disorders selectively compromise phonological processing?

Foreshadowing the conclusions, there is strong evidence that a phonological network does, in fact, exist. Several brain sites are systematically engaged in the computation of various grammatical phonological structures, these sites mediate phonological computations across individuals and languages, they are partly invariant across modalities – spoken and signed language – and they are further linked to several candidate genes. Nonetheless, none of these brain sites or genes is exclusively implicated in phonology. I will conclude the discussion by considering whether these facts are consistent with the specialization of core phonological knowledge at the functional level.

11.2   The phonological network of spoken language

In view of the very large literature examining the brain mechanisms mediating speech processing (for reviews, see Hickok & Poeppel, 2007; Poeppel et al., 2008), it is striking to see how few have examined grammatical phonological computation. Lacking a concrete model of grammatical phonological computation, specifically, we will thus use speech perception models as a point of departure. Obviously, speech perception and phonology are quite distinct – while speech perception may well be constrained by the phonological grammar, it is also shaped by several other processes, ranging from low-level spectrotemporal analysis to lexical access. Nonetheless, models of speech perception might offer a reasonable first estimate of the grammatical phonological network in spoken language, so we will use them to guide the present discussion.

An influential model by Gregory Hickok and David Poeppel (2007; see Figure 11.2) suggests that the speech stream first undergoes spectrotemporal

The phonological network of spoken language 255

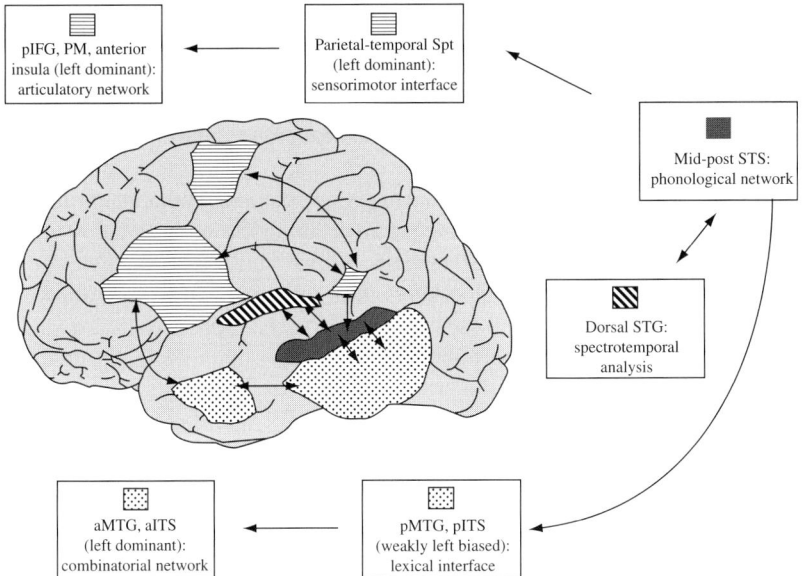

Figure 11.2 Functional anatomy of left hemisphere areas engaged in the phonological processing in spoken language and their interconnectivity (from Hickok & Poeppel, 2007)

analysis through a bilateral activation of the dorsal regions of the superior temporal gyrus (STG) as well as the middle and posterior regions of the superior temporal sulcus. Subsequent processing proceeds along two major neural pathways: ventral and dorsal. The ventral pathway maps auditory inputs onto words stored in the mental lexicon. This stream engages the posterior regions of the middle and inferior temporal gyrus (pMTG/pIFG) as well as anterior and posterior regions of the inferior temporal sulcus and anterior regions of the left middle and inferior temporal gyrus. A second, dorsal stream achieves sensorimotor integration along a left-lateralized pathway involving the left Sylvian parietal-temporal junction (SPT), as well as posterior regions of the inferior frontal gyrus (IFG), including Broca's area, as well as premotor sites and the insula.

Given the behavioral evidence (reviewed shortly), it is clear that some of these sites must be involved in processing grammatical phonological structure, but which sites, specifically, play a role and how they combine to compute phonological structure is rarely addressed. Existing research has typically adopted a rather narrow definition of phonological knowledge. Too often, phonological knowledge is defined by language-particular principles (e.g., knowledge that contrasts English and Russian), and consequently, the phonological network is equated with mechanisms that mediate language variation (e.g., lexical and phonotactic differences between languages). The behavioral evidence, however, suggests that phonological knowledge might be far broader in scope. Not only does phonological knowledge include grammatical principles that are irreducible to the statistical properties of the lexicon, but there is growing evidence that it might be universal. What brain networks might be involved in grammatical (as opposed to lexical) computation remains largely unknown, and the role of grammatical universals in the brain remains unexplored. To begin addressing these questions, we will thus review the literature in a targeted manner. Rather than asking what is known about phonological computation, generally, we will identify those aspects of phonological computation that are likely to form part of core phonology – phonological primitives and the putatively universal markedness restrictions that might govern their combinations. Our goal for now is to simply identify the regions that mediate the processing of this information – whether those mechanisms are specialized is a question that we defer to subsequent sections.

11.2.1 Phonological primitives

The behavioral evidence reviewed in previous sections suggests that all phonological systems might include several types of representational primitives. All systems apparently represent phonemes, they contrast consonants and vowels, and they encode syllables. Our question here is whether the encoding of those primitives is likewise evident in the brain – in both typical and disordered systems. The next section specifically focuses on segments and the consonant–vowel distinction; the role of syllables is explored along with markedness restrictions in the following section.

11.2.1.1 Phonemes

All phonological systems encode phonemes as discrete elements that are contrasted with each other (e.g., b vs. p). The processing of such contrasts engages several regions in the STG, and these contrasts have been observed using both functional magnetic resonance imaging (fMRI, e.g., Wilson & Iacoboni, 2006), electrophysiology (e.g., Naatanen et al., 1997; Sharma et al., 1993), and magnetoencephalography (MEG, e.g., Tervaniemi et al., 1999). But while the distinct brain responses to /b/ and /p/, for instance, could reflect phonological processing, these patterns are also amenable to alternative explanations. Stimuli such as /b/ vs. /p/ not only contrast at the phonological level but they also differ on their acoustic and phonetic properties. Accordingly, the observed brain responses could reflect sensitivity to the auditory or phonetic contrast, a possibility that is further supported by the fact that the relevant regions do, in fact, form part of the auditory cortex (broadly defined).

Unlike auditory and phonetic contrasts, however, phonemic categories promote not only differences between speech sounds but also similarities. While members of a single category – say the category of /b/ sounds – can vary in systematic ways across talkers (e.g., in their voice onset time, VOT; e.g., Theodore et al., 2009; Theodore & Miller, 2010), once these distinct members are represented at the phonological levels, the differences between them are erased. A /b/ is a /b/ irrespective of whether its VOT is short (5 ms) or long (e.g., 20 ms), and, within any given language, every phonological generalization true of a short [b] will also apply to a long one. Because the capacity to encode such classes is uniquely phonological, it offers the means to adjudicate between phonological and phonetic/acoustic accounts of discrimination. If the brain encodes speech sounds only at the auditory/phonetic level, then people should respond not only to contrasts between categories (e.g., /b/ vs. /p/) but also to differences among members of each such class (e.g., differentiate [p] with short vs. long VOT). A phonological contrast, however, should register differences between categories, but ignore within-category differences.

The predictions of the phonological account are supported by an MEG study conducted by Colin Phillips and colleagues (2000). In this study, participants were presented with multiple tokens of a single category (e.g., numerous instances of /d/) – the standard – followed by instances of a different category (/t/) – called the deviant. If the brain distinguishes between these two categories, then one would expect the deviant to elicit a change in the brain's magnetic field – a mismatch response. The change associated with the mismatch thus signals the detection of a contrast. By the same logic, however, the mismatch can also gauge the similarity among class members (e.g., of various tokens of a /d/). Indeed, an event can only be considered "deviant" when compared to a "standard," that is, if all other sounds are encoded as members of a single class. In this experiment, however, standard stimuli (e.g., various [d] sounds) were not physically identical, but rather, they slightly differed from each other on their voice onset time. If people encode only the acoustic or phonetic properties of these stimuli, then those various /d/ stimuli should not form a single class (see the top left panel of Figure 11.3), and consequently, no standard should be in place. In the absence of a standard, /t/ should not be recognized as a deviant, so no mismatch response is expected. If, however, people encode those various stimuli phonologically, as members of the /d/ class, then these stimuli should all be considered alike, and the perceived "standard" should give rise to the detection of /t/ as a deviant (see the top right panel of Figure 11.3). This is precisely what was observed: A mismatch response occurred approximately 200 ms after the onset of the auditory stimulus, and it was localized at the superior temporal plane of the left hemisphere (see Figure 11.4).

A second control condition ruled out the possibility that participants relied on a non-phonological acoustic category (see the bottom of Figure 11.3). It is

258   The phonological brain

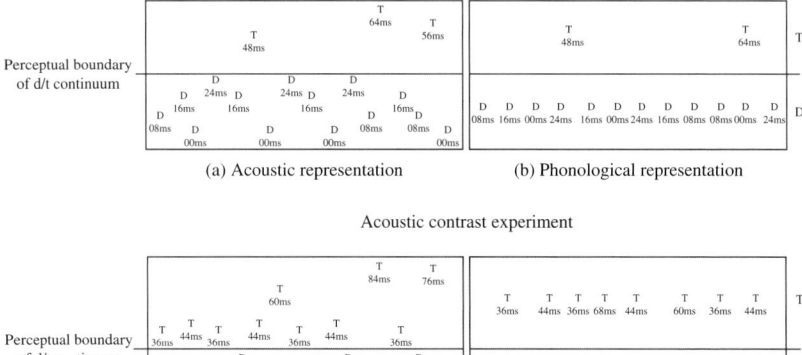

Figure 11.3 The design of Phillips et al.'s experiments (2000)

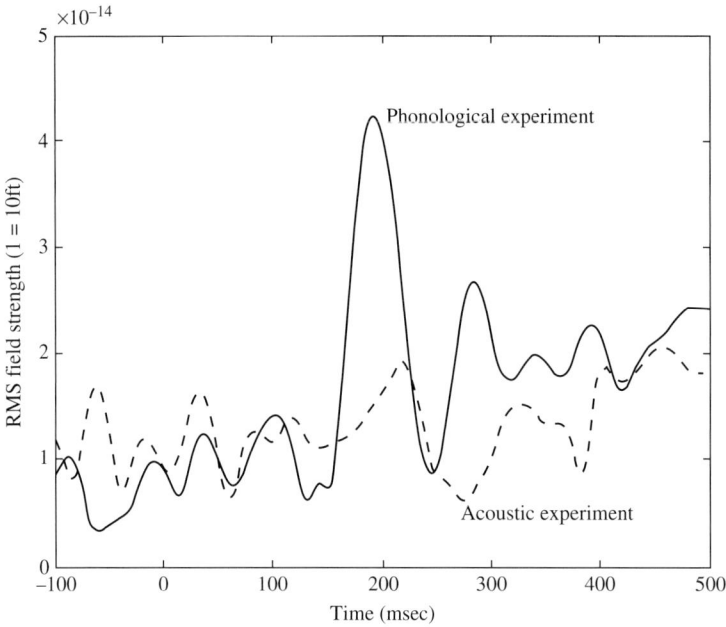

Figure 11.4 Brain responses to the phonological and acoustic control conditions in Phillips et al.'s (2000) experiments

indeed possible that people identified all "long" and "short" VOT stimuli as distinct classes of equivalent members, but the distinction between the classes was based on their mean VOT, rather than by the phonemic contrast between /d/ and /t/, specifically. To address this possibility, Phillips and colleagues increased the VOT of all items by 20 ms. This manipulation still maintained the same VOT difference between "short" and "long" stimuli, but altered their phonological structure. While in the original experiment (at the top of Figure 11.3), "standard" stimuli were identified as the voiced /d/ and the "deviant" was a /t/, most edited items (see the bottom panels) were identified as the voiceless /t/ – either shorter /t/ (for the "standard") or a long one (for the "deviant"). An auditory account would predict a similar outcome in both experiments (see the bottom left panel of Figure 11.3), but if people rely on a phonological representation, then they should now be unable to distinguish the standard from the deviant, and the mismatch effect should be eliminated (see the bottom right panel of Figure 11.3). The outcomes, once again, supported the phonological explanation (see Figure 11.4). Together, these two experiments suggest that regions of the left superior temporal plane mediate phonological processing.

Converging evidence for this conclusion is presented by transcranial magnetic stimulation. While imaging studies reflect typical brain functioning, transcranial stimulation temporarily disrupts activity in select brain areas by transmitting low-level electrical current to electrodes placed on the surface of the cortex. To the extent that that region mediates the function of interest (e.g., the identification of phonemes), then its stimulation should disrupt that function. Results from this procedure converge with the MEG findings to suggest that the middle-posterior region of the STG is involved in both phonetic and phonological processing of consonants (for review, see Boatman, 2004). Other phonological tasks that require phonemic awareness and explicit segmentation (e.g., do *pat* and *bat* share the initial consonant?) implicate additional adjacent regions in the left STG (anterior middle STG, ventral and dorsal portions of posterior STG), and the left inferior frontal lobe (Boatman, 2004) – regions that overlap with Hickok and Poeppel's proposal. And while the right hemisphere does support some limited discrimination between minimal pairs (in both healthy individuals and patients with left hemisphere damage), its effect is gradient and confined to word pairs (e.g., *beach–peach*, but not nonwords, e.g., *beesh–peesh*), whereas discrimination in the left hemisphere is categorical and extends to both words and nonwords (Wolmetz et al., 2011). These observations suggest that the right hemisphere can engage in acoustic-phonetic analysis that supports phoneme discrimination, but it is the left hemisphere that mediates the categorical perception of phonemes (Liebenthal et al., 2005; Wolmetz et al., 2011). The capacity of the left hemisphere to engage in categorical perception might be due to its preferential tuning to the sampling of acoustic information at a temporal window that matches the duration of individual phonemes (20–50 ms; Morillon et al. 2010; Poeppel, 2003), and this, in turn, might explain the advantage of the left hemisphere in the perception of phonemes.

11.2.1.2 Consonant–vowel distinctions

Consonants and vowels are distinct phonological animals. Many phonological processes specifically target one category while ignoring the other – consonant co-occurrence restrictions apply to consonants across intermediate vowels, whereas vowel harmony is opaque to intervening consonants. Furthermore, consonants and vowels carry distinct roles in the grammar and in language processing (see Chapter 4). These observations suggest that consonants and vowels form distinct functional categories in the phonological mind. In what follows, we show that the processing of consonants and vowels likewise dissociates at the neural level.

One line of evidence comes from an fMRI study of typical individuals (Obleser et al., 2010). This study compared brain activation for consonants (e.g., *da* vs. *ga*) and vowels (e.g., *da* vs. *di*). The findings revealed dissociation between consonants and vowels, but the distinction between them was rather subtle. When the results were examined using a typical subtraction methodology (by comparing brain activation to speech with noise), consonants and vowels did not differ in their pattern of activation, and they both elicited stronger activation in the left anterior-lateral superior temporal cortex. But a more sensitive analysis of the data using a classifier algorithm revealed different patches dedicated for the processing of consonants and vowels that were largely non-overlapping.

If consonants and vowels recruit different brain substrates, then it should be further possible to selectively disrupt the processing of one of these substrates (e.g., consonants) without disrupting the other. This prediction is borne out by the findings from transcranial stimulation. Earlier, we noted the role of the middle posterior region of the STG in processing the distinction between consonants. As it turns out, the same region does not respond to distinctions among vowels (Boatman, 2004). Moreover, the engagement of the STG by consonants is categorical, occurring irrespective of their specification for various phonological features, such as voicing (e.g., *b* vs. *p*) and place of articulation (e.g., *b* vs. *g*). Although the precise location (the middle posterior STG) is more posterior than the fMRI findings (Obleser et al., 2010), they nonetheless converge with other fMRI results (Wolmetz et al., 2011). Together, these two methods reflect a distinction between the neural substrates representing consonants and vowels.

Consonants and vowels likewise dissociate in naturally occurring phonological disorders. Recall (from Chapter 4) that consonants and vowels can be selectively impaired in aphasia. Alfonso Caramazza and colleagues (2000) describe two conduction aphasic patients who exhibited fluent speech and no motor or articulatory deficits, but their performance on oral repetition and naming was poor. In one patient (IFA), errors concerned mostly consonants, whereas for another (AS) the errors concerned mostly vowels. In accordance with the evidence from transcranial stimulations, the susceptibility of consonants and vowels to errors was independent of their feature constitution, as gauged by their sonority level. Moreover, one of IFA's damaged loci – the left STG – broadly overlaps with the sites implicated by transcranial stimulation and fMRI findings. In contrast, AS showed (unspecified) regions to the left parietal and temporal lobes, as well as a small region to the right parietal lobe.

Consonants and vowels likewise dissociate in reading and spelling (e.g., Cotelli et al., 2003; Miceli et al., 2004). In one extreme case, an Italian patient with a deficit to the left parietal lobe produced spelling responses that omitted all vowels while leaving consonants intact (Cubelli, 1991). The patient, Fondacaro Ciro, spelled his name as FNDCR CR; the city Bologna was spelled as BLG, and the word 'table,' *tavolino*, was spelled TVLN. The most curious case of such dissociations is presented by a healthy woman who manifests strong synesthesia that systematically associates different vowel letters with specific colors (e.g., A is associated with red; E with green, etc.; Rapp et al., 2009). These synesthetic associations are so specific and automatic that they even promote Stroop interference. In Stroop experiments, people are typically presented with words printed in color, and they are asked to name the color while ignoring the word's meaning. People typically experience difficulty when the word spells an incongruent color name (e.g., given the word GREEN printed in red). Remarkably, this synesthetic individual exhibited Stroop interference for vowels: She experienced difficulty when a vowel letter (e.g., A, which is synesthetically associated with red) was presented in a color that was incongruent with the synesthetic percept (A presented in green). Moreover, fMRI scans demonstrated that vowel letters activated brain areas that typically mediate color processing (V4 and V8). Crucially, however, these effects were only present for vowels, not consonants, and as such, they present dissociation between the brain mechanisms dedicated to these two categories.

11.2.2 Markedness restrictions

The findings reviewed in the previous section have identified various brain substrates that are engaged in the representation of phonological primitives. Core phonology, however, includes not only shared representational primitives but also broad, perhaps universal, well-formedness constraints governing their combinations. Accordingly, structures that violate those constraints (i.e., marked structures) are systematically underrepresented across languages and disfavored by individual speakers – both adults and young children. In what follows, we examine what brain substrates might mediate markedness preferences. Our case study concerns the markedness restrictions on syllable structure.

All languages constrain syllable structure: Onsets are preferred to codas (e.g., *ba*≻*ab*), simple onsets are preferred to complex ones (e.g., *ba*≻*bla*), and onsets with large sonority distances are preferred to those with smaller ones (e.g., *bla*≻*lba*). Our interest here is in the hallmarks of markedness in the brains of individual speakers – both typical participants, and those afflicted with aphasia.

11.2.2.1 Syllable markedness in typical individuals
While syllable-structure preferences have been widely documented in typology and behavioral experiments, the brain mechanisms mediating such computations remain largely unexplored in the imaging literature. The most relevant findings come from an fMRI study by Charlotte Jacquemot and colleagues (2003) that compares syllable-structure preferences of Japanese and French speakers. These two languages differ on their syllable structure: Japanese strictly disallows syllables like *eb*, so inputs like *ebzo* are systematically repaired as *ebuzo*. While the contrast between *eb.zo* and *e.bu.zo* is absent in Japanese, both structures are allowed in French (see 2). French, however, does not contrast vowels in terms of their length (e.g., it does not distinguish *ebuza* and *ebuuza*), whereas this contrast is present in Japanese. Prior research has further shown that these phonotactic preferences modulate the responses of Japanese and French speakers in behavioral experiments (Dupoux et al., 1999). The goal of the present study was to identify the correlates of those phonotactic preferences in the brain. To this end, this study compared the brain responses of Japanese and French speakers to these two contrasts – the presence/absence of a coda (e.g., *eb.za* vs. *e.bu.za*) and vowel length (e.g., *ebuza* and *ebuuza*).
(2) Phonological contrasts in Japanese and French

		Is the contrast attested?	
		Japanese	French
Contrast	*eb.za-e.bu.za*	No	Yes
	ebuza-ebuuza	Yes	No

To distinguish the brain network that mediates phonological computation from purely phonetic substrates, these authors relied on phonological attestation as a guide. Specifically, they reasoned that attested contrasts elicit phonological processing whereas unattested contrasts can only elicit phonetic processing. Following this rationale, they identified phonological areas as those that selectively mediate the processing of attested contrasts, but not unattested ones. These areas included several left perisylvian regions (the IFG, the STG, supramarginal and angular gyri, and the intraparietal sulcus), bilaterally activated regions including the cingulate cortex, insula, and precentral gyrus, and right hemisphere regions, in the frontal, superior, and middle temporal gyri. While these sites agree with Hickok and Poeppel's model of speech perception, it is unclear that they demarcate a phonological network. Because these sites were defined by contrasting attested and unattested structures, the outcome only narrowly indicates the sites that mediate knowledge of one's own language-particular phonotactics. Behavioral findings, however, suggest that speakers' phonological knowledge may well extend to unattested structures. By equating the phonological network with language-particular phonotactics, one runs the risk of over-emphasizing the contribution of lexical phonological computations to the exclusion of sites supporting core grammatical knowledge. Nonetheless, the conclusions of this pioneering work can guide future research into markedness manifestations in the brain.

11.2.2.2 The role of syllable markedness in aphasia

While imaging studies with typical individuals have not explored markedness restrictions directly, aphasia research has long been interested in the possibility that markedness might play a role in aphasia. Roman Jakobson (1968) has famously asserted that unmarked structures are more likely to be preserved in aphasia. Accordingly, other things being equal, aphasic patients should be more likely to preserve unmarked segments in their speech, and when they produce an error, they should be more likely to produce unmarked structures. This prediction is borne out by the results of several studies.

One recent demonstration of syllable-structure restrictions is presented by the case of BON, reported by Matt Goldrick and Brenda Rapp (2007). BON is an English-speaking female with left hemisphere damage affecting the left superior posterior frontal regions and the parietal lobe. While BON produced many errors in her speech, her production showed a strong effect of syllable markedness. BON was reliably more likely to produce a consonant (e.g., *p*) correctly when it occurred in the onset (e.g., *pat*) than in the coda (e.g., *up*). Goldrick and Rapp further demonstrate that the advantage of onsets is not due to various extraneous factors, including the status of the consonant as a singleton vs. cluster (e.g., *play* vs. *pay*), the frequency of the consonant in these two positions, or even the frequency of those syllable structures themselves (e.g., the frequency of CV vs. VC syllables). In fact, words including an onset are *less* frequent than those including a coda.

Another indication of the preference for unmarked syllable structures concerns complex onsets (e.g., *clip*). It has long been noticed that aphasic patients avert complex onsets – such onsets are frequently simplified by either segment deletion (e.g., *sky*→*ky*) or the epenthesis of a schwa (e.g., clip→[kəlɪp]; for review, see Blumstein, 1973; 1995; Rapp & Goldrick,

2006). One concern, however, is that these simplifications reflect not markedness pressures but rather non-grammatical sources – either the patient's inability to encode the auditory input presented to him or her or to plan and execute the relevant articulatory motor commands. A recent study by Adam Buchwald and colleagues, however, addresses this possibility (Buchwald et al., 2007).

Buchwald and colleagues discuss the case of VBR – an English-speaking female who suffered from a large fronto-parietal infarct to the left hemisphere due to a cerebral vascular accident. VBR's word production was characterized by frequent distortions of word-initial onset clusters. Given an input such as *bleed*, VBR would typically produce [bəlid] – separating the onset consonants by appending an epenthetic schwa. A detailed investigation ruled out receptive factors as the source of these errors – VBR was not simply unable to encode the acoustic input or incapable of accessing her mental lexicon. Likewise, detailed acoustic and ultrasound analyses of her production demonstrated that the errors did not result from articulatory failures. By elimination, then, these analyses suggest that epenthetic errors have a specific phonological origin, which potentially concerns the markedness of complex onsets.

Further evidence that the simplification errors in aphasia have a phonological origin is presented by their sensitivity to sonority-sequencing restrictions. Analyzing the case of DB, an Italian patient with a left fronto-parietal lesion, Cristina Romani and Andrea Calabrese (1998a) demonstrated that production errors were constrained by sonority sequencing. Moreover, sonority sequencing constrained both the intended target as well as its erroneous rendition. DB was reliably more likely to simplify onset clusters (i.e., targets) with small sonority rises (e.g., liquid-glide combinations, *rya*) compared to less marked onsets with larger distances (e.g., obstruent-liquid combinations, e.g., *tra*). Moreover, when a complex onset was simplified, DB was reliably more likely to opt for outputs that maximize sonority distance (e.g., *tra*→*ta* rather than *ra*). These errors were inexplicable by auditory failures (DB's ability to discriminate such auditory syllables fell within the normal range), nor were they due to an inability to produce certain segments (e.g., difficulty with the liquid *r*). Indeed, the errors associated with any given segment depended on its position in the syllable. For example, while DB tended to delete liquids in onset clusters (e.g., *tra*→*ta*), he did maintain them in codas (e.g., *art*→*ar*), a pattern consistent with the cross-linguistic asymmetry between onsets and codas (onsets favor a steep rise in sonority, whereas codas manifest moderate falls). Similar cases have been observed in German- (Stenneken et al., 2005) and English-speaking (e.g., Buchwald, 2009) patients. Together, those results suggest that marked structures are more likely to be impaired in aphasia.

11.2.3 Conclusions

Our discussion so far has identified several brain regions mediating phonological processing in normal individuals. These areas (including Broca's area, the posterior STG, superior temporal sulcus, and planum temporale) are implicated in the representation of both phonological primitives (segments, and consonant/vowel categories) and some markedness restrictions on syllable phonotactics. Disruptions of these areas – either transitory disruptions due to transcranial stimulation, or more permanent ones, in the case of aphasia – can selectively impair specific phonological elements (e.g., consonants). Moreover, brain injuries are more likely to spare phonological structures that are universally unmarked.

While these findings do not specifically address the connectivity between these regions, they are certainly in line with the possibility that these loci form part of a network that mediates the computation of core phonological knowledge. Nonetheless, several limitations of these conclusions are noteworthy. First, the available evidence concerning the phonological grammar is scarce. Only a handful of studies with normal individuals have explicitly examined substrates mediating the encoding of grammatical phonological primitives and constraints, so the reliability of these findings and their generality across languages requires further research. Some aphasia studies have concerned themselves with grammatical representations and constraints, and the results are generally consistent with the imaging findings with typical individuals, but in many cases, the localization of the lesions is rather coarse. Finally, the existing evidence is limited inasmuch as it focuses exclusively on the structures attested in one's language. Accordingly, the available data do not address the central question of how the brain encodes core phonological knowledge, including grammatical principles that are potentially universal.

11.3 Is the phonological network dedicated to phonological computation?

Finding that the brain is involved in grammatical phonological computation is hardly surprising given the very large corpus of behavioral data demonstrating that people constrain the phonological structure of their language. Our main question here is not *whether* the brain computes grammatical phonological structure, but rather *how* – is the relevant network specialized for phonological computation, or does it consist of domain-general mechanisms that subserve the processing of auditory sequences and vocal motor control, generally?

An answer to this question would critically depend on one's account of specialization. Earlier in this chapter, we defined two views on how functional specialization should be mirrored in the organization of the brain. A weaker version merely requires that the cognitive function of interest be associated with a cohesive brain network that is relatively fixed across individuals. A stronger view further mandates that this network and each of its components should all be *exclusively* dedicated for that function alone. The evidence reviewed so far is certainly consistent with the weaker view, but it is moot with respect to the stronger alternative.

In what follows, I evaluate this strong hypothesis from two perspectives. One way to gauge the specialization of phonology is to compare the neural networks that mediate phonological computation across modalities. If the substrates involved in the processing of spoken language phonology are general-purpose engines of auditory sequencing, for instance, then one should not expect it to mediate the computation of phonological structure in sign languages. Specialization, however, should be evident not only by the range of functions that the network mediates but also by the ones it doesn't. A brain network dedicated to phonology should be demonstrably distinct from the systems mediating the processing of non-phonological auditory sequences, such as the ones found in music. The role of the phonological network in non-phonological computations presents a second test for its specialization. Finally, I will examine whether disorders affecting the phonological system are dissociable from non-phonological deficits.

11.3.1 *An amodal phonological network?*

The discussion in previous chapters has shown that phonological patterning is not confined to speech. Like spoken languages, sign languages have a phonological structure that imposes constraints on the sequencing of meaningless sign elements. We have identified several representational primitives that are shared across modalities, including features and syllables, and reviewed markedness constraints on syllable structure that are possibly amodal. In view of those structural similarities, one wonders whether some of the brain substrates involved in phonological computation might be shared across modalities. Although numerous studies have compared the brain networks in signed and spoken language (for reviews, see Emmorey, 2002), few comparisons specifically concern phonology, and none targets phonological primitives and markedness constraints, in particular. While these crucial questions remain unanswered, there are several indications of common brain regions mediating phonological processing across modalities. Here, I review evidence from two sources: imaging findings from sign language phonology and evidence for phonological transfer across modalities.

11.3.1.1 *Brain mechanisms mediating phonological processing in sign language*

One region implicated in processing the phonological structure of spoken language is the left planum temporale – a region of the superior temporal gyrus that forms part of the classical Wernicke's receptive language area. Our question here is why is the planum temporale engaged – does it mediate sensory auditory processing, generally, or phonological patterning, specifically? To address this question, Laura Ann Petitto and colleagues (2000) examined whether this region might support phonological processing in sign language. Using a PET methodology, these researchers compared the brain activity of signers fluent in two distinct sign languages (American Sign Language and Quebec Sign Language) while viewing non-signs (phonotactically legal combinations that are not attested in participants' sign language). To control for the nonlinguistic demands associated with processing such complex visual displays, these signers were compared to non-signer controls. Results showed that signers activated the planum temporale bilaterally in processing sign language. In contrast, non-signers who viewed the same signs showed no engagement of the planum temporale. These results suggest that the planum temporale mediates phonological processing, specifically, irrespective of modality – for either spoken or signed language.

Further evidence for the role of the superior temporal gyrus in phonological processing is presented by another PET study of a deaf individual who was about to undergo surgery for the insertion of cochlear implants (Nishimura et al., 1999). Results showed that, prior to implant, sign language words elicited a bilateral activation of the supratemporal gyri relative to still-frame controls, but once this individual had undergone the implant, auditory inputs triggered activation in the primary auditory cortex (Heschl's gyrus), contra-lateral to the auditory input. These results demonstrate that phonological processing is clearly dissociable from auditory processing.

The amodal phonological network, however, also includes frontal and parietal regions. In a study comparing phonological processing in English and British Sign Language (MacSweeney et al., 2008), bilingual deaf participants were presented with pictures of two objects, and they were asked to perform two judgments regarding their phonological forms. One task required participants to determine whether the signs of those objects in British Sign Language share the same location (e.g., whether both signs are located at the forehead). A second task required participants to judge whether the English names for two objects rhyme, and the performance of these deaf signers was compared to English-speaking monolinguals. Results showed that deaf participants activated common regions in the phonological processing of the two languages (British Sign Language and English), and those regions overlapped with those used by hearing participants. Those regions included medial portions of the superior frontal gyrus, the left superior parietal lobule, the left superior portions of the supramarginal gyrus, and the left posterior inferior frontal gyrus. British Sign Language, however, resulted in greater activation of the left parietal lobule. The greater role of the parietal lobe in sign language is consistent with cortical stimulation mapping results, showing that stimulation of the supramarginal gyrus results in phonological errors (Corina & Knapp, 2006; Corina et al., 1999). The same study also suggests that sign language phonology recruits posterior aspects of Broca's area (BA 44) as well, as the stimulation of this area resulted in global phonetic distortions. Surprisingly, however, stimulation of the posterior and anterior temporal lobe, including the superior temporal gyrus, resulted in no disruptions of sign repetition (Corina et al., 1999). These results, however, were obtained from an individual with a history of complex parietal seizures, so this factor might explain the discrepancy with the imaging results of phonological processing in neurologically typical signers, where the superior temporal gyrus is often implicated.

11.3.1.2 Cross-modal phonological transfer

Further evidence for an amodal phonological network is presented by the transfer of phonological processing across modalities. It is well known that language acquisition is optimal early in development; later language learners manifest various deficits which are particularly noticeable in the area of phonology – far more so than in lexical access (Newport, 2002). Eric Lenneberg (1967) famously attributed this early window of opportunity to biological constraints on the plasticity of the language system. If such constraints, however, are amodal in nature, then exposure to phonological structure might transfer across modalities. The findings reported by Rachel Mayberry (Mayberry & Witcher, 2005; 2007) are in line with this prediction. The study examined phonological processing using the priming methodology: Participants were asked to determine whether a given "target" display is a real sign in American Sign Language (ASL). Each such target was preceded by a "prime" – a display consisting of another sign that was either unrelated to the target, or phonologically similar, such that the prime and target differed by a single phonological feature. To use an English analogy (see 3), one would compare the processing of *bee* preceded by either the phonologically similar prime *pea* (*bee* and *pea* differ by a single feature – voicing) or the unrelated control *too*. Two questions are of interest. First, are participants sensitive to the phonological properties of the target? If they are, then phonological primes should yield phonological priming – they should facilitate the processing of the target relative to controls. To the extent that people are sensitive to phonological structure, one might further inquire whether these phonological effects depend on the age of acquiring ASL.

(3) Phonological priming in English
        Target: *bee*
        Phonological prime: *pea*
        Unrelated control: *too*

Results with participants who acquired ASL early in life indeed showed phonological priming even when the interval between the prime and target was relatively short (330 ms), whereas participants who acquired ASL later in life did not benefit from the phonological prime; in fact, they showed phonological inhibition – their responses were slower in the presence of the phonologically related prime relative to the control. These results are indeed expected in light of the large body of research showing that the phonological system of late language learners differs from native speakers. The interesting twist comes from a second condition in which the interval between the prime and target was increased (to 1 second), allowing the prime additional time to exert its effect. Remarkably, the effect of this manipulation on late ASL learners critically depended on their prior experience with English. Late ASL learners who had no prior experience with English still showed no benefit from the phonologically related prime; indeed, they continued to show phonological inhibition. In contrast, late ASL learners who were previously exposed to English showed quite a different pattern. Once the long prime-target interval allowed for sufficient processing time, their results now showed a benefit from the phonologically related prime, approximating the performance of the native ASL speakers in the shorter duration. These findings suggest that early exposure to the phonological system of spoken language leads to long-term benefits that transfer to sign language phonology.

Further evidence for cross-modal transfer is presented by imaging results, showing that participants' age of acquiring their first language – British Sign Language – determined not only the regions mediating phonological processing of their native language but also affected the regions engaged in processing English – their second language (MacSweeney et al., 2008). Compared to late learners, native learners of British Sign Language exhibited stronger activation of the left posterior inferior frontal cortex in making rhyme judgments regarding English words (e.g., *chair–bear* vs. *hat–bear*). These results suggest that phonological experience in one modality may transfer to determine both the functional and the neuroanatomical characteristics of phonological processing in the other modality.

11.3.1.3 Conclusion

To summarize, the comparison of signed and spoken languages suggests that several brain substrates might mediate phonological processing across modalities, and that the early engagement of those areas in one modality transfers to the other. These results, however, do not necessarily demonstrate a shared, amodal network dedicated to phonological computation. First, there are several differences between the phonological networks in the two modalities. Such differences, however, are only expected given the functional differences between the two phonological systems. A second, more significant limitation of the present results is that they do not necessarily show that these shared substrates are, in fact, dedicated to the computation of phonological structure. Indeed, these common regions might mediate processing demands that are shared across modalities – functions such as the formation of feature-categories, the segmentation of the input stream into smaller chunks, the ability to map those chunks into meaning, etc. To assess this possibility, it is necessary to examine not only what the phonological system does, but also what it doesn't do. A strong view of specialization requires that the substrates involved in phonological computation be exclusively dedicated to this purpose. The following sections evaluate this strong hypothesis. We first examine whether any phonological region is specialized for the purpose of phonological processing; next, we examine whether the network as a whole is dedicated for phonological computation.

11.3.2 Do phonological regions mediate musical computations?

A more stringent test for the specialization of the phonological network seeks to dissociate it from non-phonological brain substrates. Strong neuroanatomical specialization would be demonstrated if at least some of the components of this network are uniquely dedicated for this purpose. The comparison of phonological processing and musical processing presents an interesting case study.

Like phonological systems in natural language, music is universally present in all cultures, and the distinct musical systems across the world share some common organizational principles that are quite distinct from the ones governing phonological structure (see Chapter 2; Jackendoff & Lerdahl, 2006; Lerdahl & Jackendoff, 1983; Patel, 2008; Peretz, 2006). The universality of music, its common design, and the emergence of musical abilities early in development render music a good candidate for a specialized knowledge domain, distinct from core phonological knowledge. Given the strong evidence for functional specialization of the two domains – phonology and music – we can now turn to examine whether any of the brain substrates mediating phonological computation are distinct from the ones implicated in music processing. Surprisingly, however, no known component of the phonological network is uniquely phonological.

The strongest candidates for specialized phonological areas are the left hemisphere sites that mediate phonological processing across modalities, spoken and signed. These sites include Broca's area (Corina et al., 1999; Gough et al., 2005; Jacquemot et al., 2003; MacSweeney et al., 2008; Petitto et al., 2000; Sahin et al., 2009); the posterior superior temporal gyrus and superior temporal sulcus (Boatman, 2004; Desai et al., 2008; Gow & Segawa, 2009; Graves et al., 2008; Liebenthal et al., 2003; Liebenthal et al., 2005; Okada & Hickok, 2006; Vouloumanos et al., 2001); and the planum temporale (Jacquemot et al., 2003;

Petitto et al., 2000). Each of these regions, however, is implicated in tasks involving nonlinguistic tonal material. Broca's area is engaged in processing unexpected harmonic progressions (Koelsch et al., 2002; Maess et al., 2001); the left STG and left superior temporal sulcus have been linked to various aspects of pitch and harmonic processing (Koelsch, 2006; Mandell et al., 2007; Tillmann et al., 2006; Wilson et al., 2009), and the left planum temporale has been likewise associated with the perception of absolute pitch (e.g., Schlaug et al., 1995) and singing (Jeffries et al., 2003; Suarez et al., 2010).

Further evidence for the sharing of neural resources across domains is presented by the strong functional links between musical and phonological abilities. Absolute pitch, for example, is more frequent among musicians who speak tonal languages (Deutsch et al., 2006), whereas musicians are more sensitive to linguistic pitch than non-musicians (Bidelman et al., 2009; Wong et al., 2007). Similarly, about a third of the people who suffer from amusia (a disorder affecting the processing of musical pitch) also manifest difficulties in the processing of linguistic pitch information (e.g., in discriminating linguistic declarative statements from questions; Patel et al., 2008). Summarizing, then, no known brain region can be linked to the computation of any specific phonological structure (e.g., the computation of syllable structure; Blumstein, 1995), and, as shown above, each of the key regions mediating segmental phonological computation is shared with musical processing.

11.3.3 The regulation of the phonological network: evidence from hereditary phonological disorders

Although we have so far failed to identify any specialized phonological regions, further evidence for specialization could conceivably come from their genetic regulation. If humans are genetically predisposed to engage in phonological patterning, then phonological ability could be linked to specific genes that have undergone changes in the human lineage, and the disruption of those genes should result in phonological disorders. But as in the case of neuroanatomical specialization, however, the interpretation of those findings depends on one's definition of specialization. A strong view would infer genetic predisposition for phonological computation only if some genetic mutation could be shown to selectively affect phonological competence; a weaker version might require that phonological competence exhibit a one-to-one correspondence with specific genes, but it would not insist on those genes being exclusively implicated in phonological functions.

Existing genetic research on the phonological competence of healthy individuals is extremely limited. The only available findings associate the prevalence of linguistic tones with the frequency of two genes related to brain growth and development in the population (*ASPM* and *Microcephalin*; Dediu & Ladd,

2007; Dediu, 2011). But the specific function of these genes and their role in phonological competence remain unknown. A large literature, however, has examined the genetic mechanisms of phonological disorders. Here, we consider the findings from individuals with various forms of Specific Language Impairment (SLI).

One of the most exciting advancements in the genetics of language has been the discovery of the *FOXP2* gene, and its involvement in speech and language disorders (Lai et al., 2001). The genetic basis of the disorder was strongly suggested by its inheritance pattern across three generations of a single British family – the KE family. About half of the family members (8/15) are affected, and affected individuals manifest a variety of linguistic disorders ranging from morphosyntax to phonology and phonetics. Imaging studies have revealed structural (Vargha-Khadem et al., 1998) and functional (Liegeois et al., 2003) abnormalities in the brain of affected members, including abnormalities to Broca's and Wernicke's areas. Subsequent analyses have linked the disorder to a mutation to a single gene – *FOXP2*, a transcription factor gene located on the long arm of chromosome 7. The mutation was transmitted in a dominant fashion, such that every family member with the mutated gene exhibited the disorder, and this mutation was only present in affected members (for reviews, see Fisher & Marcus, 2006; Fisher & Scharff, 2009; Marcus & Fisher, 2003; Vargha-Khadem et al., 2005). Subsequent studies have linked the *FOXP2* gene to the learning of the motor skills necessary for vocal communication in various other species, including birds (Haesler et al., 2004) and mice (Fischer & Hammerschmidt 2011; French et al., 2007; Shu et al., 2005). The human FOXP2 protein, however, differs from the versions found in mice by three amino acids, two of these changes occurred after the human evolutionary branch split from the chimpanzee, and a comparison of the rate of this change in nonfunctional changes (changes that do not alter amino acids) suggested that the mutation of the human *FOXP2* allele was due to evolutionary selective pressure on the human lineage, occurring within the past 200,000 years – a time that is broadly consistent with the estimated emergence of language in humans (Enard et al., 2002). The link between the *FOXP2* gene and language evolution, on the one hand, and its role in speech and language impairments, on the other, suggest that the gene regulates the assembly of brain networks that mediate linguistic computation.

Behavioral analyses of affected family members further revealed deficits to tasks that require phonemic awareness, including rhyme production (e.g., say what word rhymes with *name*?), phoneme addition (e.g., add *v* to *arg*), and phoneme deletion (e.g., say *varg* without the first sound; Alcock et al.,, 2000; Vargha-Khadem et al., 1995). Affected members also exhibit various types of phonological errors, including epenthesis (e.g., *statistics*→*sastistics*), metathesis (e.g., *cinnamon*→*cimenim*), assimilation (*parallel*→*pararrel*; examples from Shriberg et al., 2006), and simplification of complex onsets (e.g., *blue*→*bu;* Hurst et al., 1990). Although the simplification of complex onsets could reflect markedness pressures, this explanation is countered by the documentation of (phonologically unmotivated) omission errors even with simple onsets (*table*→*able*). Indeed, affected members also exhibit severe verbal dyspraxia – an impairment in the performance of the movements necessary for the production of speech – as well as marked deficits to oral facial movements of all kinds, including the production of meaningless noises (e.g., "click your tongue"), singing (e.g., "hum a tune") and various complex movements (e.g., "open your mouth wide, stick out your tongue, and say *ah*"). The production difficulties of individuals with *FOXP2* mutations could thus stem from their apraxia, rather than from a grammatical deficit.

While the grammatical competence of individuals with *FOXP2* mutations has not been fully evaluated, their phonological difficulties mirror some of the impairments seen in the broader group of individuals with linguistic difficulties that are unexpected by their overall intelligence, neurological development and environment, a disorder broadly categorized as SLI. Individuals with this milder and more common disorder typically do not exhibit mutations to the *FOXP2* gene (Balaban, 2006), but SLI is highly heritable (Bishop, 2009; Bishop & Snowling, 2004), and it has been associated with several other candidate genes (Fisher & Marcus, 2006; Newbury & Monaco, 2010). The phonological competence of such individuals has been evaluated far more extensively.

It has been well established that individuals with SLI exhibit a host of phonological disorders, ranging from the categorical perception of phonemes (e.g., van der Lely et al., 2004; Ziegler et al., 2005) to phonemic awareness (i.e., the comparison, discrimination, and segmentation of phonological forms), the use of prosodic information (Marshall et al., 2009), and word production (for reviews, see Bishop & Snowling, 2004). Production errors affect both words (e.g., Bortolini & Leonard, 2000) and nonwords (Bishop, 2006), and they are exacerbated as the number of syllables increases (e.g., Kavitskaya et al., 2011). An interesting study by Nichola Gallon and colleagues specifically links the phonological difficulties of individuals with SLI to phonological markedness (Gallon et al., 2007). While the SLI group did not differ from typical individuals (matched for their linguistic development) on the production of unmarked CVC syllables, individuals with SLI were significantly impaired on the production of more marked syllables (e.g., CCVC and CCVCC). Similar effects of markedness obtained concerning prosodic structure: SLI individuals and controls did not differ on the production of the unmarked trochee (i.e., strong–weak metrical pattern, such as drɛ-pə), but SLI individuals were impaired on the production of marked iambic (i.e., a weak–strong pattern, as in bə-drɛp). The similarity between this pattern and the one observed with younger, typically developing individuals (see Chapter 9) could suggest that the difficulty of SLI individuals might reflect developmental delays.

Other studies, however, observed no selective impairment with marked syllable structures (Kavitskaya & Babyonyshev, 2011; Kavitskaya et al., 2011), and several aspects of the error patterns point to a non-grammatical explanation for the impairment of SLI children. A markedness account would predict distinct asymmetry in the pattern of errors: Marked, complex onsets should be simplified (e.g., *block→bock*), but unmarked, simple onsets should be produced correctly (e.g., *book→book*). But as it turns out, children with SLI frequently generate complex onsets from simple ones *(book→blook)*, and they are significantly more likely to do so than typically developing children (Marshall et al., 2009). Moreover, the simplification of complex onsets and codas by SLI children is unaffected by their sonority profile (Kavitskaya & Babyonyshev, 2011). In view of these findings, one must either assume that the phonological grammar of certain SLI children is impaired (rather than merely delayed), or postulate a secondary extra-grammatical source for the errors.

These putative extra-grammatical deficits of individuals with SLI are not confined to the production system. SLI has long been linked to a series of auditory deficits (Tallal, 2004; Tallal & Piercy, 1973). To be sure, auditory deficits are not present in all affected individuals (Bishop, 2007), nor can they account for the host of grammatical problems seen in individuals at the time of testing (e.g., Marcus & Fisher, 2003; Marshall & van der Lely, 2005). But it is conceivable that individuals with developmental language disorders could manifest those deficits early in development and overcome them later in life (Galaburda et al., 2006). This possibility is indeed consistent with the observation that infants at risk of language disorders manifest deficits in the processing of auditory stimuli presented in fast succession (Benasich & Tallal, 2002; Benasich et al., 2006; Choudhury et al., 2007). Further converging evidence is presented from animal models of dyslexia (discussed in detail in the next chapter). Although dyslexia is defined as an unexplained difficulty in reading, rather than language specifically, this disorder shares important characteristics with SLI, including a strong deficit to phonological processing and a high rate of association (40–80 percent of the individuals with SLI manifest developmental dyslexia; Scerri & Schulte-Körne, 2010). Like SLI, dyslexia has been linked to auditory processing difficulties (Tallal, 2004), and a rodent model has shown that mutations to genes implicated in dyslexia produce deficits to rapid auditory processing only in juvenile animals, but not in adults (Galaburda et al., 2006; Peiffer et al., 2004).

Whether such early auditory deficits are indeed robust in individuals who are later diagnosed with SLI and whether they can specifically predict grammatical phonological development remains to be seen. But in view of the paucity of evidence regarding the phonological system in SLI, the controversy surrounding its association with auditory processing problems, and the absence of a specific genetic model, it is presently impossible to determine whether the phonological problems in SLI can be linked to any specific genes, let alone genes that are exclusively "phonological."

11.4 Minds, and brains, and core phonology

In this chapter, we have examined the brain mechanisms mediating grammatical phonological computations and their genetic control. Although very few studies have addressed grammatical phonological knowledge specifically, the available evidence implicates several left-perisylvian sites in various phonological computations, ranging from the identification of phonemes to the distinction between consonants and vowels and the sensitivity to markedness restrictions. Not only are these sites implicated in the processing of phonological structure in spoken language, but several of them are also linked to the phonology of sign language, suggesting the possibility of an amodal phonological network.

Whether this network is dedicated to phonological computation, however, is far less clear. None of the implicated areas is uniquely linked to phonology, and several sites have been shown to mediate musical processing. Similarly, phonological disorders are highly co-morbid with nonlinguistic deficits, including articulatory and auditory impairments, and no known gene, including *FOXP2*, is exclusively dedicated to language or phonology (Marcus & Fisher, 2003).

On the face of it, the failure to individuate a phonological system at the hardware level might appear to preclude any functional specialization for phonology. Many authors indeed adopt a strong hypothesis regarding specialization (see 11.1a). In this view, a cognitive system is specialized only if it can be segregated from other systems at the hardware level – if it cannot be linked to a separate piece of hardware that is exclusively dedicated to that function, then this system does not exist at the functional level. And since no known hardware is exclusively dedicated to phonological computation, the view of phonology as a system of core knowledge must be wrong.

Rather than accepting this conclusion, however, I believe one should question its premise – the requirement of "hardware segregation." Hardware segregation is indeed implausible on both biological and cognitive grounds. Considering, first, the neural level, it is well known that human brains manifest a fair degree of plasticity that allows for the recruitment of existing neural circuits at the service of novel functions. Occipital regions, for example, can be temporarily reallocated for the processing of auditory and tactile information in healthy-sighted individuals who were blindfolded for a period of five days (Pascual-Leone & Hamilton, 2001). These observations, however, do not undermine the fact that visual computations are specialized and distinct in kind from auditory and tactile ones. Clearly, the segregation of neural hardware is not a *sine qua non* for specialization even at the sensory level.

Similar problems afflict the requirement of hardware segregation at the level of genes. Most complex cognitive functions are regulated by multiple genes (Fisher & Marcus, 2006; Fisher & Scharff, 2009). Few, if any, genes are expressed in a single site, linked to a single phenotypic trait, and many significant evolutionary changes to phenotypic traits can be traced to modifications to the regulation of existing genes (Lowe et al., 2011). Consequently, hardware segregation is not merely *generally* unlikely – it is unlikely even for functions that are demonstrably specialized at the functional level. Perhaps the strongest evidence for the innateness of a functional trait is its natural selection – a proof that the trait has a distinct functional advantage that improves replicability. But traits that undergo natural selection are rarely discrete physically (Anderson, 2010; Marcus, 2006). The well-known propensity of natural selection to tinker with existing genes would suggest just the contrary: Novel neurocognitive systems should share most of their components with older systems. So hardware overlap (in both brain substrates and genes) is not merely consistent with functional specialization – it might be, in fact, its defining feature (Marcus, 2004; Marcus, 2006). While we do not know how core phonology has evolved, it is certainly conceivable that it modified gene networks regulating sensorimotor integration and motor-skill learning (Fisher & Scharff, 2009). The adjacency of phonological sites to the oral/aural brain regions and the comorbidity of phonological and orafacial disorders might well be the relics of the evolutionary history of the phonological system. Such overlap, however, is fully expected by the possibility that phonological hardware is innate and specialized, contrary to the segregation requirement.

But the hardware segregation hypothesis runs into a yet deeper problem at the cognitive level. One should not lose track of the fact that hardware segregation and domain-specificity are apples and oranges of sorts – they concern distinct concepts drawn from distinct levels of analysis. Hardware segregation concerns the topology of brain regions that are grossly implicated in phonological processing. Domain-specific systems, however, are functional systems, defined at the cognitive computational level. While we should certainly expect distinct cognitive systems to be implemented by distinct *computational* brain networks, hardware segregation does not effectively test this hypothesis. Computation is the manipulation of information using procedures that operate on symbols – "physical entities that carry information forward in time" (Gallistel & King, 2009, p. 309). Neuroimaging experiments and brain disorders identify brain sites that are grossly related to phonological processing.

Brain regions, however, are not symbols. Accordingly, the activation of these tells us virtually nothing about how these tissues represent phonological information. Finding that the superior temporal sulcus mediates phonological processing tells us nothing about how this tissue encodes "syllable," how the structure of the physical entities that encodes syllables indicates the fact that "syllable" is a complex symbol (e.g., it includes an onset and a rhyme), and what procedures are used by the brain to operate on such symbols such that the structure of input symbols will determine the structure of the output.

The problem, of course, is not specific to phonology. As Randy Gallistel and Adam Phillip detail (2009), we do not currently know how the brain computes, primarily because we know of no neural mechanisms that allow for long-term representation of symbols in a manner that would support the decoding of information later in time. Since the hypothesis that the mind has distinct systems (distinct systems of core knowledge, or "mental organs," in Noam Chomsky's words) concerns networks that effect computation, these hypotheses can be currently best evaluated at the functional level, not by the localization of its hardware (see also Barrett & Kurzban, 2006; Poeppel, 2011). Accordingly, the strong hypothesis that infers distinct functions (e.g., mental organs) only if these functions are linked to a discrete anatomical site (11.1b) is not only biologically implausible – it is also cognitively untenable. As Randy Gallistel notes (2007: 2):

Whether this organ resides in a highly localized part of the brain or arises from a language-specific interconnection of diverse data-processing modules in the brain is irrelevant to whether it constitutes a distinct organ or not. Some organs are localized (for example, the kidney) while others ramify everywhere (for example, the circulatory system). The essential feature of an organ is that it has a function distinct from the function of other organs and a structure suited to that function, a structure that makes it possible for it to do the job.

Some readers might find these conclusions disturbing. One might worry that the approach taken here licenses cognitive theory to ignore findings from neuroscience and genetics. Moreover, by releasing neuroscience from the status of an ultimate arbiter on mental architecture, one loses any hope of finding out whether cognitive systems are "real." And if one worries about individuating cognitive systems, such worries would only multiply for systems of core knowledge – systems that come with the additional conceptual baggage of being innately specified: If such systems cannot be linked to any genes that exclusively regulate these functions, then how can we determine whether the functions are innate?

I do not believe such worries are justified. The inability to reduce cognitive explanations, generally, and cognitive specialization, specifically, to the level of neuroscience and genetics illustrates a well-known difficulty in reducing scientific explanations couched at one level of explanation to lower levels of analysis. This problem is neither new nor unique to cognitive science (Chomsky, 2002; Fodor & Pylyshyn, 1988). But when it comes to the study of the mind, the problem somehow becomes more pressing. Indeed, mentalistic explanations, in general, and nativist mentalistic accounts, specifically, are subject to great mistrust that is deeply rooted in broader philosophical convictions and sociological factors (Pinker, 2002). Many people, laymen and scientists alike, consider claims about mental architecture as quite tentative. Although we rationally understand that mental capacities are intimately linked to brain functioning, we are nonetheless "surprised" to learn that plain mental characteristics can be identified in the activity of our brains. We can easily hear the difference between a concert violinist and an amateur, but we marvel at the finding that the brains of musicians and amateurs function differently – the high visibility of such findings on the pages of both the scientific and the popular press attests to this fact. Our insistence on the hardware segregation of cognitive functions (e.g., music) is the flip side of the same attitude.

Since our marvel at the operations of our brain is just as irrational as our suspicion of mental explanations, we might all benefit from attending to the origins of these emotional reactions and their role in directing scientific inquiry. While some people might explain their quest for "brain confirmation" by the unobservable nature of cognitive constructs, I suspect that this reaction is much deeper, rooted in our strongly dualist view of the world, and most notably, ourselves (Bloom, 2004). It is precisely because we are at pains to reconcile minds and brains that we find such obvious convergences reassuring. But unfortunately, when it comes to specific functional architecture, neuroscience and genetics cannot currently provide us with decisive answers. So if we are to pursue our quest to unveil the design of the phonological mind, then brain and genes cannot serve as the ultimate arbiter.

Relinquishing the decisive status of neuroanatomical evidence, however, does not mean that questions of mental architecture are unfalsifiable or impervious to external evidence. The large body of research reviewed in previous chapters demonstrates how one can evaluate the status of phonological primitives and constraints against a host of behavioral evidence, ranging from the distribution of such structures across languages, their role in linguistic processes, and their status in psychological experiments with humans – both infants and adults. Moreover, findings from neuroscience and genetics can certainly inform our understanding of how cognitive systems are implemented. Consider, for example, the role of markedness scales, such as sonority, or place of articulation. Although such proposals make no specific claims on how markedness is represented in the brain, they do predict that changes in markedness along any given markedness scale should result in a monotonic change in the activation of a single network (as opposed to non-monotonic changes across multiple circuits). Similarly, while core phonological knowledge may be linked to genes that are involved in multiple functions, one would expect to find hereditary conditions that would affect phonological competence. While there are numerous ways in which predictions associated with domain-specificity can be evaluated against neuroanatomical and genetic data, the final verdict on whether the system is specialized must rely primarily on internal evidence. Ultimately, domain-specificity in cognition is a functional question, and functional questions can only have functional answers.

12 Phonological technologies: reading and writing

> Core knowledge systems outline not only our early, instinctive, and universal understanding of the world but also provide scaffolds for subsequent learning. Like the core systems of number, physics, and social knowledge, our instinctive phonological knowledge sets the stage for the cultural invention of reading and writing. This chapter outlines the intimate link between early phonological competence and those later "phonological technologies." We will see that all writing systems – both conventional orthographies and the ones invented spontaneously by children – are based on phonological principles. Reading, in turn, entails the automatic decoding of phonological structure from print. Skilled reading recruits the phonological brain network that mediates spoken language processing. Moreover, dyslexia is typically marked by hereditary deficits to phonological processing and phonological awareness. The role of instinctive phonology as a scaffold for reading and writing is in line with its being viewed as a system of core knowledge.

12.1 Core knowledge as a scaffold for mature knowledge systems

In previous chapters, we have seen that phonological systems manifest a unique, potentially universal design that is evident already in early development. The special design of the phonological system is in line with the characteristics of core knowledge systems documented in numerous other domains, including knowledge of number, agency, space, and morality (Bloom, 2010; Carey, 2009; Carey & Spelke, 1996; Hamlin, Wynn & Bloom, 2010; Hamlin et al., 2007; Hauser & Spelke, 2004; Spelke, 2000). These early knowledge systems each includes distinct representational primitives and combinatorial principles that are innate, universal and domain specific. For example, infants as young as 4 months of age manifest rudimentary knowledge of number – they can represent the precise number of up to four objects (larger numbers are encoded approximately), and they can perform addition and subtraction operations on such small sets. In the domain of physics, young infants possess intuitive knowledge that leads them to expect objects to move cohesively (without disintegrating) and continuously (without jumping from one point to another and without intersecting other objects) as a result of contact with other objects. Other principles of morality might underlie 3-month-old infants' preference for social "helpers" (a character helping another climb up the hill) to "hinderers" (a character who interferes with the climber's efforts).

While these early, intuitive knowledge systems continue to play a role throughout development, as the child matures they gradually give rise to new bodies of knowledge that differ from their predecessors in their contents and expressive power (Carey, 2009). For example, the core number system available to infants is limited in size – it can implicitly encode precise numerosity of sets of up to four objects. Adults, in contrast, can compute the numerosity of any set by relying on a later-emerging system of recursive number that develops on the heels of the primitive number systems available to infants and animals. In a similar manner, the child's early concepts of object and motion eventually give rise to elaborate scientific theories of physics (Spelke, 1994), and infants' basic intuitive moral sense lays the foundation for moral systems that apply generally, to both kin and stranger (Bloom, 2010).

Unlike their intuitive innate predecessors, those later theories and inventions are by no means instinctive or universal, as different cultures vary in their scientific and technological advance as well as their moral codes. While the early core number systems, for example, are present universally, in any infant, the later system of recursive number depends on specific linguistic experience with number words and quantifiers, and consequently, people deprived of such linguistic devices – either because their language lacks them (Gordon, 2004) or because they lack access to a language (Spaepen et al., 2011) – do not develop the recursive number system. Likewise, scientific discoveries and moral theories are the product of deliberate reasoning and the intense research of a select few individuals, rather than the outcome of biological maturation available universally. But although the elaborate cultural discoveries, theories, and technologies of adult communities clearly differ from the intuitive, universal, and innately based early systems of core knowledge, these two kinds of knowledge are nonetheless linked inasmuch as several of those later discoveries develop on the heels of their ontogenetic predecessors.

Just as our intuitive core knowledge of number and object gives rise to mature scientific theories of mathematics and physics, so does the core system of phonology form the scaffold for a cultural technological invention – the invention of reading and writing. Indeed, reading and writing are intimately linked to phonological knowledge (Liberman, 1973; Perfetti, 1985). As I next demonstrate, all fully developed writing systems encode concepts using phonological means. The precise link between the writing system and phonological patterns varies – some writing systems encode phonemes, whereas others represent syllables. But the reliance on a phonological system is common to them all. And since writing systems are only part inventions and mostly discoveries – the discovery of spoken phonological patterns – the designs of writing systems tend to converge across cultures. In fact, such phonologically based designs reemerge spontaneously in the rudimentary writing systems that are routinely invented by many children.

Not only does core phonology form the basis for writing, but it also constrains reading. Like many skilled readers, you, the reader of this book, might decode the words printed on this page automatically and effortlessly, with no awareness of their phonological structure. For this reason, most people believe that they identify printed words in much the same way they identify traffic signs – by directly mapping visual symbols onto concepts. But a large experimental literature shows that this popular belief is in fact mistaken. All readers, both beginners and skilled, routinely go through the extra step of mapping graphemes onto phonological representations. As in the case of writing, the phonological representations decoded in reading vary in grain size depending on multiple factors (e.g., the writing system, the familiarity with the word, and the specific experimental conditions). Nonetheless, some level of phonological decoding is always present (Perfetti et al., 1992), and it is demonstrably shaped by the same phonological principles documented in spoken language. Skilled reading is thus strictly constrained by phonological competence. Conversely, when people are unable to effectively encode the phonological structures of spoken language, dyslexia typically ensues.

Why would a cultural invention such as reading pay its debt to phonology? Unlike math and physics, reading and writing are just invented codes, not theories of the physical world. While an arbitrary theory of physics cannot be maintained in the face of conflicting empirical evidence, and physically improbable technologies are bound to go extinct, phonologically arbitrary writing systems are amply feasible, and direct "visual" encoding of such systems is certainly conceivable. In fact, on some accounts, it is even likely. But precisely because phonologically arbitrary writing and reading is logically possible, the fact that such systems are disfavored suggests internal constraints on their design. The hypothesis of core phonological knowledge accounts for these facts. Like its sister systems of core knowledge – number, physics, and morality – reading and writing might be grounded in core knowledge. The link between the instinctive, early phonological system and reading and writing, in turn, provides converging evidence for the existence of a core system in phonology. This chapter reviews the evidence linking reading and writing to core phonology. These observations demonstrate that writing systems – both conventional and invented ones – are all based on phonological principles, that their decoding in reading recovers phonological structure from print, and that deficits to core phonology are linked to reading disability.

12.2 Writing systems recapitulate core phonology

12.2.1 Conventional writing systems

Writing systems are inventions that have independently emerged at least three times in the history of humanity. The first writing system was devised by the Sumerians roughly 5,000 years ago, about 1,500 years later, a second system was invented by the Chinese, and a third system was independently devised about 2,000 years ago by the Mayans (Rogers, 2005). While most inventions are judged for their originality, in the case of writing, the similarity among those independent inventions is even more striking than their differences. Each of these ancient systems includes some method of phonological organization, and the reliance on phonological encoding has since been preserved in practically every fully developed writing system (for one possible exception, Bliss, see Rogers, 2005).

Chinese characters, for example, encode syllable-size phonological units. Indeed, the number of characters expressing a single Chinese word depends on the number of syllables: Monosyllabic words are conveyed by a single character and disyllabic words by two, and this link holds irrespective of whether any given syllable is mono-morphemic or bi-morphemic (Rogers, 2005; see 1). Thus, the disyllabic words *shān hú* ('coral') and *tiě lù* ('railway') are each expressed by two characters, even though the former is mono-morphemic whereas the latter is morphemically complex.

(1) Chinese characters correspond to syllable-size units (not morphemes; examples from Rogers, 2005; *Chinese Character Dictionary*, 2010)
 a. Monosyllabic words are expressed by a single character:
   wǒ ('I') 我
   hǎo ('good') 好
 b. Disyllabic words are expressed by two characters:
   (i) monomorphemic
     hú dié ('butterfly') 蝴蝶
     shān hú ('coral') 珊瑚
   (ii) di-morphemic
     tiě lù ('railway'; = *tiě* 'iron' + *lù* 'road') 鐵路
     zì diǎn ('dictionary'; = *zì* 'character' + *diǎn* 'standard') 字典

Not only does each Chinese character correspond to a single syllable, but segments that form a single syllable may be encoded by the same character even when meaning varies (see 2). For example, the words for 'horse' and 'mother' both share the same syllable, *ma*, encoded by a common orthographic character (馬). This is not to say that the phonological representation of Chinese characters is fully predictable from print. Although some characters (about 25 percent of all characters) convey phonological information (segmental and tone) fully and reliably, others carry only partial phonological information, and some characters (estimated at 33 percent of all characters) carry no useful phonological information at all (DeFrancis, 1989: 113). But the fact that such links nonetheless exist demonstrates that phonological principles play a role in the design of this orthography.

(2) Segments that form a single syllable are encoded by the same character (examples from DeFrancis, 1989; Rogers, 2005; *Chinese Character Dictionary*, 2010)
   a. /ma/ homophones:
      媽 mā 'mother'
      馬 mǎ 'horse'
   b. /jiao/ homophones:
      僥 jiǎo 'lucky'
      澆 jiāo 'to water'

While Chinese encodes syllable-size characters, other writing systems contrast among finer-grained phonological units – moras, segments, and even feature-size units. Moras are units of prosodic weight, and weight, in turn, typically depends on the structure of the rhyme: a CV syllable counts for a single mora, whereas CVV and CVC units count for two. In the Japanese Kana systems (the 'plain' Hiragana and 'side Kana' Katakana), most characters correspond to monomoraic syllables. Accordingly, Katakana indicates monomoraic, CV syllables by a single symbol whereas bimoraic CVV and CVN (N=nasal) syllables are indicated by two symbols. For example, the initial monomoraic syllable in Toyota (see 3) is transcribed by a single Hiragana character, と, whereas the bimoraic word /too/, 'ten' comprises two symbols とお – the initial と symbol from Toyota, followed by an additional symbol for the second mora (see 3).

(3) The expression of moraic contrasts in Hiargana
   とよた Toyota /toyota/ <to.yo.ta>
   とお ten /too/ <to.o>

Moving down the inventory of phonological primitives to the level of the phonemes, we arrive at the familiar alphabetic writing systems – systems that use graphemes to encode phonemes. Some alphabets, such as English, encode both consonants and vowels, whereas consonantal orthographies such as Hebrew encode mostly consonants to the exclusion of most vowels. The examples in (4) are all morphologically relatives, derived form the root /ktv/, 'writing,' a fact depicted in the orthography by their common consonant letters (כתב).

(4) The Hebrew consonantal orthography
   כתב /katav/ 'he wrote'
   כתב /ktav/ 'handwriting'
   כתיב /ktiv/ 'spelling'

Finally, several orthographies use symbols to encode sub-phonemic feature distinctions. Korean phonology, for example, contrasts obstruents in terms of their aspiration and the tense-lax dimension. The Hangul orthography, in turn, expresses these distinctions: aspiration is marked by adding a stroke to unaspirated counterparts, and tenseness is expressed by reduplicating the corresponding lax consonant (Rogers, 2005; see 5). Moreover, Hangul conveys the syllabic constituency of the phonemes in terms of their systematic spatial arrangement: onsets are obligatorily encoded either by a consonant, or, in the case of vowel-initial syllables, by a dummy character ㅇ, vowels (including on-glides, y and w) are encoded either to the right of the onset or below it (depending on the direction of their main stroke – vertical or horizontal), and coda consonants are indicated at the bottom of the cluster (see 6).

(5) The representation of aspiration and tense-lax contrasts in Hangul
t ㄷ, t^h ㅌ, tt ㄸ
p ㅂ, p^h ㅍ, pp ㅃ
k ㄱ, k^h ㅋ, kk ㄲ

(6) The spatial depiction of syllable structure in Hangul (from Simpson & Kang, 2004) 교실 /kyo.sil/ 'classroom'
교 /kyo/
  onset: k ㄱ
  nucleus: yo ㅛ
실/sil/
  onset: s ㅅ
  nucleus: i ㅣ
  coda: lㄹ

These examples make it plain that conventional writing systems are based on phonological principles: they encode the same set of phonological primitives attested universally in phonological systems – syllables, moras, segments, and features – and, in some cases, writing even expresses their structural roles in the syllable.

12.2.2 Invented spelling systems

The link between writing systems and phonology is not merely a diachronic fact about the evolution of writing systems. Rather, it is a vital synchronic phenomenon that reemerges time and time again in the spellings invented by children. Children tend to spontaneously invent spelling systems of their own based on rudimentary familiarity with the letters of their adult community (Chomsky, 1976; Read, 1971; Treiman & Cassar, 1997). Being invented systems, these spellings are bound to differ from conventional systems. But precisely because such "misspellings" diverge from the adult model, they provide a window into the productive knowledge that shapes the formation of those systems. As it turns out, the relevant knowledge is phonological.

The role of phonological knowledge in shaping the child's invented spelling results in two broad contrasts compared to the adult's system: under-specification – cases in which the child omits some of the phonological distinctions present in the adult's system, and over-specification – cases in which children specify some phonological contrasts that adults fail to express in their mature spelling system.

Under-specification takes multiple forms. Children's early spellings of English, for example, often conflate the contrast between tense and lax vowels: Children first use tense vowels (whose sounds are familiar to them from the letters' names) to express their corresponding lax counterparts (e.g., they use *E* to spell both /i/ and /ɪ/), and once they learn the proper way to express lax vowels, they subsequently over-generalize those spellings to express tense vowels (they use *i* to spell /i/). Another common error is the omission of vowels before sonorant segments (e.g., *tiger*→ TIGR).

(7) Phonological under-specification in invented spelling (from Read, 1971)
 a. Failure to contrast tense and lax vowels:
  FEL (feel)
  FES (fish)
  SIKE (seek)
 b. Failures to specify the vowel:
  TIGR (tiger)
  DIKTR (doctor)

But these two types of errors are neither careless nor arbitrary: Both patterns reflect productive phonological knowledge. Indeed, children do not randomly confuse any two-vowel phonemes, but they specifically conflate vowels that are matched for height and differ only on their tenseness. In so doing, they demonstrate that they know that these phoneme pairs share a feature. Moreover, the conflation of the tense-lax dimension in the child's spelling mirrors a phonological abstraction present in conventional English orthography (e.g., *extreme–extremity*; *divine–divinity*), but this convergence most likely emerges independently – it is unlikely that children simply imitate the conventional adult spelling, as their invented spellings precede spelling instruction. Similarly, children's tendency to omit vowels before syllabic consonants (e.g., TIGR, DOCTR) are systematic – children are reliably more likely to omit such vowels compared to ones preceding nonsonorant consonants (e.g., *salad, basket*; Treiman, 2004). Although such spellings happen to counter the obligatory encoding of vowels in all spelled syllables, they are perfectly consistent with the phonology of English, where sonorant consonants can form a syllable of their own (e.g., the *r* in *tiger*, pronounced [tajgr̩]).

(8) Phonological over-specification in invented spellings (from Read, 1971)
   a. Regular suffix:
      MARED (married)
      HALPT (helped)
   b. Affrication:
      CHRIE (try)
      JRGAIN (dragon)
   c. Flaps:
      LADAR (letter)
      PREDE (pretty)

The systematic phonological basis of children's "misspellings" is also evident in cases where their spellings specify phonological distinctions that are attested in English phonology but are unspecified in conventional adult spelling system (i.e., over-specification). English spelling conflates the voicing distinction between the suffix in *married* and its voiceless counterpart in *walked*, but children obey this phonological contrast in their invented spellings. Similarly, by expressing the intervocalic *t* (e.g., in *letter*) by a D, children approximate its realization as a flap, rather than a [t]. In all these cases, children's spellings reflect accurate renditions of the phonology that are absent in the adult's systems. Although these examples are limited inasmuch as they are all confined to a single language, English (Share, 2008), there is some evidence that the spontaneous extraction of phonological organizational principles also extends to nonalphabetic writing systems (Nag et al., 2010).

Summarizing, then, all full writing systems deploy some phonological organizational principles. The intimate link between phonology and spelling is present in conventional orthographies that evolved from three independent phonological writing traditions, and it recapitulates in the spelling systems that are routinely and spontaneously invented by young children. Although all writing systems are ultimately inventions that utilize visual symbols, at their core, they are discoveries – the discovery of one's phonological system. Thus, writing systems are systems of visible speech (DeFrancis, 1989).

12.3 Reading recovers phonological form from print

That writing recapitulates phonology is intriguing, but perhaps not entirely surprising. After all, writing systems encode language, and all human languages are known to exhibit phonological patterns that routinely generalize to novel forms. If writing systems are to keep up with the vast richness of spoken phonological forms and their constant expansion by innovations and borrowings, then some productive phonological principles must be incorporated in the writing system itself. Remarkably, however, the phonological reflexes of written language are

288 Phonological technologies: reading and writing

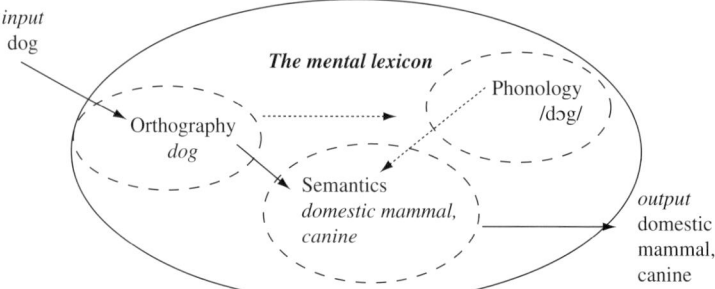

Figure 12.1 Lexical access from print via orthography vs. phonology (indicated by continuous and broken lines, respectively)

evident not only at its encoding, in writing, but even during its online decoding – in reading.

Unlike the phonological encoding of print, the reliance of reading on phonology is hardly expected. While one must rely on phonological principles to decode novel words – the *ipads*, *faxes*, and the countless other gadgets that are constantly added to our lives – the grand majority of words are highly familiar. For such words, readers have securely stored in their memory representations that link the word's spelling with its meaning. To recognize a familiar printed word – that is, to associate it with some word's meaning, stored in the mental lexicon – readers could simply retrieve the meaning directly from its spelling (see the heavy continuous arrows in Figure 12.1). Decoding the word *dog* would essentially proceed along the same lines as any other nonlinguistic sign – traffic signs, faces, and scenes.

The possibility of non-phonological decoding of familiar English words is difficult to grasp precisely because we automatically retrieve the phonological form of words from print. But a brief example from a foreign orthography reminds us that phonological decoding is by no means necessary. English readers unfamiliar with the Hebrew orthography can easily learn the meaning of כלב by associating it with an image (see Figure 12.2), oblivious to the fact that its phonological form happens to be /kelev/. In a similar manner, readers could decode most printed text by associating visual symbols directly with their meaning. But countless studies have shown that this is not what readers typically do. Rather than directly mapping graphemes and meanings, readers of all languages routinely rely on some method of phonological decoding. The precise method varies across orthographies, but the reliance on some form of phonological decoding appears to be universal. The conclusion that reading entails obligatory phonological processing underscores the intimate link between reading and the core system of phonology. In what follows, I briefly review some of the evidence for phonological decoding at both the sentence and single-word levels.

Figure 12.2 Reading without phonology

12.3.1 Phonological decoding in the silent reading of sentences

For skilled readers, the silent decoding of printed sentences typically appears just that – silent. But appearances can be misleading, and tongue-twister sentences make this fact patently clear. Tongue twisters are painfully hard to utter aloud. But remarkably, this challenge persists even when reading is silent. Numerous studies have shown that tongue-twister sentences (see 9) are harder – they take longer to (silently) read and they are subsequently harder to recall compared to control sentences matched for syntactic and semantic structure (e.g., Keller et al., 2003; Kennison et al., 2003; McCutchen & Perfetti, 1982; McCutchen et al., 1991; Zhang & Perfetti, 1993).
(9) Tongue-twister sentences and controls:

A tongue-twister sentence:	the taxis delivered the tourist directly to the tavern
Control:	the cabs hauled the visitor straight to the restaurant

Several observations suggest that the difficulties with tongue twisters are not visual confusions due to letter repetitions. First, the same difficulties obtain even when the repeated phonological elements (phonemes, features) are expressed by different letters (e.g., *t* vs. *d*, in English; McCutchen & Perfetti, 1982). Second, the interference from tongue twisters interacts with a secondary phonological task of digit recall (McCutchen et al., 1991). In the experiment, people are first asked to memorize a set of five numbers (e.g., 2, 12, 25, 22, 29), they are next presented with a sentence – either tongue twister or control which they are asked to judge for meaning – and they are finally asked to recall the set of digits in the order they were presented. Results show that the tongue-twister effect depends on the phonological similarity between the sentence and the digits. For example, tongue twisters repeating an alveolar stop (*the taxis delivered the tourist* ...) are harder to read in the context of phonologically similar digits (e.g., the initial alveolar stop in 2, 12, 25, 22) compared to less similar controls (e.g., the initial alveolar fricatives in 6, 7, 66), and the similarity between the sentences and numbers impairs number recall as well. A third piece of evidence demonstrating that the tongue-twister effect cannot be due to visual confusions comes from the demonstration of this effect in Chinese, a nonalphabetic orthography (Zhang & Perfetti, 1993). Like their English-speaker counterparts, Chinese readers take longer to silently read tongue-twister stories compared to controls, and they make more errors in their recall. Finally, a functional MRI study of the silent reading of tongue twisters demonstrates that such sentences engage various phonological sites (e.g., Broca's area, the left angular/supramarginal gyri, and the areas along the left superior temporal sulcus) compared to control sentences (Keller et al., 2003). As these authors conclude, tongue twisters twist not only the tongue but also the brain – specifically, the regions involved in phonological processing and maintenance.

Why do people bother to decode the phonological forms of sentences, even though doing so interferes with the experimental task of digit recall? The answer to this puzzle becomes immediately apparent once we consider the properties of working memory. To escape oblivion, words must be maintained in a memory buffer called working memory (Baddeley, 1986). Working memory maintenance, however, is executed using a phonological format – try to memorize a phone number, and this will immediately become evident. So the phonological decoding of printed materials is mandated by the format of our short-term memory system. And because digit recall puts additional demands on working memory, it interferes with the phonological maintenance of sentences, an interference that is further exacerbated by the phonological similarity between words and digits.

12.3.2 Phonological activation in single-word recognition

The phonological format of working memory explains why all linguistic materials – printed or spoken – must ultimately undergo phonological encoding. Phonological decoding, however, begins immediately upon the recognition of single isolated words, even in tasks that impose only the slightest demands on working-memory maintenance. We now turn to examine the mechanisms mediating the phonological decoding of isolated words.

Alphabetic orthographies such as English allow readers to obtain phonological representations in two ways (see Figure 12.3). One method obtains the word's phonological form even before it is retrieved from the lexicon (i.e., pre-lexically) by mapping its graphemes to phonemes – a process known as *phonology assembly* (marked by the non-continuous line in Figure 12.3). English speakers, for example, know that the letter *d* signals the phoneme /d/, o can signal /ɔ/, and g corresponds to /g/. By relying on such regularities, readers can assemble the

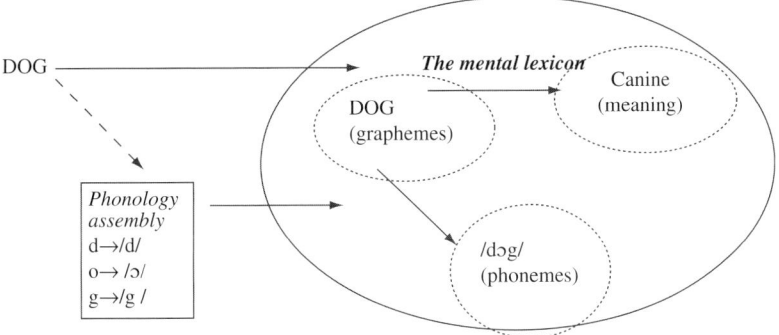

Figure 12.3 Two routes to phonology from print: assembled and addressed phonology (indicated by the broken and continuous lines, respectively)

phonological representations of many printed words – familiar and novel. Familiar words such as *dog*, however, are also stored in the mental lexicon, linked to their phonological forms. Such words can thus be decoded along a second lexical route. The lexical route retrieves the word's lexical phonological representation directly from its graphemic form, a process known as *addressed phonology* (marked by the continuous line in Figure 12.3). While alphabetic systems such as English allow readers to obtain phonology from print along either the assembled or addressed route, in other writing systems – both alphabetic (e.g., Hebrew) and nonalphabetic (e.g., Chinese) – a word's phonological form cannot be fully generated by phonology assembly, so readers must rely on lexical retrieval to a greater extent. While the precise origins of a word's phonological form – assembled or addressed phonology – vary depending on the orthography, reading skill, and subtle properties of task demands, the reliance on some form of phonological encoding appears to occur universally, in all writing systems. This conclusion is supported by literally hundreds of published studies in many languages. Here, we will illustrate some of the main findings from behavioral methods. Additional insights from neuroimaging studies are discussed in the next section.

12.3.2.1 Phonological predictability effects
A common method to gauge the contribution of phonology assembly in reading concerns predictability effects. This method exploits the well-known fact that, too often, phonological forms are only partly predictable from print. Consider English, for instance. While words like *dog* and *cat* can be reliably decoded by mapping their graphemes to phonemes, other words, like *come* and *put*, are not fully predictable, as the assembly of their phonology would yield incorrect

over-regularized forms, rhyming with *home* and *mute*. Such unpredictability can be annoying for readers, but it comes in handy to researchers who wish to determine how reading works. The rationale is simple: If English readers rely on phonology assembly, then phonologically unpredictable words (e.g., *come*) should exert a cost compared to predictable controls (e.g., *home*) – they should take longer to read and produce more errors. Such costs have indeed been detected by many studies (e.g., Andrews, 1982; Baron & Strawson, 1976; Glushko, 1979; Jared et al., 1990). While early research obtained these effects only for unfamiliar words, or unskilled readers (e.g., Seidenberg, 1985; Seidenberg et al., 1984), subsequent studies (Jared, 2002) documented predictability effects even for highly familiar words (provided that their pronunciation is highly unpredictable, i.e., that their incorrect pronunciation rhymes with many more familiar words than the correct one). The generality of these effects is important because it demonstrates that skilled readers assemble phonological forms for all words – rare or familiar.

Nonetheless, these findings are limited in several ways. First, predictability effects are typically observed only when people are asked to read the word aloud, but not in the silent lexical decision tasks (e.g., Berent, 1997; Seidenberg et al., 1984), a finding that is sometimes interpreted to suggest that silent reading is not mediated by phonology assembly. A second limitation of predictability effects is that they gauge reliance on phonology only when a word's pronunciation is unpredictable (e.g., for *come*). Accordingly, this method cannot determine the role of phonology when a word's phonological form is predictable – for phonologically "regular" words (e.g., for *home*) or in "transparent" alphabetic orthographies (e.g., Spanish, Italian) – and it is altogether useless in nonalphabetic orthographies (e.g., Chinese). To address these limitations, we now turn to a second marker of phonological decoding – homophony effects.

12.3.2.2 Homophony effects

Homophony in phonological priming/masking

Homophony effects gauge readers' sensitivity to phonological similarity – either the similarity among words (e.g., *rose, rows*) or nonwords (e.g., *rose–roze*). If phonological representations are employed in reading, then letter strings that share their phonology (e.g., *rose–rows*) should be perceived as similar, and this similarity should facilitate lexical access. To see why, let's think of lexical access as the opening of a door to the mental lexicon. If the lexical door has a phonological entry code, then words sharing the same phonological form (e.g., *rose, rows, roze*) should all be equally able to open the door. So once two homophones are presented in succession (e.g., *roze–rose*), *roze* will crack the lexical door of *rose*, and as *rose* follows, its identification (i.e., access to its lexical entry) should be easier.

It is of course possible, however, that *roze* is helpful because it shares some of *rose*'s letters, rather than its sound. To control for this possibility, we can compare the facilitation from *roze* to *roge*, for instance. These two nonwords – *roze* and *roge* – are matched for their spelling similarity to *rose*, but differ on their phonological overlap. If the lexical entry to *rose* is specifically mediated phonology, then *roze* should have an advantage over *roge* – a case of phonological priming. If the process of extracting phonological representation from print occurs rapidly, then phonological priming should emerge even when the first word (a prime) is presented subliminally (see 10). In a similar fashion, presenting *roze* after *rose* should reinstate its phonological form, and consequently, it should facilitate its identification compared to spelling controls. As with priming, these effects, known as phonological masking, should obtain even when both words are displayed extremely briefly, masked by visual shapes.

(10) Phonological priming effects

Target: rose
Phonological prime: roze
Spelling control: roge

These predictions have been amply supported in numerous orthographies. Research by Charles Perfetti and colleagues has demonstrated that the identification of English words benefits from a brief phonological prime or mask, presented for as little as 35 ms under heavy visual masking (Perfetti & Bell, 1991; Perfetti et al., 1988). The finding that phonological similarity can affect word identification even after a brief encounter suggests that these representations become available rather quickly. Moreover, these effects obtain for all words, both infrequent and highly familiar, suggesting once again that the contribution of phonological representations to reading is general in scope.

Phonological priming and masking effects have since been detected in numerous orthographies and writing systems. Among the alphabetic writing systems, phonological masking and priming have been reported not only in orthographies that readily support the assembly of phonology from print, such as French (e.g., Ferrand & Grainger, 1992) and Spanish (e.g., Carreiras et al., 2009), but also the opaque consonantal script of Hebrew (e.g., Berent & Frost, 1997; Frost et al., 2003). Moreover, phonological priming has been demonstrated even in nonalphabetic writing systems – in Japanese (e.g., Buchanan & Besner, 1993; Chen et al., 2007) and Chinese (e.g., Perfetti & Zhang, 1991; Tan & Perfetti, 1997). Unlike English, however, Chinese does not allow one to assemble phonology by mapping graphemes to phonemes on the fly, so phonological priming in Chinese reflects the retrieval of stored phonological forms from the lexicon (addressed phonology), rather than its online assembly.

Although the phonological representations available to Chinese and English readers are obtained by different mechanisms, both orthographies are decoded by reliance on phonology.

Phonological interference effects

The findings reviewed so far suggest that the reflex of phonological decoding is quite robust. A true reflex, however, is judged by its automaticity. Once its triggering conditions are present, a reflex will proceed to completion, irrespective of whether its immediate effect is desired. The ultimate test of the phonological decoding reflex thus concerns not circumstances in which it is expected to help reading but those in which it is potentially deleterious. Our question is whether phonological decoding takes place under such conditions.

Semantic categorization experiments present one such case. Participants are asked to judge whether a word forms part of a semantic category (e.g., is *rose* a flower?) – a task that clearly does not require that phonology be decoded from print. In fact, it tacitly discourages participants from doing so. Because many trials include foils that are phonologically similar to a flower exemplar (e.g., *rows*; *roze*), reliance on phonology is detrimental. But results show that readers nonetheless rely on phonology. Consequently, people tend to incorrectly accept homophone and pseudohomophone foils (e.g., categorize *roze* as a flower) relative to spelling controls, and these effects obtain in both alphabetic (English: Jared & Seidenberg, 1991; Van Orden, 1987; Van Orden et al., 1988; French: Ziegler et al., 1999) and nonalphabetic orthographies (Chinese: Perfetti & Zhang, 1995; Tan & Perfetti, 1997; Japanese Kanji: Wydell et al., 1993). People in these experiments do not fall for phonological foils on every trial (e.g., they do not invariably categorize *roze* as a flower) because they can put their phonological instincts in check using a spelling-verification mechanism. But if phonology is nonetheless active, homophones should be more likely to slip in, which is precisely what is observed.

In fact, people demonstrably rely on phonological representations even when they are explicitly required to avoid reading altogether, in the Stroop task. The Stroop task, discussed in previous chapters, presents participants with letter strings printed in colors (e.g., the word *green* printed in the color red), and participants are asked to name the color of the ink (*red*) while ignoring the printed letter. The classical finding is that readers automatically decode the words, and consequently, color naming is impaired when the printed word spells the name of an incongruent color (Stroop, 1935). Subsequent research, however, showed that Stroop interference obtains not only from the conventional spelling of color names but also by their homophones (e.g., the word *bloo* printed in red). These homophonic Stroop effects have been reported in various alphabetic systems (English: Naish, 1980; French: Ziegler et al., 1999; Hebrew: Berent et al., 2006; Tzelgov et al., 1996) as well as in Chinese (Guo et al., 2005; Spinks et al., 2000; Wang et al., 2010). Such demonstrations suggest that the encounter with a printed word automatically triggers phonological decoding even when people attempt to avoid doing so.

12.4 Reading recruits the phonological brain network

Another test of the role of phonological reflexes in reading is presented by the brain networks that mediate silent reading. If silent reading recruits the phonological system, then printed words should engage brain mechanisms implicated in the phonological processing of spoken language. The previous section has already offered one illustration of the role of phonological brain networks in the silent reading of tongue-twister sentences. In what follows, we further evaluate this prediction at the level of single words.

In one study (Buchsbaum et al., 2005), participants were presented with pronounceable monosyllabic pseudowords across two modalities – either in print or aurally. Results revealed numerous brain regions that were common to the two conditions, including the left planum temporale, left STG, and the left middle temporal gyrus, as well as the left IFG – regions that correspond to the phonological brain network discussed in Chapter 10.

Further evidence for the phonological functions of these temporal-lobe sites is offered by their differential contribution to the naming of novel letter strings – either meaningless letter strings (i.e., pseudowords, e.g., *bleg*) or ones homophonous to real words (e.g., *burth*) – as compared with irregular words (e.g., *pint*). Because novel words are not lexically stored, their phonological form can be obtained productively only by mapping graphemes to phonemes (i.e., phonology assembly). Irregular words, by contrast, have a phonological form that is unpredictable from print, so their pronunciation requires lexical retrieval (i.e., addressed phonology). To the extent that assembled and addressed phonology engage different brain regions, one would thus expect novel words and irregular words to elicit different patterns of activation. An fMRI experiment by Simos and colleagues supports these predictions (Simos et al., 2002). Irregular words activated the posterior middle temporal gyrus and the middle temporal lobe to a greater extent than novel words, suggesting that those regions specifically mediate the retrieval of addressed phonological forms. In contrast, activation in the posterior STG correlated with the pronunciation time of nonwords (but not exception words), suggesting that this region subserves the assembly of phonology from print. Interestingly, however, the posterior STG also contributed to the processing of irregular words, indicating that naming invariably triggers phonology assembly of all words – regular or irregular. A subsequent magnetoencephalography study (Simos et al., 2009) observed bilateral activation of the posterior superior temporal gyri and inferior frontal gyri using the lexical decision task. These findings are significant because they suggest that the phonological network mediates silent reading, and it is active quite generally, irrespective of the frequency of those words. Moreover, a meta-analysis of imaging studies of reading across Western alphabetic and Eastern orthographies (Bolger et al., 2005) concluded that the left posterior STG (Brodmann Area 22) is active in all orthographies, including the nonalphabetic Chinese and Japanese Kanji systems (but see Liu et al., 2009; Siok et al., 2008, for different conclusions). These studies illustrate the conclusions emerging from a very large literature that links the phonological network of spoken language to reading. These findings underscore an interesting link between mind and brain. Just as the cognitive architecture of reading is based on older phonological principles, so does the phonological brain assemble the reading networks by recycling older substrates that mediate core phonology (Dehaene et al., 2010; Dehaene & Cohen, 2011).

12.5   Grammatical phonological reflexes in reading

The behavioral and neural evidence reviewed in previous sections make it clear that reading entails phonological decoding. It is the recovery of phonological structure that allows readers to determine that two printed words are homophonous (e.g., *rose–rows–roze*) and leads to the over-regularization errors of irregular words (e.g., *come*). While these results firmly establish that readers extract some phonological representations, they do not determine whether these representations are in fact identical to the ones extracted in spoken language processing. To determine that two words are homophonous, for example, readers could rely on simple phonological codes that list only phonemes and their linear order. This possibility would entail a rather superficial link between reading and the putative system of core phonology – while both systems might encode phonological information, the representations computed by the two systems would be qualitatively different. On an alternative account, the representations computed to visual and spoken language are isomorphic: They encode the same set of primitives, and they are constrained by the same set of grammatical constraints. The link between reading and core phonology is thus intimate. The neural evidence summarized in the previous section is consistent with this latter possibility. Further support for this view is presented by behavioral evidence showing that silent reading yields structured phonological representations. These representations specify the same phonological primitives implicated in spoken language processing, and they are subject to some of the same grammatical constraints – both language-particular and universal.

12.5.1   *Phonological primitives in silent reading*

The review of spoken language phonology suggests that phonological features, CV frames, and syllables form the representational primitives of core phonology. Each of these elements has also been implicated in reading.

Consider, for example, the role of phonological features in silent reading. If the representations assembled in reading encode the feature composition of segments, then words that share most of their features should be perceived as phonologically similar even if they do not share phonemes. In line with this prediction, printed words that share most of their phonological features (e.g., *sea–zea*, which differ only on the voicing) have been shown to prime each other to a greater extent than controls (e.g., *sea–vea*, which differ by voicing and place of articulation; Lukatela et al., 2001; see 11). Remarkably, readers are sensitive to phonological features even when the task, lexical decision (e.g., is *sea* a real English word?), does not require phonological encoding or articulatory response. Additional auxiliary analyses rule out the possibility that the similarity between *sea* and *zea* reflects visual features. Together, these findings suggest that the computation of sub-segmental phonological detail might be quite general.

(11) Priming by phonological features in silent reading

Target: sea
Phonologically similar prime: zea
Control prime: vea

Other findings indicate that such sub-segmental information includes even non-contrastive features. It is well known that the acoustic duration of the vowel in CVC words differs as a function of the following coda consonant. Vowels followed by voiced codas (e.g., *plead*) are longer than ones preceding voiceless codas (e.g., *pleat*). This acoustic difference is analog and non-contrastive (i.e., no English words differ solely on this dimension), but it nonetheless reflects a regular characteristic of phonological systems. Strikingly, however, this subtle phonetic contrast has been shown to affect silent reading: Readers take longer to classify *plead* as an English word compared to its counterpart *pleat*, even though the materials are printed and the task (lexical decision) requires no articulatory response. This effect, first demonstrated by Abramson and Goldinger (1997), was subsequently replicated in both behavioral (Ashby et al., 2009; Lukatela et al., 2004) and EEG measures (Ashby et al., 2009).

Moving up in the hierarchy of phonological primitives, readers encode various prosodic constituents. Readers in different orthographies are sensitive to the syllable structure of multisyllabic words. Participants in these experiments are presented with a printed multisyllabic target word beginning with either a CV or a CVC syllable (e.g., *ba.sin* vs. *bas.ket*; see 12), and each such target is preceded by either a CV prime (e.g., *ba*) or a CVC prime (e.g., *bas*). Results show that readers are sensitive to the congruency between the syllable structure of the target and prime. For example, people are faster to classify *basket* as an English word when preceded by *bas* compared to *ba* despite not being required to articulate either the target or prime. Syllable congruency effects, moreover, obtain not only in so-called syllable-timed languages – languages like Spanish and French (Carreiras et al., 2005; Chetail & Mathey, 2009; Colé et al., 1999), where syllable structure is well defined – but even in English, a stress-timed language whose syllable structure is relatively opaque. In a series of examples, Jane Ashby and colleagues demonstrated that the brief priming of the target by a syllable-congruent prime that is presented parafoveally (i.e., in the area surrounding the fovea, such that readers are unaware of the prime) facilitates target recognition (Ashby & Rayner, 2004) and decreases the negative N1 evoked-potential brain response to the target word (Ashby, 2010; Ashby & Martin, 2008).

(12) Syllable priming effects in silent reading

		Target	
		CV (e.g., *ba.sin*)	CVC (e.g., *bas.ket*)
Prime	Congruent	*ba*	*bas*
	Incongruent	*bas*	*ba*

12.5.2 Grammatical constraints in silent reading

Not only do readers decode phonological primitives from print, but they further subject these representations to the same grammatical constraints operative in spoken language. The evidence is particularly strong when it comes from novel words. Unlike familiar words, the phonological forms of novel words cannot be retrieved from the lexicon, so such phonological effects imply a productive grammatical mechanism that is operative online in reading. Previous chapters have discussed various examples of such grammatical constraints in detail, so here we will review some of these examples rather briefly.

Recall, for example, that Hebrew constrains the location of identical segments in the root – roots with initial identical consonants are ill formed (e.g., *ssm*) whereas roots with identical consonants at their end are quite frequent (e.g., *smm*). A second restriction bans non-identical segments that share the same place of articulation (i.e., homorganic consonants, such as the two labials in *smb*). Results show that Hebrew speakers are sensitive to both constraints in silent reading (Berent et al., 2001b; Berent et al., 2004). Because ill-formed strings are less wordlike, such strings should be easier to identify as nonwords. And indeed, words derived from ill-formed roots (e.g., *ssm, smb*) are identified more readily than well-formed controls (e.g., *smm*), matched for segment co-occurrence. In fact, the restriction on root structure modulates reading even when people are explicitly asked to avoid reading altogether, in a modified Stroop task. In these experiments, Hebrew speakers were presented with words printed in color. As in the typical Stroop task, the task was to name the color of the ink while ignoring the printed letters. These letters, however, corresponded to novel words constructed with novel roots – either well-formed roots (e.g., *smm*) or ill-formed controls (e.g., *ssm*). Because ill-formed roots engage the grammar to a lesser extent than better-formed controls, readers should find it easier to ignore ill-formed structure, a fact that might leave them free to quickly name the color and ignore the root. Results indeed showed that the structure of the root affected color naming: People named the color faster with ill-formed *ssm*-type roots compared to well-formed controls (Berent et al., 2006). These results indicate that grammatical phonological knowledge concerning Hebrew phonotactics constrains reading automatically, in a reflex-like manner.

Additional results from English suggest that these grammatical constraints might include principles that are potentially universal. One such case concerns the markedness of syllable frames. Across languages, syllables that manifest an onset are preferred to onsetless ones, and simple syllable margins are preferred to complex ones. For these reasons, syllables manifesting an onset and simple margins (e.g., CVC) are unmarked compared to syllables with no onset and complex margins (e.g., VCC). If marked structures are less likely to engage the language system, then VCC structures should be easier to ignore than CVC ones. The results from modified Stroop experiments are consistent with this prediction (Marom & Berent, 2010). In these studies, people were presented with novel words printed in color. None of these words shared phonemes or graphemes with the color name, but in some cases, the color name and the word shared the same CV skeletal frame (e.g., the word TROP printed in the color *black* – a CCVC frame), whereas in others they were incongruent. Crucially, incongruent frames were either relatively unmarked (e.g., the CVC frame in GUF) or more marked (e.g., VCC, as in OCP). As expected, people were sensitive to the skeletal congruency between the color and the word, resulting in faster naming time in the CV-congruent condition (see also Berent & Marom, 2005). But crucially, responses to the incongruent condition were modulated by markedness, such that incongruent frames were easier to ignore when they were marked (e.g., for VCC compared to CVC).

Other, putatively universal markedness restrictions concern sonority sequencing. Recall that speakers constrain the sonority distance of auditory onsets that are unattested in their language – the smaller the distance, the more likely is the onset to be epenthetically repaired. In particular, people tend to misidentify ill-formed onsets of falling sonority as identical to their epenthetic counterparts (e.g., *lbif=ləbif*), they are less likely to do so for sonority plateaus (e.g., *bdif*), and they are least likely to misidentify well-formed onsets of rising sonority (e.g., *bnif*). Interestingly, these sonority restrictions extend to printed words: CCVC syllables with ill-formed onsets are harder to distinguish from their CəCVC counterparts (Berent & Lennertz, 2010; Berent et al., 2009). Although the findings from printed materials are less robust than those with auditory stimuli, and they obtain only when the task allows for sufficiently detailed processing of the printed words (Berent, 2008), the available results nonetheless show that, given sufficient processing time, silent reading is shaped by grammatical constraints and yields phonological forms that are largely isomorphic to the ones extracted from spoken language.

12.5.3 Dyslexia as a phonological disorder

The previous sections have demonstrated that reading ability is intimately linked to phonological competence. Not only does phonology form the basis for the design of writing systems, but it is routinely recruited in its online decoding, in reading single words and texts. The close link between reading ability and core phonology carries some direct implication for reading disability as well. If skilled reading relies on core phonology, then deficits to core phonology are expected to impair the acquisition of reading skill. In what follows, I evaluate this prediction by examining the phonological competence of individuals with developmental dyslexia and evaluate the etiology of this disorder.

12.5.3.1 Phonological deficits in dyslexia

Developmental dyslexia is a deficit in the acquisition of reading skill that is unexplained by intelligence or emotional, motivational, and social factors, and it affects between 5 and 17 percent of the population (Shaywitz, 1998). Dyslexia is a complex disorder with multiple causes, including visual deficits, working memory and attention limitations. Many dyslexic individuals, however, also demonstrate subtle phonological impairments (Dehaene, 2009; Ramus, 2001; Shaywitz, 1998).

One aspect of this impairment is evident in reading itself, especially when it comes to the decoding of novel words (e.g., *blig*). Novel words indeed exert far greater phonological demands than existing words. Existing words (e.g., *block*) have memorized phonological forms that can be retrieved by association with their spellings, akin to the retrieval of a person's name from the sight of his or her face. By contrast, novel words' pronunciations can only be obtained "from scratch" – by a productive process that maps each of their graphemes to phonemes (e.g., b→/b/). And before a child can even learn the mapping, he or she must first become aware that words comprise phonemes (Liberman, 1989). A failure to encode the phonological structure of spoken language, to gain awareness of its constituent phonemes, and to automatically link them to graphemes is bound to impair nonword naming. And indeed, many studies have shown that unskilled readers of alphabetic orthographies manifest difficulties in the decoding of nonwords (e.g., Paulesu et al., 2001; Rack et al., 1992; Ziegler et al., 2009).

But the phonological difficulties of dyslexics are not limited to reading tasks nor are they confined to alphabetic writing systems. Numerous studies have shown that dyslexics experience greater difficulties in the processing of spoken language. For example, dyslexics are impaired in tasks that elicit explicit analysis and awareness of the phonological structure of spoken language – tasks such as rhyming (e.g., does *bat* rhyme with *hat*?) phoneme deletion (say *block* without the first sound), and spoonerisms (e.g., *bed+lot*➔*led+bot*). Such deficits have been repeatedly demonstrated not only in alphabetic writing systems (Bishop & Snowling, 2004; Manis et al., 1997) but also in Chinese (Siok et al., 2008). Although the failure to develop phonemic awareness could also result from illiteracy (Morais et al., 1979), and as such, its impairment in dyslexia could reflect a symptom of the disorder, rather than its cause, reading skill alone is insufficient to explain the phonological delays of dyslexics. Indeed, the phonological deficits of dyslexics are evident even when compared to controls matched for reading skill (Ziegler et al., 2009), and they even extend to tasks that do not require awareness of phonological structure – in repeating orally presented words, rapidly naming objects, and the maintenance of words and digits in memory (Paulesu et al., 2001; Ziegler et al., 2009). These deficits, moreover, can be traced to difficulties in extracting phonological features from speech signals, such as place (e.g., *ba* vs. *da*) and manner of articulation (e.g., *ba* vs. *fa*; e.g., Mody et al., 1997; Serniclaes et al., 2001; Ziegler et al., 2009). Because many dyslexics do not manifest frank linguistic deficits in either comprehension or production, such subtle speech perception deficits are typically detectable only when the perception of speech is made more challenging, by degrading the speech signal, presenting it masked in noise (Ziegler et al., 2009), or eliciting discrimination among sounds presented in rapid succession (Mody et al., 1997). Nonetheless, these abnormalities in the perception of speech sounds have been documented in several studies. These findings indicate that the phonetic representations extracted by dyslexic individuals from speech are fragile. In fact, these deficits in speech perception can be detected already in infancy.

Longitudinal studies have shown that event-related brain potential responses to speech stimuli obtained from an individual at infancy can be linked to that individual's reading scores at 8 years of age (Molfese, 2000). Moreover, infants from families with high incidence of dyslexia differ from controls in their perception of phonetic contrasts, such as the Finnish contrast in consonant duration (e.g., the contrast between *ata* and *atta*; Leppänen et al., 2002) and the Dutch contrast between /bak/ and /dak/ (van Leeuwen et al., 2007). Specifically, while left hemisphere brain responses of control infants differentiated /bAk/ and /dAk/, the brain responses of 2-month-old infants at high risk of dyslexia failed to differentiate among these exemplars, and their brain activity originated predominantly in the right hemisphere.

The fact that familial pedigree presents a risk factor for dyslexia also underscores the strong hereditary basis of this disorder. It has long been noted that dyslexia runs in families. Families, of course, share both genes and environment, so familial patterns do not uniquely demonstrate a genetic link. Twin studies, however, allow one to dissociate genetic from environmental factors by comparing the prevalence of the disorder among monozygotic twins – twins who share all their genes – and dizygotic twins who share on average 50 percent of their genes. To the extent that dyslexia has a genetic basis, one would expect it to be more prevalent among monozygotic twins. Twin studies indeed show that if one sibling suffers from dyslexia, his or her twin is significantly more likely to exhibit this disorder when the twins are monozygotic. The precise heritability of dyslexia (i.e., the amount of variance associated with genetic factors) varies across studies (Bishop & Snowling, 2004), but according to one estimate (Pennington & Bishop, 2009) the overall heritability of dyslexia is 0.58 (Pennington & Bishop, 2009), and the heritability that is specifically associated with phonological awareness reaches 0.53 (Byrne et al., 2002). Subsequent genetic studies have identified several candidate genes, including *DYX1C1*, *DCDC2*, *KIAA0319*, and *ROBO1* (Galaburda et al., 1985; Pennington & Bishop, 2009; Shastry, 2007).

Although there is strong evidence that individuals with dyslexia suffer from a highly heritable deficit to speech processing, these data do not establish its source – whether it originates from lower-level impairment to phonetic processing or whether it extends to the phonological grammar. Although there is a large literature on the perception of phonetic categories, we know very little about the sensitivity of dyslexics to phonological structure, and the existing findings are inconsistent. Some researchers found that dyslexics exhibit phonotactic deficits. For example, dyslexics are less sensitive to phonotactic probability (Bonte et al., 2007), and they experience difficulties in the production of consonant clusters in unstressed syllables (Marshall & Van Der Lely, 2009). Other researchers, however, failed to find any phonotactic deficit (Szenkovits et al., 2011). Like normal (French) controls, dyslexic individuals were sensitive to the contrast between onsets such as *bl* (attested in French) and *dl* (which is unattested), and both groups showed a similar tendency to misidentify unattested onsets as their licit counterparts (e.g., *dl*➔ *bl*). Similarly, dyslexics distinguished between phonological processes that are obligatory in their language and ones that are unattested. They correctly produced voicing assimilation, a process that is obligatory in their language (e.g., *cape gris* ➔ [kabgʁiz]), but did not assimilate place of articulation – a process that is unattested in French (*zone portuaire*➔* [zompoʁtyɛʁ]), and like their typical counterparts, they tended to perceptually compensate for assimilation (i.e., they failed to detect assimilation in legal contexts) only when the assimilatory process was legal in their language – for voicing, but not place assimilation.

The small number of studies makes it difficult to draw any firm conclusions regarding the status of the phonological grammar in dyslexia. One possibility is that at least some dyslexics manifest deficit in the phonological grammar, even if they do not otherwise show any frank speech or language disorder. The failure to detect such impairments consistently might be due to individual differences in the severity of the deficit, task sensitivity, or the specific aspect of phonological competence under investigation. Indeed, existing studies have tested the sensitivity of dyslexics only to structures attested in their language, so in such cases, individuals could compensate for their grammatical deficit by relying on lexical memory. Unattested structures, however, might be less likely to benefit from lexical compensation, so it is conceivable that future investigations of such cases might reveal a grammatical deficit. On an alternative account, the phonological deficit of dyslexics originates from low-level difficulties in the extraction of phonetic structure, rather than the phonological grammar per se. Because imprecise phonetic representations can lead to imprecise phonological forms, such low-level deficits could occasionally interfere with tasks that require phonological judgment (e.g., phonotactic sensitivity) even if the phonological grammar is otherwise intact. It is indeed well known that dyslexic individuals manifest deficits in the extraction of phonetic features, and existing studies have not ruled out the possibility that phonological errors might result from phonetic processing impairments (Bonte et al., 2007; Marshall & Van Der Lely, 2009). The precise locus of the phonological deficits in dyslexia awaits further research.

12.5.3.2 *The etiology of dyslexia*

While existing research makes it clear that many individuals with dyslexia manifest phonological deficits, the nature of these impairments remains unknown. These outstanding questions regarding the phonological deficit in dyslexia also make it difficult to evaluate the etiology of the disorder. As with other language developmental disorders, such as Specific Language Impairment and Speech Sound Disorder – hereditary disorders that exhibit high comorbidity with dyslexia (Pennington & Bishop, 2009) – the class of potential explanations for dyslexia ranges from domain-specific accounts to domain-generalist explanations. Domain-specific accounts attribute these disorders to a specialized language system, either a specialized speech perception mechanism (e.g., Mody et al., 1997) or components of the grammar (van der Lely, 2005; van der Lely et al., 2004). Alternative explanations view those linguistic deficits as secondary to basic impairments in either nonlinguistic systems (e.g., the magnocellular system; Stein & Walsh, 1997) or domain-general mechanisms, such as procedural learning (Ullman & Pierpont, 2005), lexical retrieval (Wolf et al., 1986), and working memory (Gathercole & Baddeley, 1990). Of these various domain-general accounts of dyslexia, low-level auditory processing deficits have received wide attention. In a series of influential studies (Benasich & Tallal, 2002; Tallal, 2004; Tallal & Piercy, 1973; Temple et al., 2000; Temple et al., 2003), Paula Tallal and colleagues proposed that dyslexia and specific language impairments result from low-level deficits in the processing of brief or rapid auditory events. In support of this proposal, Tallal and colleagues demonstrated that the processing deficits of individuals with language-learning impairments extend to nonlinguistic stimuli – to auditory tones presented either briefly or in rapid succession – that these deficits are present in infants at high risk of language disorders (Benasich & Tallal, 2002), and that training on the discrimination of such brief auditory and linguistic events (specifically, on modified speech designed to slow or amplify rapid frequency transitions) might improve reading skill (Temple et al., 2003). Although deficits to the processing of brief/rapid auditory events are not seen in all adult dyslexics (e.g., Mody et al., 1997), they appear to be more prevalent in younger children (Tallal, 2004).

The auditory origins of dyslexia and their transitory developmental nature are both captured by an influential genetic model of this disorder. Examining the brains of deceased individuals with dyslexia, Albert Galaburda has noticed subtle cortical anomalies that originate from a disruption to neural migration during embryonic development (Galaburda et al., 1985). Indeed, several of the candidate susceptibility genes for dyslexia have been linked to neural migration and growth (Galaburda et al., 2006), and a rat model (e.g., Burbridge et al., 2008) shows that disruption to these genes results in cortical and subcortical anomalies that mirror the ones found in dyslexic individuals, and it also yields similar behavioral symptoms. Like dyslexic individuals, affected rats manifest disruption to the discrimination of auditory tones presented in rapid succession (Peiffer et al., 2004), these deficits are larger in juvenile animals (Peiffer et al., 2004), and they are more prevalent in males (Galaburda et al., 2006), a finding that mirrors the larger prevalence of dyslexia in human males (Rutter et al., 2004). These results open up the possibility that the phonological deficits of dyslexics might originate from prenatal insults to brain development that result in impairments in the perception of rapid/brief auditory stimuli. While these auditory deficits might be initially present in all individuals, they might eventually ameliorate in later development (Galaburda et al., 2006).

The neural migration theory is unique in its ability to offer a comprehensive account that ranges the entire gamut from genes to behavior: It links phonological deficits to specific brain abnormalities, identifies genetic mutations that cause those neural changes, and traces the dynamic unfolding of these gene–brain–behavior interactions throughout development. It is still too early to tell whether this model can in fact capture the full range of behavioral and neural abnormalities reported in humans, but the available data are promising. At first blush, this proposal would seem to challenge the core phonology hypothesis. If hereditary phonological deficits could originate from low-level auditory disorders, then perhaps auditory mechanisms might be sufficient to account for phonological competence in healthy individuals. But the challenge is only apparent. To begin with, it is unclear whether most dyslexics exhibit a deficit to the phonological grammar, so their putative auditory processing impairment could be unrelated to phonological competence (Berent et al., 2012c; Ramus & Szenkovits, 2006). Even if the phonological grammar were impaired, this would hardly imply that auditory processing subsumes core phonology. Correlations between phonological and auditory processing could reflect some third common factor that governs the development or use of both systems. And even if these two systems share some of their hardware resources (brain sites and genes), they might still be segregated at the functional level – a conclusion defended in Chapter 10.

In summary, there is mounting evidence that dyslexia is associated with a host of phonological deficits, but many questions remain regarding the scope of these deficits and their origins. Concerning the scope of the disorder, we do not currently know whether most dyslexics manifest deficits in the organization of the phonological grammar, or whether their difficulties in the processing of spoken language are confined to the phonetic and acoustic levels. Similarly, it is unknown whether the phonological processing deficits of dyslexics are secondary to a lower-level auditory deficit, or whether they originate from injuries to the language system occurring at either the phonetic or the phonological level. So while the clear phonological deficit in dyslexia certainly underscores the link between normal reading and the phonological system, it is unclear whether the subtle deficits to spoken language processing that are characteristic of many dyslexics do in fact reside specifically in the grammatical phonological component.

12.6 Conclusion

Humans are equipped with several systems of core knowledge. Core knowledge systems, such as primitive number systems and naïve physics, each manifest a unique design that is universal, adaptive, and present early in birth. Unlike the universal instinctive core knowledge systems present in infancy, other aspects of knowledge are the domain of the select few, they are discovered only in late development or adulthood through intense reasoning or scientific methods, and transmitted by explicit instruction. But the structure of those invented systems can be traced back to their origins in core knowledge systems. The propensity of those core knowledge systems to lay the foundation for cultural inventions is one of their defining features.

Previous chapters have examined the hypothesis that the phonological grammar forms a system of core knowledge. In support of this possibility, we have demonstrated that the system manifests universal substantive constraints that appear to be adaptive and active early in life. Here, we have seen that, like the core systems of number and physics, the putative core system of phonology lays the foundation for the invented systems of readings and writing.

Considering writing, we have seen that all full writing systems encode phonological primitives, and that the link between core phonology and writing recapitulates spontaneously in the writing systems invented by young children. While writing invariably encodes phonological structure, reading universally decodes it. The precise method and grain-size of decoding varies – transparent alphabetic writing systems like Italian and Spanish allow for a rapid extraction of phonology by mapping graphemes to phonemes; at the other extreme, the Chinese syllabic orthography conveys segmental phonology only at the syllabic levels. But despite those differences, in all writing systems, skilled readers extract phonological structure from print, and at least in alphabetic orthographies, they do so by relying on brain sites that mediate the phonological processing of spoken language. But the computation of phonological structure from print goes beyond the bare minimum of segmental phonology – a closer inspection offers numerous demonstrations that the representations computed on printed language exhibit significant overlap with those computed on spoken language, including shared primitives and structural constraints. And indeed, deficits in processing of spoken language are associated with dyslexia. While the precise source of the deficits – whether they result from a deficit to the language/speech system or to auditory processing – and their extent – whether they are confined to feature extraction or extend to the organization of the grammar – remains debated, the centrality of phonology to the design of writing and to its online decoding underscores the intimate link between these cultural technologies and the phonological system at their core.

13 Conclusions, caveats, questions

> Finally, his time has come, the same angel approaches him and says to him: "Do you recognize me?" And the man tells him: "Yes." And says: "Why did you come to me today of all days?" "To remove you from this world since it is your time to abate," says the angel. Immediately he begins crying, and sounds his voice from one end of the world to the other. And no one recognizes or hears his voice, except the rooster alone. And he tells the angel, "You had already removed me from two worlds and brought me into this one." And the angel tells him, "But I had already told you, against your will you were created, and against your will you were born, and against your will you live, and against your will you die, and against your will you are to give an account before God almighty" ...
>
> (Midrash Tanhuma, pekudei: 3, translation mine)

In this Jewish tradition, innate knowledge and destiny go hand in hand. It is the same angel who had endowed the embryo with knowledge of the entire Torah that now appears to him in old age and summons him to the next world. Recognizing the angel, the man wishes to avert his demise, just as he had previously attempted to prevent his birth. But neither knowledge nor fate is in our hands. And just as man's vain resistance to his arrival on this earth only resulted in the loss of precious knowledge, so is it futile when it comes to his departure.

While modern science attributes to humans far greater control over their knowledge, the findings reviewed in this book suggest that some forms of knowledge might be nonetheless determined by our biology. But precisely what aspects of our knowledge are universally human, to what extent knowledge is predetermined, and what mechanisms are responsible for its fixation and plasticity are the subject of much debate. This final chapter summarizes my conclusions so far.

13.1 Phonological instincts: what needs to be explained

Phonological patterns are universal. All languages, both spoken and signed, organize a set of phonological primitives according to algebraic combinatorial constraints. These constraints operate on elements that are digital and discrete, and they support generalizations that apply across the board, to all members of a class. But the commonalities across phonological patterns go beyond their computational properties – the reliance on algebraic combinatorial structure. It is the shared design of phonological systems – their common representational primitives and combinatorial constraints – that presents the strongest case for the specialization of the phonological system.

These putative design universals are not arbitrary. Rather, they are intimately linked to the phonetic channel inasmuch as they typically optimize the perception and production of language. But unlike the phonetic interface, phonological systems optimize those pressures using mechanisms that are algebraic. Accordingly, phonological constraints are autonomous from the phonetic channel, and they will apply even if their consequences are phonetically undesirable in a given context.

Phonological constraints, moreover, are general across languages, and possibly universal. Putative universal constraints on phonological patterns are evident in language typology, they are partly shared across modalities – for spoken and signed language – and they are mirrored in the behavior of individual speakers. In fact, speakers converge on their phonological preferences even when the relevant structures do not exist in their own language and they exhibit their preferences already in early language development. Most strikingly, these recurrent phonological patterns emerge spontaneously, among people who have never been exposed to any phonological system at all.

Phonological patterning further manifests itself in cultural inventions – in reading and writing. Unlike natural language phonology, writing systems are by no means universal or instinctive. Nonetheless, every time such a system is invented, either in the history of humanity or in the development of individual children, it invariably exploits the phonological principles of spoken language. And each time skilled readers decode the printed word, they automatically recover some aspects of its phonological structure. The decoding of spoken and printed language indeed relies on a common brain network. While none of these brain sites is exclusive to language or phonology, the phonological network is abstract, and partly invariant across modalities – it mediates the computation of phonological structure in both signed and spoken languages – and it is demonstrably dissociable from non-phonological systems, including auditory processing, the formation of phonetic categories, and the encoding of musical inputs.

Not only is phonological patterning universal to all human languages but it might also be unique to humans. Naturally, no system of animal communication incorporates the specific phonological primitives and combinatorial principles found in human languages – this is hardly surprising given the intimate link between human phonologies and phonetics, on the other hand, and the anatomical differences separating the vocal tracts of humans from even their closest relatives, on the other. But the argument for specialization does not rest on these obvious differences. Rather, it is the general design of phonological systems that is special.

In its broadest form, the human phonological system reflects the marriage of two ingredients: algebraic mechanisms and functional adaptation. It patterns meaningless units using algebraic means, and the resulting patterns are designed to optimize phonetic pressures on natural communication. Neither of these two ingredients is uniquely human. Many animals possess algebraic mechanisms that they deploy in laboratory settings, and in the case of birdsong, algebraic patterns are apparently used spontaneously. Many forms of animal communication are likewise shaped by functional constraints. But while algebraic mechanisms and adaptation to functional pressures are each widely attested in the animal kingdom, no nonhuman species is known to combine these two ingredients together, giving rise to a system of algebraic constraints that optimize phonetic pressures.

13.2 Some explanations

What mechanisms support the unique and universal capacity of humans for phonological patterning? I believe there are currently too many open questions and unknowns to warrant a definitive answer to this question. But the available evidence nonetheless allows one to rule out two extreme positions.

13.2.1 A strong empiricist position

A strong empiricist view attributes the phonological talents of our species solely to domain-general mechanisms of inductive learning – either associative or algebraic. I believe this extreme position is untenable. It fails to explain the similar designs seen in phonological systems across different languages, the sensitivity of individual speakers to phonological restrictions that are unattested in their language, and the spontaneous emergence of universal design principles in the phonological systems invented by deaf signers.

Proponents of the empiricist position deny the role of universal grammatical constraints in shaping these patterns. Phonological universals are attributed solely to cultural evolution, whose outcomes are molded by the limitations of subsidiary domain-general systems – audition and motor control. But cultural evolution cannot explain why individual speakers exhibit grammatical (rather than merely phonetic) preferences concerning structures that they have never heard before, nor can it account for the spontaneous emergence of phonological patterns. The similarities across languages and their convergence with the preferences of individual speakers also cannot be explained away as artifacts of audition and articulation. Unlike the analog, continuous, and modality-specific nature of acoustic inputs and motor plans, phonological primitives are digital, discrete, and amodal, and the restrictions governing their combinations are algebraic. Experimental research suggests that these putatively universal phonological preferences are dissociable from these phonetic pressures, and brain research shows that the phonological brain network is distinct from phonetic substrates in normal individuals, and it can be selectively impaired, either temporarily, by transcranial stimulation, or more permanently, in aphasia.

At this point, one might conjecture that the shared design of the phonological mind could emerge spontaneously, by the interaction of inductive algebraic computational principles and domain-general functional pressures. The challenge facing this proposal is to explain what allows these distinct mechanisms to interact in such predictable ways: why speakers of different languages converge on similar phonological systems despite wide variations in their linguistic experience, and why some aspects of phonological patterns are shared across modalities. Proponents of domain-general explanations might attribute the convergence to dynamical principles of self-organization. This possibility requires further research, but it is not a priori clear that self-organization will suffice. Indeed, the two ingredients of phonological patterns – the ability for algebraic combinations and functionally constrained communication – do not invariably give rise to a human-like phonological system. Nonhuman animals that possess the separate capacities for algebraic computation and functionally grounded vocal communication do not spontaneously merge them in their natural communication. And even when these two characteristics are each deployed in the species' own vocal communication, in birdsong, there is still no evidence for algebraic combinatorial constraints that are phonetically grounded. While these observations obviously do not deny the contribution of dynamical self-organization, they do question the possibility that self-organization, experience, and domain-general mechanisms are sufficient to explain the human capacity for phonological patterns.

13.2.2   A strong nativist account

Although the properties of phonological systems defy a strong empiricist account, they are not easily amenable to a radical nativist approach either. In that nativist view, phonological primitives and constraints are all specified at birth, irrespective of linguistic experience, generally, and experience with the phonetic characteristics of linguistic inputs, specifically. The strong link between phonological and phonetic facts is solely due to constraints that are shaped only in phylogeny, not ontogeny.

There are several reasons to question this proposal. If grammatical primitives and constraints were all pre-specified irrespective of experience, ready for use at birth, then one would expect all newborns to develop a full-fledged universal phonological grammar, irrespective of whether they have been exposed to any phonological patterns. This, however, is clearly not the case. While deaf individuals who are deprived of phonological input exhibit a remarkable capacity for spontaneous phonological patterning, their initial phonological systems are rudimentary. The home signs of isolated deaf individuals show only kernels of phonological structure (Brentari et al., 2012), and initial generations of the Al-Sayyid Bedouin Sign Language reportedly lacked any phonological system at all (Sandler et al., 2011). This situation is quite similar to the impoverished species-specific song of isolate birds (Fehér et al., 2009). But while, absent critical experience, neither bird nor man exhibits intact sound patterns, across generations, both species spontaneously converge on species-specific patterns that they have never fully heard before. These observations suggest that, contrary to the extreme nativist position, grammatical universals are not fully specified at birth. Rather than providing the newborn with a readymade blueprint of universal grammar, innate knowledge specifies a recipe for the assembly of some universal grammatical properties over time, guided by some minimal experience.

Further evidence for the critical role of experience in triggering innate grammatical universals is presented by cross-linguistic variations in the organization of phonological grammars. The contrast between signed and spoken language is one obvious example. Although signed and spoken languages share primitives (e.g., syllables) and constraints (e.g., sonority restrictions), several discrepancies are noteworthy. While the requirement for an onset, for example, is among the most highly ranked constraints in many spoken languages, no parallel constraint has been identified on signed syllables. Signed and spoken languages further differ on their feature inventories, their internal organization (e.g., the choice of the articulator and place of articulation are largely redundant in spoken language, but not so in signed languages; van der Hulst, 2000), and the degree of simultaneity of signing (Sandler & Lillo-Martin, 2006). Likewise, the organization of spoken language phonology correlates with the phonetic characteristics of the language (Hayes et al., 2004b). Correlations do not necessarily indicate causations, but the numerous links between phonetics and phonology, both within and across modalities, suggest that some phonological constraints might emerge in ontogeny, informed by phonetic experience.

13.3 The core phonology hypothesis: some open questions

While the observations discussed in previous chapters do not lend themselves to either a radical nativist or an extreme empiricist explanation, I believe they are in line with an intermediate position, identified as the core knowledge hypothesis. This view attributes the design of the phonological mind to an innate, specialized knowledge system whose structure is shaped by multiple sources operating in phylogeny and ontogeny. Like the core systems of number, naïve physics, and social cognition, the phonological systems of different languages are a priori biased to converge on a universal set of primitives and principles that are adaptive, active very early in life, and offer scaffolds for later learning and cultural inventions.

The hypothesis of putative innate constraints on the development of phonological systems obviously sets this account apart from the empiricist position. At the same time, the core knowledge hypothesis also differs from the radical nativist explanation. While in the radical proposal, innate knowledge is hardwired and independent of experience, the core knowledge hypothesis, as portrayed here, presents a more nuanced approach that assumes an interaction of genetic factors and experience. Innate mental structures, in this view, are genotypes, not phenotypes. The phonological genotype outlines the functional properties of human phonological systems. Phonological phenotypes, namely, the phonological systems of distinct human languages, are the expressions of that genotype. Like any other phenotype (e.g., the human visual system), however, the expression of the phonological genotype is shaped by multiple sources, including genetic constraints, experience, interactions with nonlinguistic systems (e.g., the phonetic interface, memory, attention), self-organization, and chance. And just as changes in the quality and timing of critical triggering conditions can result in profound changes to the visual system (for review, see Marcus, 2004), so does the expression of the phonological genotype depend on experiential triggers. Accordingly, the primitives and combinatorial principles of phonological systems might vary within limits. While certain grammatical constraints, such as "ONSET," might emerge in many grammars, ONSET is probably not hardwired in the human genome. Rather, the constraints active in any particular grammar (e.g., English) are jointly shaped by both genetic endowment and experiential triggers, including triggers that are phonetic. For this reason, the grammars of signed and spoken phonologies are likely to differ, and some limited variation might likewise emerge among the grammars of spoken languages as well. The sensitivity to triggers, however, does not mean that grammatical primitives and constraints are learned. And indeed, speakers manifest sensitivity to grammatical principles that are unattested in their language, and such principles emerge spontaneously in nascent languages. The core knowledge hypothesis explains the regenesis of phonological universals and their active role in the brains of individual speakers whose language lacks the relevant structures.

While the core phonology proposal seems to presently offer the best explanation for the wide range of evidence considered in this book, the available evidence is insufficient to fully evaluate this hypothesis. These open questions,

outlined next, await further testing and examination. A resolution of these issues is critical for determining the adequacy of the core knowledge proposal.
- *The scope of cross-linguistic preferences.* One crucial question that is still fiercely debated in the literature concerns the scope of the similarities between phonological grammars – whether some primitives and constraints are truly universal across all languages, or whether the undeniable cross-linguistic preferences for certain grammatical structures is annulled in certain grammars. A complete answer to this question will require a concerted effort across multiple disciplines.

At the formal front, it is crucial to further explore putative counterexamples to grammatical universals. The role of syllable primitives and the preference for onsets are a case in point. Syllables have been widely implicated in many languages, but in some languages, syllables appear less central (Hyman, 1985), and some researchers view them as obsolete (Steriade, 1999). Similarly, while many languages require an onset, some Australian languages have been cited as counterexamples (Breen & Pensalfini, 2001). In both cases, however, the debate on the proper linguistic analysis of these phenomena is ongoing (e.g., Berry, 1998; Hyman, 2011; McCarthy & Prince, 1986; Nevins, 2009; Smith, 2005).

Unveiling grammatical phonological universals requires not only formal linguistic analysis but also experimental investigation and computational analyses that gauge the learnability of specific grammars given speakers' linguistic experience. While the experimental cases reviewed in previous chapters certainly show that speakers' phonological preferences are quite broad, these select examples by no means prove that grammatical preferences are all universal or innate.

- *The role of phonetic experience.* Putative counterexamples to language universals are significant for reasons that go beyond the question of universality per se. Indeed, in several such cases, variations in the organization of the phonological grammar have been linked to phonetic factors. But whether phonetic factors do, in fact, shape the organization of the grammar in ontogeny, and what mechanisms mediate such effects, is unknown. We do not know whether the manipulation of phonetic factors could lead language learners to alter their grammatical preferences, nor do we know what is the window of opportunity for such plasticity to take place in human development. A resolution of these questions presents important challenges for future research.
- *Development: early and late.* The developmental trajectory of universal markedness restrictions is another crucial question for which we still lack the necessary evidence. While the review of language acquisition in Chapter 9 suggests that a precursor of the phonological grammar might be in place within the first years of life, the phonological preferences of young

children are gleaned mostly from language production. It is thus unclear whether those preferences reflect algebraic and amodal grammatical constraints or articulatory preferences.

We also know very little about the factors governing the development of reading and writing – the late emerging phonologically based "technologies" – from the putative system of core phonology. While reading research has made it clear that the development of reading ability requires awareness of the phonological structure of spoken language (Liberman, 1973), we do not know what allows children to gain this awareness, and why some children gain phonemic awareness spontaneously whereas others struggle to do so despite intense instruction. Another important question is whether some of the mechanisms that support the transition from early core knowledge to later developing systems are partly shared across domains. For example, is the transition from core phonology to reading linked to the transition from core number systems to the later-emerging knowledge of recursive number? The development of the phonological mind and its link to other putative core systems present many outstanding challenges.

- *Mind, brain, genes and evolution.* While the study of grammatical phonological universals has been the topic of much research in linguistics, and recently, also in experimental phonology, there has been very little research on the representation of grammatical markedness restrictions in the brain. We know practically nothing on how the activity in phonological brain networks results in the grammatical computations evident functionally, what genes are expressed in these areas, and how networks are assembled in ontogeny or phylogeny.

These neural and genetic mechanisms could also shed light on the nature of grammatical universals themselves. It could reveal, for example, whether markedness hierarchies (e.g., sonority, place of articulation) are represented as continuous scales, or whether they are in fact the aggregate of distinct systems. Unveiling the genes that are expressed in these regions is critical to explain not only typical phonological systems but also a host of phonology-related disorders, ranging from Specific Language Impairment and dyslexia to autism. Finally, while most comparative animal research has focused on syntax, there is a great lacuna in the area of phonology. A comparative research program is acutely needed to unveil the precursors of the human phonological mind and explain why our phonological talents seem to lack any parallels in the great apes while being partly shared with birds.

As the answers to these questions begin to arrive, they might require numerous revisions to the account of phonology outlined in this book. At this point, all we know for sure is that the right theory will have to explain a host of facts, including how members of our species spontaneously converge on similar phonological preferences despite having minimal evidence or none at all, what allows them to broadly generalize these preferences to novel instances, and why this capacity for algebraic patterning is rarely found in animal communication. In learning more about this instinctive capacity of our species, we might also recognize the origins of human knowledge and its scope.

References

Abler, W. L. (1989). On the particulate principle of self-diversifying systems. *Journal of Social and Biological Systems*, 12, 1–13.
Abramson, M. & Goldinger, S. D. (1997). What the reader's eye tells the mind's ear: silent reading activates inner speech. *Perception and Psychophysics*, 59, 1059–1068.
Adam, G. & Bat-El, O. (2009). When do universal preferences emerge in language development? The acquisition of Hebrew stress. *Brill's Annual of Afroasiatic Language and Linguistics*, 1, 255–282.
Adriaans, F. & Kager, R. (2010). Adding generalization to statistical learning: the induction of phonotactics from continuous speech. *Journal of Memory and Language*, 62, 311–331.
Albright, A. (2007). Natural classes are not enough: biased generalization in novel onset clusters. Manuscript, Massachusetts Institute of Technology.
  (2009). Feature-based generalisation as a source of gradient acceptability. *Phonology*, 26, 9–41.
Alcock, K. J., Passingham, R. E., Watkins, K. E. & Vargha-Khadem, F. (2000). Oral dyspraxia in inherited speech and language impairment and acquired dysphasia. *Brain and Language*, 75, 17–33.
Allen, G. D. (1983). Some suprasegmental contours in French two-year-old children's speech. *Phonetica: International Journal of Speech Science*, 40, 269–292.
Allen, J. S. & Miller, J. L. (2004). Listener sensitivity to individual talker differences in voice-onset-time. *The Journal of the Acoustical Society of America*, 115, 3171–3183.
Alonzo, A. & Taft, M. (2002). Sonority constraints on onset-rime cohesion: evidence from native and bilingual Filipino readers of English. *Brain and Language*, 81, 368–383.
Anderson, M. (2010). Neural re-use as a fundamental organizational principle of the brain. *Behavioral and Brain Sciences*, 33, 245–266.
Andrews, S. (1982). Phonological recoding: is the regularity effect consistent? *Memory and Cognition*, 10, 565–575.
Anttila, A. (1997). Deriving variation from grammar. In F. Hinskens, R. van Hout & L. Wetzels (eds.), *Variation, Change and Phonological Theory* (pp. 35–68). Amsterdam: John Benjamins.
Arnold, K. & Zuberbuhler, K. (2006). Language evolution: semantic combinations in primate calls. *Nature*, 441, 303.
Ashby, J. (2010). Phonology is fundamental in skilled reading: evidence from ERPs. *Psychonomic Bulletin & Review*, 17, 95–100.

References

Ashby, J. & Martin, A. E. (2008). Prosodic phonological representations early in visual word recognition. *Journal of Experimental Psychology: Human Perception and Performance*, 34, 224–236.

Ashby, J. & Rayner, K. (2004). Representing syllable information during silent reading: evidence from eye movements. *Language and Cognitive Processes*, 19, 391–426.

Ashby, J., Sanders, L. D. & Kingston, J. (2009). Skilled readers begin processing subphonemic features by 80 ms during visual word recognition: evidence from ERPs. *Biological Psychology*, 80, 84–94.

Baddeley, A. D. (1986). *Working Memory*. Oxford University Press.

Balaban, E. (1988a). Bird song syntax: learned intraspecific variation is meaningful. *Proceedings of the National Academy of Sciences of the United States of America*, 85, 3657–3660.

(1988b). Cultural and genetic variation in Swamp Sparrows (Melospiza georgiana): II. Behavioral salience of geographic song variants. *Behaviour*, 105, 292–322.

(2006). Cognitive developmental biology: history, process and fortune's wheel. *Cognition*, 101, 298–332.

Barlow, J. A. (2001). The structure of /s/-sequences: evidence from a disordered system. *Jounal of Child Language*, 28, 291–324.

(2005). Sonority effects in the production of consonant clusters by Spanish-speaking children. In D. Eddington (ed.), *Selected Proceedings of the Sixth Conference on the Acquisition of Spanish and Portuguese as First and Second Languages* (pp. 1–14). Somerville, MA: Cascadilla Proceedings Project.

Barner, D., Wood, J., Hauser, M. & Carey, S. (2008). Evidence for a non-linguistic distinction between singular and plural sets in rhesus monkeys. *Cognition*, 107, 603–622.

Baron, J. & Strawson, C. (1976). Use of orthographic and word-specific knowledge in reading words aloud. *Journal of Experimental Psychology: Human Perception and Performance*, 2, 386–393.

Barrett, H. C. & Kurzban, R. (2006). Modularity in cognition: framing the debate. *Psychological Review*, 113, 628–647.

Bat-El, O. (1994). Stem modification and cluster transfer in modern Hebrew. *Natural Language and Linguistic Theory*, 12, 571–596.

(2003). The fate of the consonantal root and the binyan in Optimality Theory. *Recherches Linguistiques de Vincennes*, 32, 31–60.

(2004). Parsing forms with identical consonants: Hebrew reduplication. In D. Ravid & H. B. Z. Shyldkrot (eds.), *Perspectives on Language and Language Development* (pp. 25–34). Dordrecht: Kluwer.

(2006). Consonant identity and consonant copy: the segmental and prosodic structure of Hebrew reduplication. *Linguistic Inquiry*, 37, 179–210.

Becker, M., Ketrez, N. & Nevins, A. (2011). The surfeit of the stimulus: analytic biases filter lexical statistics in Turkish laryngeal alternations. *Language*, 87, 84–125.

Becker, M., Nevins, A. & Levine, J. (2011). Asymmetries in generalizing alternations to and from initial syllables. Manuscript, University of Massachusetts, Amherst, UCL, Harvard.

Benasich, A. A. & Tallal, P. (2002). Infant discrimination of rapid auditory cues predicts later language impairment. *Behavioural Brain Research*, 136, 31–49.

Benasich, A. A., Choudhury, N., Friedman, J. T., Realpe-Bonilla, T., Chojnowska, C. & Gou, Z. (2006). The infant as a prelinguistic model for language learning impairments: predicting from event-related potentials to behavior. *Neuropsychologia*, 44, 396–411.

Berent, I. (1997). Phonological priming in the lexical decision task: regularity effects are not necessary evidence for assembly. *Journal of Experimental Psychology: Human Perception and Performance*, 23, 1727–1742.

(2008). Are phonological representations of printed and spoken language isomorphic? Evidence from the restrictions on unattested onsets. *Journal of Experimental Psychology: Human Perception and Performance*, 34, 1288–1304.

Berent, I. & Frost, R. (1997). The inhibition of polygraphic consonants in spelling Hebrew: evidence for a recurrent assembly of spelling and phonology in visual word recognition. In C. Perfetti, M. Fayol & L. Rieben (eds.), *Learning to Spell: Research, Theory, and Practice across Languages* (pp. 195–219). Hillsdale, NJ: Lawrence Erlbaum.

Berent, I. & Lennertz, T. (2010). Universal constraints on the sound structure of language: phonological or acoustic? *Journal of Experimental Psychology: Human Perception and Performance*, 36, 212–223.

Berent, I., Lennertz, T. & Rosselli, M. (2012). Universal phonological restrictions and language-specific repairs: evidence from Spanish. *The Mental Lexicon*, 13.2.

Berent, I. & Marom, M. (2005). The skeletal structure of printed words: evidence from the Stroop task. *Journal of Experimental Psychology: Human Perception and Performance*, 31, 328–338.

Berent, I. & Perfetti, C. A. (1995). A rose is a REEZ: the two cycles model of phonology assembly in reading English. *Psychological Review*, 102, 146–184.

Berent, I. & Shimron, J. (1997). The representation of Hebrew words: Evidence from the Obligatory Contour Principle. *Cognition*, 64, 39–72.

(2003). Co-occurrence restrictions on identical consonants in the Hebrew lexicon: are they due to similarity? *Journal of Linguistics*, 39, 31–55.

Berent, I., Everett, D. L. & Shimron, J. (2001a). Do phonological representations specify variables? Evidence from the obligatory contour principle. *Cognitive Psychology*, 42, 1–60.

Berent, I., Shimron, J. & Vaknin, V. (2001b). Phonological constraints on reading: evidence from the Obligatory Contour Principle. *Journal of Memory and Language*, 44, 644–665.

Berent, I., Marcus, G. F., Shimron, J. & Gafos, A. I. (2002). The scope of linguistic generalizations: evidence from Hebrew word formation. *Cognition*, 83, 113–139.

Berent, I., Vaknin, V. & Shimron, J. (2004). Does a theory of language need a grammar? Evidence from Hebrew root structure. *Brain and Language*, 90, 170–182.

Berent, I., Tzelgov, J. & Bibi, U. (2006). The autonomous computation of morphophonological structure in reading: evidence from the Stroop task. *The Mental Lexicon*, 1–2, 201–230.

Berent, I., Steriade, D., Lennertz, T. & Vaknin, V. (2007a). What we know about what we have never heard: evidence from perceptual illusions. *Cognition*, 104, 591–630.

Berent, I., Vaknin, V. & Marcus, G. (2007b). Roots, stems, and the universality of lexical representations: evidence from Hebrew. *Cognition*, 104, 254–286.

Berent, I., Lennertz, T., Jun, J., Moreno, M. A. & Smolensky, P. (2008). Language universals in human brains. *Proceedings of the National Academy of Sciences*, 105, 5321–5325.

Berent, I., Lennertz, T., Smolensky, P. & Vaknin-Nusbaum, V. (2009). Listeners' knowledge of phonological universals: evidence from nasal clusters. *Phonology* 26, 75–108.

Berent, I., Balaban, E., Lennertz, T. & Vaknin-Nusbaum, V. (2010). Phonological universals constrain the processing of nonspeech. *Journal of Experimental Psychology: General*, 139, 418–435.

Berent, I., Harder, K. & Lennertz, T. (2011a). Phonological universals in early childhood: evidence from sonority restrictions. *Language Acquisition*, 18, 281–293.

Berent, I., Lennertz, T. & Smolensky, P. (2011b). Markedness and misperception: it's a two-way street. In C. E. Cairns & E. Raimy (eds.), *Handbook of the Syllable* (pp. 373–394), Leiden: E. J. Brill.

Berent, I., Lennertz, T. & Balaban, E. (2012a). Language universals and misidentification: a two way street. *Language and Speech*, 1–20.

Berent, I., Wilson, C., Marcus, G. & Bemis, D. (2012b). On the role of variables in phonology: remarks on Hayes and Wilson. *Linguistic Inquiry*, 43, 97–119.

Berent, I., Dupuis, A., & Brentari, D. (forthcoming a). Amodal aspects of linguistic design. Manuscript submitted for publication.

Berent, I., Lennertz, T. & Rosselli, M. (forthcoming b). Universal phonological restrictions and language-specific repairs: evidence from Spanish. Manuscript submitted for publication.

Berent, I., Vaknin-Nusbaum, V., Balaban, E., & Galaburda, A. M. (2012). Dyslexia impairs speech recognition but can spare phonological competence. *PLOS One*, 7(9), e44875.doi:10.1371/journal.pone0044875.

Berkley, D. M. (2000). *Gradient obligatory contour principle effects*. Ph.D. thesis, Northwestern University, Dept. of Linguistics.

Bernhardt, B. H. & Stemberger, J. (1998). *Handbook of Phonological Development*. San Diego: Academic Press.

(2007). Phonological impairment in children and adults. In P. de Lacy (ed.), *The Cambridge Handbook of Phonology* (pp. 575–593). Cambridge University Press.

Berry, L. (1998). *Alignment and Adjacency in Optimality Theory: Evidence from Walpiri and Arrernte*. University of Sydney.

Bertoncini, J. & Mehler, J. (1981). Syllables as units in infant speech perception. *Infant Behavior and Development*, 4, 247–260.

Berwick, R. C., Okanoya, K., Beckers, G. J. L. & Bolhuis, J. J. (2011). Songs to syntax: the linguistics of birdsong. *Trends in Cognitive Sciences*, 15, 113–121.

Bidelman, G. M., Gandour, J. T. & Krishnan, A. (2009). Cross-domain effects of music and language experience on the representation of pitch in the human auditory brainstem. *Journal of Cognitive Neuroscience*, 23, 425–434.

Bishop, D. V. (2006). What causes specific language impairment in children? *Current Directions in Psychological Sciences*, 15, 217–221.

(2007). Using mismatch negativity to study central auditory processing in developmental language and literacy impairments: where are we, and where should we be going? *Psychological Bulletin*, 133, 651–672.

(2009). Genes, cognition, and communication: insights from neurodevelopmental disorders. *Annals of the New York Academy of Sciences*, 1156, 1–18.

Bishop, D. V. & Snowling, M. J. (2004). Developmental dyslexia and specific language impairment: same or different? *Psychological Bulletin*, 130, 858–886.

Blevins, J. (2004). *Evolutionary Phonology*. Cambridge University Press.

(2006). A theoretical synopsis of evolutionary phonology *Theoretical Linguistics*, 32, 117–165.

(2008). Consonant epenthesis: natural and unnatural histories. In J. Good (ed.), *Linguistic Universals and Language Change* (pp. 79–107). Oxford; New York: Oxford University Press.

Bloom, P. (2004). *Descartes' Baby: How the Science of Child Development Explains What Makes Us Human.* New York: Basic Books.

(2010). How do morals change? *Nature*, 464, 490.

Blumstein, S. E. (1973). *A Phonological Investigation of Aphasic Speech.* The Hague: Mouton.

(1995). The neurobiology of the sound structure of language. In M. Gazzaniga (ed.), *The Cognitive Neurosciences.* (pp. 915–929). Cambridge, MA: MIT Press.

Blust, R. (2004). *t to k: an Austronesian sound change revisited. *Oceanic Linguistics*, 43, 365.

Boatman, D. (2004). Cortical bases of speech perception: evidence from functional lesion studies. *Cognition*, 92, 47–65.

Boersma, P. & Hamann, S. (2008). The evolution of auditory dispersion in bidirectional constraint grammar. *Phonology*, 25, 217–270.

Bolger, D. J., Perfetti, C. A. & Schneider, W. (2005). Cross-cultural effect on the brain revisited: universal structures plus writing system variation. *Human Brain Mapping*, 25, 92–104.

Bonatti, L. L., Peña, M., Nespor, M. & Mehler, J. (2005). Linguistic constraints on statistical computations: the role of consonants and vowels in continuous speech processing. *Psychological Science*, 16, 451–459.

Bongard, S. & Nieder, A. (2010). Basic mathematical rules are encoded by primate prefrontal cortex neurons. *Proceedings of the National Academy of Sciences of the United States of America*, 107, 2277–2282.

Bonte, M. L., Poelmans, H. & Blomert, L. (2007). Deviant neurophysiological responses to phonological regularities in speech in dyslexic children. *Neuropsychologia*, 45, 1427–1437.

Bortolini, U. & Leonard, L. B. (2000). Phonology and children with specific language impairment: status of structural constraints in two languages. *Journal of Communication Disorders*, 33, 131–149.

Bowling, D. L., Gill, K., Choi, J. D., Prinz, J. & Purves, D. (2010). Major and minor music compared to excited and subdued speech. *Journal of the Acoustical Society of America*, 127, 491–503.

Breen, G. & Pensalfini, R. (2001). A language with no syllable onsets. *Linguistic Inquiry*, 30, 1–25.

Brentari, D. (1998). *A Prosodic Model of Sign Language Phonology.* Cambridge, MA: MIT Press.

Brentari, D., Coppola, M., Mazzoni, L. & Goldin-Meadow, S. (2012). When does a system become phonological? Handshape production in gestures, signers and homesigners. *Natural Language & Linguistic Theory*, 30, 1–31.

Bromberger, S. & Halle, M. (1989). Why phonology is different. *Linguistic Inquiry*, 20, 51–70.

Broselow, E. & Finer, D. (1991). Parameter setting in second language phonology and syntax. *Second Language Research*, 7, 35–59.

Broselow, E., Chen, S.-I. & Wang, C. (1998). The emergence of the unmarked in second language phonology. *Studies in Second Language Acquisition*, 20, 261–280.

Broselow, E. & Xu, Z. (2004). Differential difficulty in the acquisition of second language phonology. *International Journal of English Studies*, 4, 135–163.
Buchanan, L. & Besner, D. (1993). Reading aloud: evidence for the use of a whole word nonsemantic pathway. *Canadian Journal of Experimental Psychology*, 47, 133–152.
Buchsbaum, B. R., Olsen, R. K., Koch, P. F., Kohn, P., Kippenhan, J. S. & Berman, K. F. (2005). Reading, hearing, and the planum temporale. *Neuroimage*, 24, 444–454.
Buchwald, A. B. (2009). Minimizing and optimizing structure in phonology: evidence from aphasia. *Lingua*, 119, 1380–1395.
Buchwald, A. B., Rapp, B. & Stone, M. (2007). Insertion of discrete phonological units: an articulatory and acoustic investigation of aphasic speech. *Language and Cognitive Processes*, 22, 910–948.
Burbridge, T. J., Wang, Y., Volz, A. J., Peschansky, V. J., Lisann, L., Galaburda, A. M., et al. (2008). Postnatal analysis of the effect of embryonic knockdown and overexpression of candidate dyslexia susceptibility gene homolog Dcdc2 in the rat. *Neuroscience*, 152, 723–733.
Burns, E. M. & Ward, W. D. (1978). Categorical perception – phenomenon or epiphenomenon: evidence from experiments in the perception of melodic musical intervals. *Journal of the Acoustical Society of America*, 63, 456–468.
Bybee, J. & McClelland, J. L. (2005). Alternatives to the combinatorial paradigm of linguistic theory based on domain general principles of human cognition. *Linguistic Review*, 22, 381–410.
Bybee, J. L. (2008). Linguistic universals and language change. In J. Good (ed.), *Linguistic Universals and Language Change* (pp. 108–121). Oxford; New York: Oxford University Press.
Byrne, B., Delaland, C., Fielding-Barnsley, R., Quain, P., Samuelsson, S., Hoien, T., Corle, R., DeFries, J. C., Wadsworth, S., Willcutt, E. & Olson, R. K. (2002). Longitudinal twin study of early reading development in three countries: Preliminary results. *Annals of Dyslexia*, 52, 49–73.
Cantlon, J. F. & Brannon, E. M. (2006). Shared system for ordering small and large numbers in monkeys and humans. *Psychological Science*, 17, 401–406.
Caramazza, A., Chialant, D., Capasso, R. & Miceli, G. (2000). Separable processing of consonants and vowels. *Nature*, 403, 428–430.
Carey, S. (2009). *The Origin of Concepts*. Oxford; New York: Oxford University Press.
Carey, S. & Spelke, E. (1996). Science and core knowledge. *Philosophy of Science*, 63, 515–533.
Carreiras, M., Alvarez, C. J. & de Vega, M. (1993). Syllable frequency and visual word recognition in Spanish. *Journal of Memory and Language*, 32, 766–780.
Carreiras, M., Ferrand, L., Grainger, J. & Perea, M. (2005). Sequential effects of phonological priming in visual word recognition. *Psychological Science: A Journal of the American Psychological Society / APS* 16(8), 585–589.
Carreiras, M., Perea, M., Vergara, M. & Pollatsek, A. (2009). The time course of orthography and phonology: ERP correlates of masked priming effects in Spanish. *Psychophysiology*, 46, 1113–1122.
Chater, N., Reali, F. & Christiansen, M. H. (2009). Restrictions on biological adaptation in language evolution. *Proceedings of the National Academy of Sciences of the United States of America*, 106, 1015–1020.

Chen, H.-C., Yamauchi, T., Tamaoka, K. & Vaid, J. (2007). Homophonic and semantic priming of Japanese Kanji words: a time course study. *Psychonomic Bulletin & Review*, 14, 64–69.

Chen, S., Swartz, K. B. & Terrace, H. S. (1997). Knowledge of the ordinal position of list items in rhesus monkeys. *Psychological Science*, 8, 80–86.

Chetail, F. & Mathey, S. (2009). Syllabic priming in lexical decision and naming tasks: the syllable congruency effect re-examined in French. *Canadian Journal of Experimental Psychology* 63(1), 40–48.

Chinese Character Dictionary. Retrieved May, 21, 2010, from /www.mandarintools.com/chardict.html.

Cholin, J. (2011). Do syllables exist? Psycholinguistic evidence of the retrieval of syllabic units in speech production. In C. E. Cairns & E. Raimy (eds.), *Handbook of the Syllable* (pp. 225–248). Leiden: E. J. Brill.

Chomsky, N. (1957). *Syntactic Structures*. Gravenhage: Mouton.

(1965). *Aspects of the Theory of Syntax*. Cambridge: MIT Press.

(1972). *Language and Mind* (Enl. edn.). New York: Harcourt Brace Jovanovich.

(1976). Approaching reading through invented spelling, Paper presented at the Conference on *Theory and Practice of Beginning Reading*. University of Pittsburgh, Learning Research and Development Center.

(1980). *Rules and Representations*. New York: Columbia University Press.

(2002). *On the Nature of Language*. Cambridge University Press.

(2005). Three factors in language design. *Linguistic Inquiry*, 36, 1–22.

Chomsky, N. & Halle, M. (1968). *The Sound Pattern of English*. New York: Harper & Row.

Choudhury, N., Leppanen, P. H. T., Leevers, H. J. & Benasich, A. A. (2007). Infant information processing and family history of specific language impairment: converging evidence for RAP deficits from two paradigms. *Developmental Science*, 10, 213–236.

Christiansen, M. H., Reali, F. & Chater, N. (2011). Biological adaptations for functional features of language in the face of cultural evolution. *Human Biology*, 83, 247–259.

Clements, G. N. (1990). The role of the sonority cycle in core syllabification. In J. Kingston & M. Beckman (eds.), *Papers in Laboratory Phonology* (Vol. I: *Between the Grammar and Physics of Speech*, pp. 282–333). Cambridge University Press.

(2005). The role of features in phonological inventories. In E. Raimy & C. Cairns (eds.), *Contemporary Views on Architecture and Representations in Phonological Theory* (pp. 19–68). Cambridge, MA: MIT Press.

Clements, G. N. & Keyser, S. J. (1983). *CV Phonology*. Cambridge, MA: MIT Press.

Coetzee, A. (2008). Grammaticality and ungrammaticality in phonology. *Language*, 84, 218–257.

(2011). Syllables in speech perception: evidence from perceptual epenthesis. In C. Cairns & E. Raimy (eds.), *Handbook of the Syllable* (pp. 295–325). Leiden: E. J. Brill.

Coetzee, A., W. & Pater, J. (2008). Weighted constraints and gradient restrictions on place co-occurrence in Muna and Arabic. *Natural Language and Linguistic Theory*, 26, 289–337.

Coetzee, A. W. and Pretorius, R. (2010). Phonetically grounded phonology and sound change: the case of Tswana labial plosives. *Journal of Phonetics*, 38(3), 404–421.

Cole, J. (2009). Emergent feature structures: harmony systems in exemplar models of phonology. *Language Sciences*, 31, 144–160.
Colé, P., Magnan, A. & Grainger, J. (1999). Syllable-sized units in visual word recognition: evidence from skilled and beginning readers of French. *Applied Psycholinguistics*, 20(4), 507–532.
Coleman, J. & Pierrehumbert, J. (1997). Stochastic phonological grammars and acceptability. In J. Coleman (ed.), *Third Meeting of the ACL Special Interest Group in Computational Phonology: Proceedings of the Workshop* (pp. 49–56). East Stroudsburg, PA: Association for Computational Linguistics.
Conrad, M., Carreiras, M., Tamm, S. & Jacobs, A. M. (2009). Syllables and bigrams: orthographic redundancy and syllabic units affect visual word recognition at different processing levels. *Journal of Experimental Psychology: Human Perception and Performance*, 35, 461–479.
Corballis, M. C. (2009). Do rats learn rules? *Animal Behaviour*, 78, e1–e2.
Corina, D. P., McBurney, S. L., Dodrill, C., Hinshaw, K., Brinkley, J. & Ojemann, G. (1999). Functional roles of Broca's area and SMG: evidence from cortical stimulation mapping in a deaf signer. *Neuroimage*, 10, 570–581.
Corina, D. P. & Knapp, H. (2006). Sign language processing and the mirror neuron system. *Cortex*, 42, 529–539.
Cosmides, L. & Tooby, J. (1994). Origins of domain specificity: the evolution of functional organization. In L. A. Hirschfeld & S. A. Gelman (eds.), *Mapping the Mind: Domain Specificity in Cognition and Culture* (pp. 85–116). New York: Cambridge University Press.
Costa, A. & Sebastian-Gallés, N. (1998). Abstract structure in language production: evidence from Spanish. *Journal of Experimental Psychology: Learning, Memory, and Cognition*, 24, 886–903.
Cotelli, M., Abutalebi, J., Zorzi, M. & Cappa, S. F. (2003). Vowels in the buffer: a case study of acquired dysgraphia with selective vowel substitutions. *Cognitive Neuropsychology*, 20, 99–114.
Crain, S. & Nakayama, M. (1987). Structure dependence in grammar formation. *Language*, 63, 522–543.
Crain, S., Gualmini, A., & Pietroski, P. (2005). Brass tacks in linguistic theory: innate grammatical principles. In P. Carruthers, S. Laurence & S. Stich (eds.), *The Innate Mind: Structure and Contents* (pp. 175–197). New York: Oxford University Press.
Crockford, C. & Boesch, C. (2005). Call combinations in wild chimpanzees. *Behaviour*, 142, 397–421.
Crockford, C., Herbinger, I., Vigilant, L. & Boesch, C. (2004). Wild chimpanzees produce group-specific calls: a case for vocal learning? *Ethology*, 110, 221–243.
Cubelli, R. (1991). A selective deficit for writing vowels in acquired dysgraphia. *Nature*, 353, 258–260.
Cynx, J. (1990). Experimental determination of a unit of song production in the Zebra Finch (Taeniopygia guttata). *Journal of Comparative Psychology*, 104, 3–10.
Daland, R., Hayes, B., Garellek, M., White, J., Davis, A. & Norrmann, I. (2011). Explaining sonority projection effects. *Phonology*, 28, 197–234.
Davidson, L. (2000). Experimentally uncovering hidden strata in English phonology. In L. Gleitman and A. Joshi (eds.), *Proceedings of the 22nd Annual Conference of the Cognitive Science Society* (p. 1023). Mahwah, NJ: Lawrence Erlbaum.

(2006a). Schwa elision in fast speech: segmental deletion or gestural overlap? *Phonetica*, 63, 79–112.

(2006b). Phonotactics and articulatory coordination interact in phonology: evidence from nonnative production. *Cognitive Science*, 30, 837–862.

Davidson, L., Jusczyk, P., Smolensky, P., Kager, R., Pater, J. & Zonneveld, W. (2004). The initial and final states: theoretical implications and experimental explorations of richness of the base. In Paul Smolensky & Geraldine Legendre (eds.), *Constraints in Phonological Acquisition* (pp. 321–368). New York: Cambridge University Press.

Davidson, L., Jusczyk, P. & Smolensky, P. (2006). Optimality in language acquisition I: the initial and final state of the phonological grammar. In P. Smolensky & G. Legendre (eds.), *The Harmonic Mind: From Neural Computation to Optimality-Theoretic Grammar* (pp. 231–278). Cambridge, MA: MIT Press.

Dawkins, R. (1987). *The Blind Watchmaker: Why the Evidence of Evolution Reveals a Universe without Design.* New York: W. W. Norton & Co.

de Lacy, P. (2004). Markedness conflation in Optimality Theory. *Phonology*, 21, 145–188.

(2006). *Markedness: Reduction and Preservation in Phonology.* Cambridge; New York: Cambridge University Press.

(2007). The interaction of tone, sonority, and prosodic structure. In P. de Lacy (ed.), *The Cambridge Handbook of Phonology* (pp. 281–307). Cambridge University Press.

de Lacy, P. & Kingston, J. (2006). Synchronic explanation. Unpublished manuscript, Rutgers University and the University of Massachusetts Amherst.

Dediu, D. (2011). A Bayesian phylogenetic approach to estimating the stability of linguistic features and the genetic biasing of tone. *Proceedings. Biological Sciences / The Royal Society*, 278, 474–479.

Dediu, D. & Ladd, D. R. (2007). Linguistic tone is related to the population frequency of the adaptive haplogroups of two brain size genes, ASPM and Microcephalin. *Proceedings of the National Academy of Sciences of the United States of America*, 104, 10944–10949.

DeFrancis, J. (1989). *Visible Speech: The Diverse Oneness of Writing Systems.* Honolulu: University of Hawaii Press.

Dehaene, S. (2009). *Reading and the Brain: The Science and Evolution of a Human Invention.* New York: Viking.

Dehaene, S. & Cohen, L. (2011). The unique role of the visual word form area in reading. *Trends in Cognitive Sciences*, 15, 254–262.

Dehaene, S., Pegado, F., Braga, L. W., Ventura, P., Nunes Filho, G., Jobert, A., Dehaene-Lambertz, G., Kolinsky, R., Morais, J. & Cohen, L. (2010). How learning to read changes the cortical networks for vision and language. *Science (New York, N.Y.)*, 330(6009), 1359–1364.

Dehaene-Lambertz, G., Dupoux, E. & Gout, A. (2000). Electrophysiological correlates of phonological processing: a cross-linguistic study. *Journal of Cognitive Neuroscience*, 12, 635–647.

Dell, F. & Elmedlaoui, M. (1985). Syllabic consonants and syllabification in Imdlawn Tashlhiyt Berber. *Journal of African Languages and Linguistics*, 7, 105–130.

Demuth, K. (1995). Markedness and the development of prosodic structure. In J. Beckman (ed.), *Proceedings of the North Eastern Linguistic Society 25* (pp. 13–25). Amherst, MA: GLSA, University of Massachusetts.

(2011). The acquisition of phonology. In J. Goldsmith, J. Riggle & A. Yu (eds.), *The Handbook of Phonological Theory* (2nd edn., pp. 571–595). Malden, MA: Blackwell.
Demuth, K. & McCullough, E. (2009). The longitudinal development of clusters in French. *Journal of Child Language*, 36, 425–448.
Desai, R., Liebenthal, E., Waldron, E. & Binder, J. R. (2008). Left posterior temporal regions are sensitive to auditory categorization. *Journal of Cognitive Neuroscience*, 20, 1174–1188.
Deutsch, D., Henthorn, T., Marvin, E. & Xu, H. (2006). Absolute pitch among American and Chinese conservatory students: prevalence differences, and evidence for a speech-related critical period. *Journal of the Acoustical Society of America*, 119, 719–722.
Doignon, N. & Zagar, D. (2005). Illusory conjunctions in French: the nature of sublexical units in visual word recognition. *Language and Cognitive Processes*, 20, 443–464.
Domahs, U., Kehrein, W., Knaus, J., Wiese, R. & Schlesewsky, M. (2009). Event-related potentials reflecting the processing of phonological constraint violations. *Language and Speech*, 52, 415–435.
Dooling, R. J., Best, C. T. & Brown, S. D. (1995). Discrimination of synthetic full-formant and sinewave/ra-la/continua by budgerigars (Melopsittacus undulatus) and Zebra Finches (Taeniopygia guttata). *Journal of the Acoustical Society of America*, 97, 1839–1846.
Dupoux, E., Kakehi, K., Hirose, Y., Pallier, C. & Mehler, J. (1999). Epenthetic vowels in Japanese: a perceptual illusion? *Journal of Experimental Psychology: Human Perception and Performance*, 25, 1568–1578.
Dupoux, E., Parlato, E., Frota, S., Hirose, Y. & Peperkamp, S. (2011). Where do illusory vowels come from? *Journal of Memory and Language*, 64, 199–210.
Dyer, F. C. & Seeley, T. D. (1991). Dance dialects and foraging range in three Asian honey bee species. *Behavioral Ecology and Sociobiology*, 28, 227–233.
Dyer, F. C. & Dickinson, J. A. (1994). Development of sun compensation by honeybees: how partially experienced bees estimate the sun's course. *Proceedings of the National Academy of Sciences of the United States of America*, 91, 4471–4474.
Eimas, P. & Seidenberg, M. (1997). Do infants learn grammar with algebra or statistics? *Science*, 284, 433.
Eimas, P. D., Siqueland, E. R., Jusczyk, P. & Vigorito, J. (1971). Speech perception in infants. *Science*, 171, 303–306.
Elman, J. (1993). Learning and development in neural networks: the importance of starting small. *Cognition*, 48, 71–99.
Emlen, S. T. (1975). The stellar-orientation system of a migratory bird. *Scientific American*, 233, 102–111.
  (1976). Magnetic direction finding: evidence for its use in migratory indigo buntings. *Science*, 193, 505–508.
Emmorey, K. (2002). *Language, Cognition, and the Brain: Insights from Sign Language Research*. Mahwah, NJ: Lawrence Erlbaum.
Enard, W., Przeworski, M., Fisher, S. E., Lai, C. S., Wiebe, V., Kitano, T., et al. (2002). Molecular evolution of FOXP2, a gene involved in speech and language. *Nature*, 418, 869–872.

Endress, A. D., Cahill, D., Block, S., Watumull, J. & Hauser, M. D. (2009). Evidence of an evolutionary precursor to human language affixation in a non-human primate. *Biological Letters*, 5, 749–751.

Evans, N. & Levinson, S. (2009). The myth of language universals: language diversity and its importance for cognitive science. *Behavioral and Brain Sciences*, 32, 429–492.

Everett, D. L. (2008). *Don't Sleep, There Are Snakes: Life and Language in the Amazonian Jungle*. New York: Pantheon Books.

Fehér, O., Wang, H., Saar, S., Mitra, P. P. & Tchernichovski, O. (2009). De novo establishment of wild-type song culture in the Zebra Finch. *Nature*, 459, 564–568.

Feigenson, L., Carey, S. & Hauser, M. (2002). The representations underlying infants' choice of more: object files versus analog magnitudes. *Psychological Science*, 13, 150–156.

Ferrand, L. & Grainger, J. (1992). Phonology and orthography in visual word recognition: evidence from masked non-word priming. *The Quarterly Journal of Experimental Psychology Section A: Human Experimental Psychology*, 45, 353–372.

Fikkert, P. (2007). Acquiring phonology. In P. de Lacy (ed.), *The Cambridge Handbook of Phonology* (pp. 536–554). Cambridge University Press.

Fikkert, P. & Levelt, C. (2008). How does place fall into place? The lexicon and emergent constraints in children's developing phonological grammar. In P. Avery, B. E. Dresher & K. Rice (eds.), *Contrast in Phonology: Theory, Perception, Acquisition* (pp. 231–268). Berlin: Mouton de Gruyter.

Finley, S. & Badecker, W. (2008). Analytic biases for vowel harmony languages, *West Coast Conference of Formal Linguistics*. UCLA.

(2009). Artificial language learning and feature-based generalization. *Journal of Memory and Language*, 61, 423–437.

(2010). Linguistic and non-linguistic influences on learning biases for vowel harmony. In O. S. & R. Catrambone (eds.), *Proceedings of the 32nd Annual Conference of the Cognitive Science Society* (pp. 706–711). Austin, TX: Cognitive Science Society.

Fischer, J. & Hammerschmidt, K. (2011) Ultrasonic vocalizations in mouse models for speech and socio-cognitive disorders: insights into the evolution of vocal communication. *Genes, Brain, Behavior*, 10, 17–27.

Fisher, S. & Marcus, G. (2006). The eloquent ape: genes, brains and the evolution of language. *Nature Reviews*, 7, 9–20.

Fisher, S. E. & Scharff, C. (2009). FOXP2 as a molecular window into speech and language. *Trends in Genetics*, 25, 166–177.

Fitch, W. T. (2010). *The Evolution of Language*. Leiden: Cambridge University Press.

Fitch, W. T. & Fritz, J. B. (2006). Rhesus macaques spontaneously perceive formants in conspecific vocalizations. *Journal of the Acoustical Society of America*, 120, 2132–2141.

Fitch, W. T. & Hauser, M. D. (2004). Computational constraints on syntactic processing in a nonhuman primate. *Science*, 303, 377–380.

Fitch, W. T., Hauser, M. D. & Chomsky, N. (2005). The evolution of the language faculty: clarifications and implications. *Cognition*, 97, 179–210.

Fleishhacker, H. (2001). *Onset Transfer in Reduplication*. Los Angeles: UCLA.

Flemming, E. (2001). Scalar and categorical phenomena in a unified model of phonetics and phonology. *Phonology*, 18, 7–44.

Fodor, J. A. (1983). *The Modularity of Mind*. Cambridge, MA: MIT Press.
Fodor, J. A. (1975). *The Language of Thought*. Cambridge, MA: Harvard University Press.
Fodor, J. A. & Piattelli-Palmarini, M. (2010). *What Darwin Got Wrong* New York: Farrar, Straus and Giroux.
Fodor, J. A. & Pylyshyn, Z. (1988). Connectionism and cognitive architecture: a critical analysis. *Cognition*, 28, 3–71.
Fowler, C., Treiman, R. & Gross, J. (1993). The structure of English syllables and polysyllables. *Journal of Memory and Language*, 32, 115–140.
Franz, M. & Goller, F. (2002). Respiratory units of motor production and song imitation in the Zebra Finch. *Journal of Neurobiology*, 51, 129–141.
French, C. A., Groszer, M., Preece, C., Coupe, A. M., Rajewsky, K. & Fisher, S. E. (2007). Generation of mice with a conditional Foxp2 null allele. *Genesis*, 45, 440–446.
Friederici, A., D. & Wessels, J. M. (1993). Phonotactic knowledge of word boundaries and its use in infant speech perception. *Perception and Psychophysics*, 54, 287–295.
Friedrich, M. & Friederici, A. D. (2005). Phonotactic knowledge and lexical-semantic processing in one-year-olds: brain responses to words and nonsense words in picture contexts. *Journal of Cognitive Neuroscience*, 17, 1785–1802.
Frisch, S. A. & Zawaydeh, B. A. (2001). The psychological reality of OCP-place in Arabic. *Language*, 77, 91–106.
Frisch, S. A., Pierrehumbert, J. B. & Broe, M. B. (2004). Similarity avoidance and the OCP. *Natural Language and Linguistic Theory*, 22, 197–228.
Frost, R., Ahissar, M., Gotesman, R. & Tayeb, S. (2003). Are phonological effects fragile? The effect of luminance and exposure duration on form priming and phonological priming. *Journal of Memory and Language*, 48, 346–378.
Gafos, A., I. (1999). *The Articulatory Basis of Locality in Phonology*. New York: Garland.
Galaburda, A. M., Sherman, G. F., Rosen, G. D., Aboitiz, F. & Geschwind, N. (1985). Developmental dyslexia: four consecutive patients with cortical anomalies. *Annals of Neurology*, 18, 222–233.
Galaburda, A. M., LoTurco, J., Ramus, F., Fitch, R. H. & Rosen, G. D. (2006). From genes to behavior in developmental dyslexia. *Nature Neuroscience*, 9, 1213–1217.
Gallistel, C. R. (1990). *The Organization of Learning*. Cambridge, MA: MIT Press.
  (2007). Learning organs (L'apprentissage de matières distinctes exige des organes distincts). In J. B. J. Franck (ed.), *Cahier n° 88: Noam Chomsky* (pp. 181–187). Paris: L'Herne.
Gallistel, C. R. & King, A. P. (2009). *Memory and the Computational Brain: Why Cognitive Science Will Transform Neuroscience*. Chichester; Malden, MA: Wiley-Blackwell.
Gallon, N., Harris, J. & van der Lely, H. (2007). Non-word repetition: an investigation of phonological complexity in children with Grammatical SLI. *Clinical Linguistics & Phonetics*, 21, 435–455.
Gardner, R. A. & Gardner, B. T. (1969). Teaching sign language to a chimpanzee. *Science (New York, NY)*, 165, 664–672.
Gaskell, M. G. & Marslen-Wilson, W. D. (1998). Mechanisms of phonological inference in speech perception. *Journal of Experimental Psychology: Human Perception and Performance*, 24, 380–396.

Gaskell, M., Hare, M. & Marslen-Wilson, W. (1995). A connectionist model of phonological representation in speech perception. *Cognitive Science*, 19, 407–439.

Gathercole, S. E. & Baddeley, A. D. (1990). Phonological memory deficits in language disordered children: is there a causal connection? *Journal of Memory and Language*, 29, 336–360.

Gentner, T. Q. (2008). Temporal scales of auditory objects underlying birdsong vocal recognition. *Journal of the Acoustical Society of America*, 124, 1350–1359.

Gentner, T. Q. & Hulse, S. H. (2000). Perceptual classification based on the component structure of song in European starlings. *Journal of the Acoustical Society of America*, 107, 3369–3381.

Gentner, T. Q., Fenn, K. M., Margoliash, D. & Nusbaum, H. C. (2006). Recursive syntactic pattern learning by songbirds. *Nature*, 440, 1204–1207.

Gerken, L. (1994). A metrical template account of children's weak syllable omissions from multisyllabic words. *Journal of Child Language*, 21, 565–584.

Gervain, J., Macagno, F., Cogoi, S., Peña, M. & Mehler, J. (2008). The neonate brain detects speech structure. *Proceedings of the National Academy of Sciences of the United States of America*, 105, 14222–14227.

Gervain, J., Berent, I. & Werker, J. (2012). Binding at birth: newborns detect identity relations and sequential position in speech. *Journal of Cognitive Neuroscience*, 24, 564–574.

Gervain, J., Berent, I., Dupoux, E. & Werker, J. F. (forthcoming). Distinct networks for music and speech perception in the newborn brain.

Gibson, E. & Fedorenko, E. (2010). Weak quantitative standards in linguistics research. *Trends in Cognitive Sciences*, 14, 233–234.

Gierut, J. A. (1999). Syllable onsets: clusters and adjuncts in acquisition. *Journal of Speech, Language, and Hearing Research: JSLHR*, 42, 708–726.

Gill, K. Z. & Purves, D. (2009). A biological rationale for musical scales. *PLoS One*, 4, e8144–e8144.

Giurfa, M., Zhang, S., Jenett, A., Menzel, R. & Srinivasan, M. V. (2001). The concepts of "sameness" and "difference" in an insect. *Nature*, 410, 930.

Glushko, R. (1979). The organization and activation of orthographic knowledge in reading aloud. *Journal of Experimental Psychology: Human Perception and Performance*, 5, 674–691.

Gnanadesikan, A. (2004). Markedness and faithfulness constraints in child phonology. In R. Kager, J. Pater & W. Zonneveld (eds.), *Constraints in Phonological Acquisition* (pp. 73–108). New York: Cambridge University Press.

Goad, H. & Rose, Y. (2004). Input elaboration, head faithfulness, and evidence for representation in the acquisition of left-edge clusters in West Germanic. In R. Kager, J. Pater & W. Zonneveld (eds.), *Constraints in Phonological Acquisition* (pp. 109–157). New York: Cambridge University Press.

Goldin-Meadow, S. & Mylander, C. (1983). Gestural communication in deaf children: noneffect of parental input on language development. *Science*, 221, 372–374.

(1998). Spontaneous sign systems created by deaf children in two cultures. *Nature*, 391, 279–281.

Goldinger, S. D. (1998). Echoes of echoes? An episodic theory of lexical access. *Psychological Review*, 105, 251–279.

Goldrick, M. & Rapp, B. (2007). Lexical and post-lexical phonological representations in spoken production. *Cognition*, 102, 219–260.
Goldsmith, J. (2002) Probabilistic models of grammar: phonology as information minimization. *Phonological Studies*, 5, 21–46.
Gordon, P. (1985). Level ordering in lexical development. *Cognition*, 21, 73–93.
(2004). Numerical cognition without words: evidence from Amazonia. *Science*, 306, 496–499.
Gough, P. M., Nobre, A. C. & Devlin, J. T. (2005). Dissociating linguistic processes in the left inferior frontal cortex with transcranial magnetic stimulation. *Journal of Neuroscience*, 25, 8010–8016.
Gould, S. J. & Lewontin, R. C. (1979). The spandrels of San Marco and the Panglossian paradigm: a critique of the adaptationist programme. *Proceedings of the Royal Society of London. Series B, Containing Papers of a Biological Character*, 205, 581–598.
Gouskova, M. (2001). Falling sonority onsets, loanwords, and syllable contact. *CLS*, 37, 175–185.
Gow, D. W. J. (2001). Assimilation and anticipation in continuous spoken word recognition. *Journal of Memory and Language*, 45, 133–159.
Gow, D. W. J. & Segawa, J. A. (2009). Articulatory mediation of speech perception: a causal analysis of multi-modal imaging data. *Cognition*, 110, 222–236.
Graves, W. W., Grabowski, T. J., Mehta, S. & Gupta, P. (2008). The left posterior superior temporal gyrus participates specifically in accessing lexical phonology. *Journal of Cognitive Neuroscience*, 20, 1698–1710.
Greenberg, J. H. (1966). *Language Universals, with Special Reference to Feature Hierarchies*. The Hague: Mouton.
(1978). Some generalizations concerning initial and final consonant clusters. In J. H. Greenberg, C. A. Ferguson & E. A. Moravcsik (eds.), *Universals of Human Language* (Vol. II, pp. 243–279). Stanford University Press.
Guion, S. G. (1996). Velar palatalization: coarticulation, perception, and sound change. Ph.D. dissertation, University of Texas at Austin.
(1998). The role of perception in the sound change of velar palatalization. *Phonetica*, 55, 18–52.
Guo, T., Peng, D. & Liu, Y. (2005). The role of phonological activation in the visual semantic retrieval of Chinese characters. *Cognition*, 98, B21–B34.
Haesler, S., Wada, K., Nshdejan, A., Morrisey, E. E., Lints, T., Jarvis, E. D., et al. (2004). FoxP2 expression in avian vocal learners and non-learners. *Journal of Neuroscience*, 24, 3164–3175.
Haesler, S., Rochefort, C., Georgi, B., Licznerski, P., Osten, P. & Scharff, C. (2007). Incomplete and inaccurate vocal imitation after knockdown of FoxP2 in songbird basal ganglia nucleus Area X. *PLoS Biol*, 5(12), e321. doi: 06-PLBI-RA-1785 [pii].
Hale, M. & Reiss, C. (2008). *The Phonological Enterprise*. Oxford; New York: Oxford University Press.
Halle, M. (1971). *The Sound Pattern of Russian: A Linguistic and Acoustical Investigation*. The Hauge, Mouton.
Hallé, P. A., Segui, J., Frauenfelder, U. & Meunier, C. (1998). The processing of illegal consonant clusters: a case of perceptual assimilation? *Journal of Experimental Psychology: Human Perception and Performance*, 24, 592–608.

Hallé, P. A., Dominguez, A., Cuetos, F. & Segui, J. (2008). Phonological mediation in visual masked priming: evidence from phonotactic repair. *Journal of Experimental Psychology: Human Perception and Performance*, 34, 177–192.
Hamlin, J. K., Wynn, K. & Bloom, P. (2007). Social evaluation by preverbal infants. *Nature*, 450, 557–559.
　(2010). Three-month-olds show a negativity bias in their social evaluations. *Developmental Science*, 13, 923–929.
Harm, M. W. & Seidenberg, M. S. (1999). Phonology, reading acquisition, and dyslexia: insights from connectionist models. *Psychological Review*, 106, 491–528.
Haugeland, J. (1985). *Artificial Intelligence: The Very Idea*. Cambridge, MA: MIT Press.
Hauser, M. D. (1996). *The Evolution of Communication*. Cambridge, MA: MIT Press.
Hauser, M. D. & Glynn, D. (2009). Can free-ranging rhesus monkeys (Macaca mulatta) extract artificially created rules comprised of natural vocalizations? *Journal of Comparative Psychology*, 123, 161–167.
Hauser, M. D. & Konishi, M. (1999). *The Design of Animal Communication*. Cambridge, MA: MIT Press.
Hauser, M. D. & Spelke, E. (2004). Evolutionary and developmental foundations of human knowledge. In M. S. Gazzaniga (ed.), *The Cognitive Neurosciences* (3rd edn., pp. 853–864). Cambridge, MA: MIT Press.
Hauser, M. D., Newport, E. L. & Aslin, R. N. (2001). Segmentation of the speech stream in a non-human primate: statistical learning in cotton-top tamarins. *Cognition*, 78, B53–64.
Hauser, M. D., Chomsky, N. & Fitch, W. T. (2002). The faculty of language: what is it, who has it, and how did it evolve? *Science*, 298, 1569–1579.
Havy, M. l. & Nazzi, T. (2009). Better processing of consonantal over vocalic information in word learning at 16 months of age. *Infancy*, 14, 439–456.
Hayes, B. P. (1980). A metrical theory of stress rules. Unpublished Ph.D. thesis, Massachusetts Institute of Technology.
　(1999). Phonetically driven phonology: the role of Optimality Theory and inductive grounding. In E. A. M. Danell, F. Newmeyer, M. Noonan & K. W. Wheatley (eds.), *Formalism and Functionalism in Linguistics* (Vol. I, pp. 243–285). Amsterdam: Benjamins.
　(2009). *Introductory Phonology*. Malden, MA; Oxford: Wiley-Blackwell.
　(forthcoming). Interpreting sonority – projection experiments: the role of phonotactic modeling. Paper presented at the Proceedings of the 17th International Congress of Phonetic Sciences, Hong Kong.
Hayes, B. & Steriade, D. (2004). A review of perceptual cues and cue robustness. In B. Hayes, R. M. Kirchner & D. Steriade (eds.), *Phonetically Based Phonology* (pp. 1–33). Cambridge University Press.
Hayes, B. & Wilson, C. (2008). A maximum entropy model of phonotactics and phonotactic learning. *Linguistic Inquiry* 39, 379–440.
Hayes, B., Kirchner, R. M. & Steriade, D. (eds.). (2004). *Phonetically Based Phonology*. Cambridge University Press.
Hengeveld, K. (2006). Linguistic typology. In R. Mairal & J. Gil (eds.), *Linguistic Universals* (pp. 46–66). Cambridge University Press.
Hickok, G. & Poeppel, D. (2007). The cortical organization of speech processing. *Nature Reviews Neuroscience*, 8, 393–402.

Hillenbrand, J. (1983). Perceptual organization of speech sounds by infants. *Journal of Speech and Hearing Research*, 26, 268–282.
Hockett, C. F. (1960). The origin of speech. *Scientific American*, 203, 89–96.
Hooper, J. B. (1976). *An Introduction to Natural Generative Phonology*. New York: Academic Press.
Humboldt, W. (1997). *Essays on Language*. Frankfurt am Main; New York: P. Lang.
Hurst, J. A., Baraitser, M., Auger, E., Graham, F. & Norell, S. (1990). An extended family with a dominantly inherited speech disorder. *Developmental Medicine and Child Neurology*, 32, 352–355.
Hyman, L. (1985). *A Theory of Phonological Weight*. Dordrecht: Foris.
  (2001). On the limits of phonetic determinism in phonology: *NC revisited. In B. Hume & K. Johnson (eds.), *The Role of Speech Perception Phenomena in Phonology* (pp. 141–185). New York: Academic Press.
  (2008). Universals in phonology. *The Linguistic Review*, 25, 83–137.
  (2011). Does Gokana really have no syllables? Or: what's so great about being universal? *Phonology*, 28, 55–85.
Immelmann, K. (1969). Song development in the Zebra Finch and other estrildid finches. In R. A. Hinde (ed.), *Bird Vocalizations* (pp. 61–74). London: Cambridge University Press.
Israel, A. & Sandler, W. (2009). Phonological category resolution: a study of handshapes in younger and older sign languages. In A. Castro Caldas & A. Mineiro (eds.), *Cadernos de Saúde* (Vol. II, Special issue *Linguas Gestuais, UCP*, pp. 13–28). Lisbon: UCP.
Iverson, J., R. & Patel, A. D. (2008). Perception of rhythmic grouping depends on auditory experience. *Journal of the Acoustical Society of America*, 124, 2263–2271.
Jackendoff, R. (2002). *Foundations of Language: Brain, Meaning, Grammar, Evolution*. Oxford; New York: Oxford University Press.
Jackendoff, R. & Lerdahl, F. (2006). The capacity for music: what is it, and what's special about it. *Cognition*, 100, 33–72.
Jacquemot, C., Pallier, C., LeBihan, D., Dehaene, S. & Dupoux, E. (2003). Phonological grammar shapes the auditory cortex: a functional magnetic resonance imaging study. *Journal of Neuroscience*, 22/23, 9541–9546.
Jakobson, R. (1941). *Kindersprache, Aphasie und allgemeine Lautgesetze* [Child language, aphasia and phonological universals]. Frankfurt: Suhrkamp.
  (1962). *Selected Writings 1: Phonological Studies* (Vol. I). The Hague: Mouton.
  (1968). *Child Language Aphasia and Phonological Universals*. The Hague: Mouton.
Jared, D. (2002). Spelling-sound consistency and regularity effects in word naming. *Journal of Memory and Language*, 46, 723–750.
Jared, D. & Seidenberg, M. S. (1991). Does word identification proceed from spelling to sound to meaning? *Journal of Experimental Psychology: General*, 120, 358–394.
Jared, D., McRae, K. & Seidenberg, M. (1990). The basis of consistency effect in word naming. *Journal of Memory and Language*, 29, 687–715.
Jeffries, K., Fritz, J. & Braun, A. (2003). Words in melody: an H(2)15O PET study of brain activation during singing and speaking. *Neuroreport*, 14, 749–754.

Jun, J. (2004). Place assimilation. In B. Hayes, R. Kirchner & D. Steriade (eds.), *Phonetically Based Phonology* (pp. 58–86). Cambridge University Press.
Jusczyk, P. W., Friederici, A. D., Wessels, J., Svenkerud, V. Y. & Jusczyk, A. M. (1993). Infants' sensitivity to the sound patterns of native language words. *Journal of Memory and Language*, 32, 402–420.
Jusczyk, P. W., Luce, P. A. & Luce, C. J. (1994). Infants' sensitivity to phonotactic patterns in the native language. *Journal of Memory and Language*, 33, 630–645.
Jusczyk, P. W., Smolensky, P. & Allocco, T. (2002). How English-learning infants respond to markedness and faithfulness constraints. *Language Acquisition*, 10, 31–73.
Kager, R. & Pater, J. (2012). Phonotactics as phonology: knowledge of a complex constraint in Dutch. *Phonology*, 29, 81–111.
Kaminski, J., Call, J. & Fischer, J. (2004). Word learning in a domestic dog: evidence for "fast mapping." *Science*, 304, 1682–1683.
Kang, K.-S. (2003). The status of onglides in Korean: evidence from speech errors. *Studies in Phonetics, Phonology, and Morphology*, 9, 1–15.
Kavitskaya, D. & Babyonyshev, M. (2011). The role of syllable structure: the case of Russian-speaking children with SLI. In C. E. Cairns & E. Raimy (eds.), *Handbook of the Syllable* (pp. 353–369). Leiden: E. J. Brill.
Kavitskaya, D., Babyonyshev, M., Walls, T. & Grigorenko, E. (2011). Investigating the effects of syllable complexity in Russian-speaking children with SLI. *Journal of Child Language, FirstView*, 1–20.
Kawahara, S., Ono, H. & Sudo, K. (2006). Consonant co-occurrence restrictions in Yamato Japanese. In T. Vance & K. Jones (eds.), *Japanese/Korean Linguistics* (Vol. XIV, pp. 27–38). Stanford: CSLI Publications.
Kawasaki-Fukumori, H. (1992). An acoustical basis for universal phonotactic constraints. *Language and Speech*, 35, 73–86.
Keating, P. A. (1984). Phonetic and phonological representation of stop consonant voicing. *Language*, 60, 286–319.
  (1985). Universal phonetics and the organization of grammars. In V. Fromkin (ed.), *Phonetic Linguistics: Essays in Honor of Peter Ladefoged* (pp. 115–132). Orlando, FL: Academic.
  (1988). The phonology-phonetics interface. In F. Newmeyer (ed.), *Linguistics: The Cambridge Survey* (Vol. I, pp. 281–302). Cambridge University Press.
Keil, F. C. (1986). The acquisition of natural kind and artifact term. In W. Demopoulos & A. Marras (eds.), *Language Learning and Concept Acquisition* (pp. 133–153). Norwood, NJ: Ablex.
Keller, T. A., Carpenter, P. A. & Just, M. A. (2003). Brain imaging of tongue-twister sentence comprehension: twisting the tongue and the brain. *Brain and Language*, 84, 189–203.
Kennison, S. M., Sieck, J. P. & Briesch, K. A. (2003). Evidence for a late-occurring effect of phoneme repetition during silent reading. *Journal of Psycholinguistic Research*, 32, 297–312.
Kenstowicz, M. (1994). *Phonology in Generative Grammar*. Cambridge, MA: Blackwell.
Kim, J. W. & Kim, H. (1991). The characters of Korean glides. *Studies in the Linguistic Sciences*, 21, 113–125.
Kiparsky, P. (1979). Metrical structure assignment is cyclic. *Linguistic Inquiry*, 10, 421–442.

(2008). Universals constrain change; change results in typological generalizations. In J. Good (ed.), *Language Universals and Language Change* (pp. 23–53). Oxford University Press.

Kirchner, R. (2000). Geminate inalterability and lenition. *Language*, 76, 509–545.

Kirk, C. (2008). Substitution errors in the production of word-initial and word-final consonant clusters. *Journal of Speech, Language & Hearing Research*, 51, 35–48.

Kirk, C. & Demuth, K. (2005). Asymmetries in the acquisition of word-initial and word-final consonant clusters. *Journal of Child Language*, 32, 709–734.

Kirkham, N. Z., Slemmer, J. A. & Johnson, S. P. (2002). Visual statistical learning in infancy: evidence for a domain general learning mechanism. *Cognition*, 83, B35–42.

Kisilevsky, B. S., Hains, S. M. J., Brown, C. A., Lee, C. T., Cowperthwaite, B., Stutzman, S. S. et al. (2009). Fetal sensitivity to properties of maternal speech and language. *Infant Behavior and Development*, 32, 59–71.

Kluender, K. R., Lotto, A. J., Holt, L. L. & Bloedel, S. L. (1998). Role of experience for language-specific functional mappings of vowel sounds. *Journal of the Acoustical Society of America*, 104(6), 3568–3582.

Knobel, M. & Caramazza, A. (2007). Evaluating computational models in cognitive neuropsychology: the case from the consonant/vowel distinction. *Brain and Language*, 100, 95–100.

Koelsch, S. (2006). Significance of Broca's area and ventral premotor cortex for music-syntactic processing. *Cortex: A Journal Devoted To The Study of the Nervous System And Behavior*, 42, 518–520.

Koelsch, S., Gunter, T., Cramon, D., Zysset, S., Lohmann, G. & Friederici, A. (2002). Bach speaks: a cortical "language-network" serves the processing of music. *Neuroimage*, 17, 956–966.

Kuhl, P. K. (1991). Human adults and human infants show a "perceptual magnet effect" for the prototypes of speech categories, monkeys do not. *Perception and Psychophysics*, 50, 93–107.

Kuhl, P. K. & Miller, J. D. (1975). Speech perception by the chinchilla: voiced-voiceless distinction in alveolar plosive consonants. *Science*, 190, 69–72.

Kuhl, P. K. & Padden, D. M. (1983). Enhanced discriminability at the phonetic boundaries for the place feature in macaques. *Journal of the Acoustical Society of America*, 73, 1003–1010.

Kuhl, P. K., Williams, K. A., Lacerda, F., Stevens, K. N. & Lindblom, B. (1992). Linguistic experience alters phonetic perception in infants by 6 months of age. *Science*, 255, 606–608.

Kusumoto, K. & Moreton, E. (1997). Native language determines parsing of nonlinguistic rhythmic stimuli. *Journal of the Acoustical Society of America*, 102, 3204.

Lachlan, R. F., Verhagen, L., Peters, S. & ten Cate, C. (2010). Are there species-universal categories in bird song phonology and syntax? A comparative study of chaffinches (Fringilla coelebs), Zebra Finches (Taenopygia guttata), and Swamp Sparrows (Melospiza georgiana). *Journal of Comparative Psychology*, 124, 92–108.

Ladefoged, P. (1975). *A Course in Phonetics* (2nd edn.). New York: Harcourt Brace Jovanovich.

Laganaro, M. & Alario, F. X. (2006). On the locus of the syllable frequency effect in speech production. *Journal of Memory and Language*, 55, 178–196.

Lai, C. S., Fisher, S. E., Hurst, J. A., Vargha-Khadem, F. & Monaco, A. P. (2001). A forkhead-domain gene is mutated in a severe speech and language disorder. *Nature*, 413, 519–523.

Lenneberg, E. H. (1967). *Biological Foundations of Language*. Oxford: Wiley.

Lennertz, T. & Berent, I. (2011). People's knowledge of phonological universals: evidence from fricatives and stops. Unpublished manuscript.

Leppänen, P. H., Richardson, U., Pihko, E., Eklund, K. M., Guttorm, T. K., Aro, M., et al. (2002). Brain responses to changes in speech sound durations differ between infants with and without familial risk for dyslexia. *Developmental Neuropsychology*, 22, 407–422.

Lerdahl, F. & Jackendoff, R. (1983). *A Generative Theory of Tonal Music*. Cambridge, MA: MIT Press.

Levelt, C. C. (2009). An experimental approach to coda-omissions in early child language. Paper presented at the Boston University Conference on Language Development, Boston, MA.

Levelt, C. C., Schiller, N. O. & Levelt, W. J. (1999). A developmental grammar for syllable structure in the production of child language. *Brain and Language*, 68, 291–299.

Levitt, A., Healy, A. F. & Fendrich, D. W. (1991). Syllable-internal structure and the sonority hierarchy: differential evidence from lexical decision, naming, and reading. *Journal of Psycholinguistic Research*, 20, 337–363.

Lewkowicz, D. J. & Berent, I. (2009). Sequence learning in 4-month-old infants: do infants represent ordinal information? *Child Development*, 80, 1811–1823.

Liberman, A. M. & Mattingly, I. G. (1989). A specialization for speech perception. *Science*, 243, 489–494.

Liberman, A. M., Harris, K. S., Kinney, J. A. & Lane, H. (1961). The discrimination of relative onset-time of the components of certain speech and nonspeech patterns. *Journal of Experimental Psychology*, 61, 379–388.

Liberman, A. M., Cooper, F. S., Shankweiler, D. P. & Studdert-Kennedy, M. (1967). Perception of the speech code. *Psychological Review*, 74, 431–461.

Liberman, I. Y. (1973). Segmentation of the spoken word and reading acquisition. *Bulletin of the Orton Society*, 23, 65–77.

  (1989). Phonology and beginning reading revisited. In C. von Euler (ed.), *Wenner-Gren International Symposium Series: Brain and Reading* (pp. 207–220). Hampshire, UK: Macmillan.

Liddell, S. K. & Johnson, R. E. (1989). American sign language: the phonological base. *Sign Language Studies*, 64, 195–278.

Lidz, J., Waxman, S. & Freedman, J. (2003). What infants know about syntax but couldn't have learned: experimental evidence for syntactic structure at 18 months. *Cognition*, 89, B65–B73.

Liebenthal, E., Binder, J. R., Piorkowski, R. L. & Remez, R. E. (2003). Short-term reorganization of auditory analysis induced by phonetic experience. *Journal of Cognitive Neuroscience*, 15, 549–558.

Liebenthal, E., Binder, J. R., Spitzer, S. M., Possing, E. T. & Medler, D. A. (2005). Neural substrates of phonemic perception. *Cerebral Cortex*, 15, 1621–1631.

Lieberman, P. (2006). *Toward an Evolutionary Biology of Language*. Cambridge, MA: Belknap Press of Harvard University Press.

Liegeois, F., Baldeweg, T., Connelly, A., Gadian, D. G., Mishkin, M. & Vargha-Khadem, F. (2003). Language fMRI abnormalities associated with FOXP2 gene mutation. *Nature Neuroscience*, 6, 1230–1237.
Lindblom, B. (1998). Systemic constraints and adaptive changes in the formation of sound structure. In J. R. Hurford, M. Studdert-Kennedy & C. Knight (eds.), *Approaches to the Evolution of Language: Social and Cognitive Bases* (pp. 242–263). Cambridge University Press.
Lisker, L. & Abramson, A. (1964). A cross-language study of voicing in initial stops: acoustical measurements. *Word*, 20, 384–422.
Liu, L., Deng, X., Peng, D., Cao, F., Ding, G., Jin, Z., et al. (2009). Modality- and task-specific brain regions involved in Chinese lexical processing. *Journal of Cognitive Neuroscience*, 21, 1473–1487.
Lombardi, L. (1999). Positional faithfulness and voicing assimilation in optimality theory. *Natural Language & Linguistic Theory*, 17, 267–302.
Lotto, A. J., Kluender, K. R. & Holt, L. L. (1997). Perceptual compensation for coarticulation by Japanese quail (Coturnix coturnix japonica). *Journal of the Acoustical Society of America*, 102, 1134–1140.
Lowe, C. B., Kellis, M., Siepel, A., Raney, B. J., Clamp, M., Salama, S. R., et al. (2011). Three periods of regulatory innovation during vertebrate evolution. *Science*, 333, 1019–1024.
Lukaszewicz, B. (2007). Reduction in syllable onsets in the acquisition of Polish: deletion, coalescence, metathesis and gemination. *Journal of Child Language*, 34, 53–82.
Lukatela, G., Eaton, T. & Turvey, M. T. (2001). Does visual word identification involve a sub-phonemic level? *Cognition*, 78, B41–B52.
Lukatela, G., Eaton, T., Sabadini, L. & Turvey, M. T. (2004). Vowel duration affects visual word identification: evidence that the mediating phonology is phonetically informed. *Journal of Experimental Psychology: Human Perception and Performance*, 30, 151–162.
MacNeilage, P. F. (1998). The frame/content theory of evolution of speech production. *Behavioral and Brain Sciences*, 21, 499–511.
  (2008). *The Origin of Speech*. Oxford; New York: Oxford University Press.
MacNeilage, P. F. & Davis, B. L. (2000). On the origin of internal structure of word forms. *Science*, 288, 527–531.
MacSweeney, M., Waters, D., Brammer, M. J., Woll, B. & Goswami, U. (2008). Phonological processing in deaf signers and the impact of age of first language acquisition. *Neuroimage*, 40, 1369–1379.
Maess, B., Koelsch, S., Gunter, T. C. & Friederici, A. D. (2001). Musical syntax is processed in Broca's area: an MEG study. *Nature Neuroscience*, 4, 540.
Mandell, J., Schulze, K. & Schlaug, G. (2007). Congenital amusia: an auditory-motor feedback disorder? *Restorative Neurology and Neuroscience*, 25, 323–334.
Manis, F. R., McBride-Chang, C., Seidenberg, M. S., Keating, P., Doi, L. M., Munson, B., et al. (1997). Are speech perception deficits associated with developmental dyslexia? *Journal of Experimental Child Psychology*, 66, 211–235.
Marcus, G. F. (1998). Rethinking eliminative connectionism. *Cognitive Psychology*, 37, 243–282.
  (2001). *The Algebraic Mind: Integrating Connectionism and Cognitive Science*. Cambridge: MIT Press.

(2004). *The Birth of the Mind: How a Tiny Number of Genes Creates the Complexities of Human Thought*. New York: Basic Books.

(2006). Cognitive architecture and descent with modification. *Cognition*, 101, 443–465.

Marcus, G. F. & Fisher, S. E. (2003). FOXP2 in focus: what can genes tell us about speech and language? *Trends in Cognitive Sciences*, 7, 257–262.

Marcus, G. & Rabagliati, H. (2006). What developmental disorders can tell us about the nature and origins of language. *Nature Neuroscience*, 9, 1226–1229.

Marcus, G. F., Vijayan, S., Bandi Rao, S. & Vishton, P. M. (1999). Rule learning by seven-month-old infants. *Science*, 283, 77–80.

Marcus, G. F., Fernandes, K. J. & Johnson, S. P. (2007). Infant rule learning facilitated by speech. *Psychological Sciences*, 18, 387–391.

Marler, P. (1997). Three models of song learning: evidence from behavior. *Journal of Neurobiology*, 33, 501–516.

Marler, P. & Peters, S. (1988). The role of song phonology and syntax in vocal learning preferences in the Song Sparrow, Melospiza melodia. *Ethology*, 77, 125–149.

Marler, P. & Pickert, R. (1984). Species-universal microstructure in the learned song of the Swamp Sparrow (Melospiza georgiana). *Animal Behaviour*, 32, 673–689.

Marler, P. & Sherman, V. (1985). Innate differences in singing behaviour of sparrows reared in isolation from adult conspecific song. *Animal Behaviour*, 33, 57–71.

Marom, M. & Berent, I. (2010). Phonological constraints on the assembly of skeletal structure in reading. *Journal of Psycholinguistic Research*, 39, 67–88.

Marshall, A. J., Wrangham, R. W. & Arcadi, A. C. (1999). Does learning affect the structure of vocalizations in chimpanzees? *Animal Behaviour*, 58, 825–830.

Marshall, C. R. & van der Lely, H. K. (2005). A challenge to current models of past tense inflection: the impact of phonotactics. *Cognition*, 100, 302–20.

(2009). Effects of word position and stress on onset cluster production: evidence from typical development, specific language impairment, and dyslexia. *Language*, 85, 39–57.

Marshall, C. R., Harcourt-Brown, S., Ramus, F. & van der Lely, H. K. (2009). The link between prosody and language skills in children with specific language impairment (SLI) and/or dyslexia. *International Journal of Language and Communication Disorders*, 44, 466–488.

Massaro, D. W. & Cohen, M. M. (1983). Phonological constraints in speech perception. *Perception and Psychophysics*, 34, 338–348.

Mattingly, I. G. (1981). Phonetic representation and speech synthesis by rule. In T. Myers, J. Laver & J. Anderson (eds.), *The Cognitive Representation of Speech* (pp. 415–420). Amsterdam: North Holland.

Mattys, S. L. & Jusczyk, P. W. (2001). Phonotactic cues for segmentation of fluent speech by infants. *Cognition*, 78, 91–121.

Mattys, S. L., Jusczyk, P. W., Luce, P. A. & Morgan, J. L. (1999). Phonotactic and prosodic effects on word segmentation in infants. *Cognitive Psychology*, 38, 465–494.

Mayberry, R. I. (2007). When timing is everything: age of first-language acquisition effects on second-language learning. *Applied Psycholinguistics*, 28, 537–549.

Mayberry, R. I. & Witcher, P. (2005). *What Age of Acquisition Effects Reveal about the Nature of Phonological Processing*. San Diego, La Jolla: University of California.

Maye, J., Werker, J. F. & Gerken, L. (2002). Infant sensitivity to distributional information can affect phonetic discrimination. *Cognition*, 82, B101–111.

Mazuka, R., Cao, Y., Dupoux, E. & Christophe, A. (2012). The development of a phonological illusion: a cross-linguistic study with Japanese and French infants. *Developmental Science*, 14, 693–699.
McCarthy, J. (1979). *Formal Problems in Semitic Phonology and Morphology.* Doctoral dissertation, MIT. New York: Garland Press, 1985.
  (1981). A prosodic theory of nonconcatenative morphology. *Linguistic Inquiry*, 12, 373–418.
  (1982). Prosodic structure and expletive infixation. *Language*, 58, 574–590.
  (1994). The phonetics and phonology of Semitic pharyngeals. In P. Keating (ed.), *Papers in Laboratory Phonology III* (pp. 191–283). Cambridge University Press, Cambridge.
  (forthcoming). Autosegmental spreading in Optimality Theory. In J. Goldsmith, E. Hume & L. Wetzels (eds.), *Tones and Features (Clements Memorial Volume)*. Berlin: Mouton de Gruyter.
McCarthy, J. J. & Prince, A. (1986). *Prosodic Morphology.* Rutgers: Rutgers University Center for Cognitive Science.
  (1993). *Prosodic Morphology I: Constraint Interaction and Satisfaction* (Report no. RuCCS-TR-3). New Brunswick, NJ: Rutgers University Center for Cognitive Science.
  (1995). Prosodic morphology. In J. A. Goldsmith (ed.), *Phonological Theory* (pp. 318–366). Oxford: Blackwell.
  (1998). Prosodic morphology. In A. Spencer & A. M. Zwicky (eds.), *Handbook of Morphology* (pp. 283–305). Oxford: Blackwell.
McClelland, J. L. & Patterson, K. (2002). Rules or connections in past-tense inflections: what does the evidence rule out? *Trends in Cognitive Sciences*, 6, 465–472.
McCutchen, D. & Perfetti, C. A. (1982). The visual tongue-twister effect: phonological activation in silent reading. *Journal of Verbal Learning & Verbal Behavior*, 21, 672–687.
McCutchen, D., Bell, L. C., France, I. M. & Perfetti, C. A. (1991). Phoneme-specific interference in reading: the tongue-twister effect revisited. *Reading Research Quarterly*, 26, 87–103.
McIntosh, J. B. (1944). Huichol phonemes. *International Journal of American Linguistics*, 11, 31–35.
McMurray, B. & Aslin, R. N. (2005). Infants are sensitive to within-category variation in speech perception. *Cognition*, 95, B15–26.
Mehler, J., Jusczyk, P., Lambertz, G., Halsted, N., Bertoncini, J. & Amiel-Tison, C. (1988). A precursor of language acquisition in young infants. *Cognition*, 29, 143–178.
Mester, A. R. & Ito, J. (1989). Feature predictability and underspecification: palatal prosody in Japanese mimetics. *Language*, 65, 258–293.
Miceli, G., Capasso, R., Benvegnú, B. & Caramazza, A. (2004). The categorical distinction of vowel and consonant representations: evidence from dysgraphia. *Neurocase: Case Studies in Neuropsychology, Neuropsychiatry, and Behavioural Neurology*, 10, 109–121.
Midrash Tanhuma. [R. Tanhuma] *Exodus, parashat pekudei.* Retrieved November 24, 2010, from http://kodesh.snunit.k12.il/tan/b0023.htm/#3

Miksis-Olds, J. L., Buck, J. R., Noad, M. J., Cato, D. H. & Stokes, M. D. (2008). Information theory analysis of Australian humpback whale song. *Journal of the Acoustical Society of America*, 124, 2385–2393.

Miller, J. L. (2001). Mapping from acoustic signal to phonetic category: Internal category structure, context effects and speeded categorization. *Language and Cognitive Processes*, 16, 683–690.

Miller, J. L. & Volaitis, L. E. (1989). Effect of speaking rate on the perceptual structure of a phonetic category. *Perception and Psychophysics*, 46, 505–512.

Mody, M., Studdert-Kennedy, M. & Brady, S. (1997). Speech perception deficits in poor readers: auditory processing or phonological coding? *Journal of Experimental Child Psychology*, 64, 199–231.

Molfese, D. L. (2000). Predicting dyslexia at 8 years of age using neonatal brain response. *Brain and Language*, 72, 238–245.

Monaghan, P. & Shillcock, R. (2003). Connectionist modelling of the separable processing of consonants and vowels. *Brain and Language*, 86, 83–98.

Morais, J., Cary, L., Alegria, J. & Bertelson, P. (1979). Does awareness of speech as a sequence of phonemes arise spontaneously? *Cognition*, 7, 323–331.

Morely, R. L. (2008). Generalization, lexical statistics, and a typologically rare system. Ph.D. dissertation, Johns Hopkins University, Baltimore.

Moreton, E. (2002). Structural constraints in the perception of English stop-sonorant clusters. *Cognition*, 84, 55–71.

(2008). Analytic bias and phonological typology. *Phonology*, 25, 83–127.

Morillon, B., Lehongre, K., Frackowiak, R. S. J., Ducorps, A., Kleinschmidt, A., et al. (2010). Neurophysiological origin of human brain asymmetry for speech and language. *Proceedings of the National Academy of Sciences of the United States of America*, 107, 18688–18693.

Murphy, R. A., Mondragón, E. & Murphy, V. A. (2008). Rule learning by rats. *Science*, 319, 1849–1851.

Naatanen, R., Lehtokoski, A., Lennes, M., Cheour, M., Huotilainen, M., Iivonen, A., et al. (1997). Language-specific phoneme representations revealed by electric and magnetic brain responses. *Nature*, 385, 432–434.

Nag, S., Treiman, R. & Snowling, M. (2010). Learning to spell in an alphasyllabary: the case of Kannada. *Writing Systems Research*, 2, 41–52.

Naish, P. (1980). Phonological recoding and the Stroop effect. *British Journal of Psychology*, 71, 395–400.

Nazzi, T. (2005). Use of phonetic specificity during the acquisition of new words: differences between consonants and vowels, *Cognition*, 98(1), 13–30.

Nazzi, T. & Bertoncini, J. (2009). Phonetic specificity in early lexical acquisition: new evidence from consonants in coda positions. *Language and Speech*, 52, 463–480.

Nazzi, T. & New, B. (2007). Beyond stop consonants: consonantal specificity in early lexical acquisition. *Cognitive Development*, 22, 271–279.

Nazzi, T., Bertoncini, J. & Mehler, J. (1998). Language discrimination by newborns: toward an understanding of the role of rhythm. *Journal of Experimental Psychology: Human Perception and Performance*, 24, 756–766.

Nazzi, T., Dilley, L. C., Jusczyk, A. M., Shattuck-Hufnagel, S. & Jusczyk, P. W. (2005). English-learning infants' segmentation of verbs from fluent speech. *Language & Speech*, 48(Part 3), 279–298.

Nazzi, T., Floccia, C., Moquet, B. & Butler, J. (2009). Bias for consonantal information over vocalic information in 30-month-olds: cross-linguistic evidence from French and English. *Journal of Experimental Child Psychology*, 102, 522–537.

Nelson, D. A. & Marler, P. (1989). Categorical perception of a natural stimulus continuum: birdsong. *Science*, 244, 976.

Nespor, M., Peña, M. & Mehler, J. (2003). On the different roles of vowels and consonants in speech processing and language acquisition. *Lingue e Linguaggio*, 2, 223–229.

Nevins, A. (2009). On formal universals in phonology. *Behavioral and Brain Sciences*, 32, 461–432.

Newbury, D. F. & Monaco, A. P. (2010). Genetic advances in the study of speech and language disorders. *Neuron*, 68, 309–320.

Newport, E. L. (2002). Critical periods in language development. In L. Nadel (ed.), *Encyclopedia of Cognitive Science* (pp. 737–740). London: Macmillan Publishers Ltd. / Nature Publishing Group.

Newport, E. L., Hauser, M. D., Spaepen, G. & Aslin, R. N. (2004). Learning at a distance II: statistical learning of non-adjacent dependencies in a non-human primate. *Cognitive Psychology*, 49, 85–117.

Niddah (I. W. Slotski, trans.). (1947). In I. Epstein (ed.), *Babylonian Talmud*. London: Soncino Press.

Nishimura, H., Hashikawa, K., Doi, K., Iwaki, T., Watanabe, Y., Kusuoka, H., et al. (1999). Sign language "heard" in the auditory cortex. *Nature*, 397, 116.

Nowak, M. A. & Krakauer, D. C. (1999). The evolution of language. *Proceedings of the National Academy of Sciences of the United States of America*, 96, 8028–8033.

Obleser, J., Leaver, A., VanMeter, J. & Rauschecker, J. P. (2010). Segregation of vowels and consonants in human auditory cortex: evidence for distributed hierarchical organization. *Frontiers in Psychology*, 1. Available at: www.frontiersin.org/ Auditory_Cognitive_Neuroscience/10.3389/fpsyg.2010.00232/abstract.

Ohala, D. K. (1999). The influence of sonority on children's cluster reductions. *Journal of Communication Disorders*, 32, 397–421.

Ohala, J. J. (1975). Phonetic explanations for nasal sound patterns. In C. A. Ferguson, L. M. Hyman & J. J. Ohala (eds.), *Nasalfest: Papers from a symposium on nasals and nasalization* (pp. 289–316). Stanford: Language Universals Project.

  (1989). Sound change is drawn from a pool of synchronic variation. In L. E. Breivik & E. H. Jahr (eds.), *Language Change: Contributions to the Study of Its Causes* (pp. 173–198). Berlin: Mouton de Gruyter.

  (1990). Alternatives to the sonority hierarchy for explaining segmental sequential constraints. *Papers from the Regional Meetings, Chicago Linguistic Society*, 2, 319–338.

Ohala, J. J. & Riordan, C. J. (1979). Passive vocal tract enlargement during voiced stops. In J. J. Wolf & D. H. Klatt (eds.), *Speech Communication Papers* (pp. 89–92). New York: Acoustical Society of America.

Okada, K. & Hickok, G. (2006). Identification of lexical-phonological networks in the superior temporal sulcus using functional magnetic resonance imaging. *Neuroreport*, 17, 1293–1296.

Onishi, K. H. & Baillargeon, R. (2005). Do 15-month-old infants understand false beliefs? *Science*, 308, 255–258.

Ota, M. (2006). Input frequency and word truncation in child Japanese: structural and lexical effects. *Language and Speech*, 49, 261–295.

Ouattara, K., Lemasson, A. & Zuberbühler, K. (2009). Campbell's monkeys concatenate vocalizations into context-specific call sequences. *Proceedings of the National Academy of Sciences of the United States of America*, 106, 22026–22031.

Oudeyer, P.-Y. (2001). The origins of syllable systems: an operational model. In J. Moore & K. Stenning (eds.), *Proceedings of the 23rd Annual Conference of the Cognitive Science Society, COGSCI'2001* (pp. 744–749). London: Lawrence Erlbaum.

(2006). *Self-Organization in the Evolution of Speech*. Oxford; New York: Oxford University Press.

Padden, C. A. & Perlmutter, D. M. (1987). American Sign Language and the architecture of phonological theory. *Natural Language & Linguistic Theory*, 5, 335–375.

Padgett, J. (1995). *Structure in Feature Geometry*. Stanford: CSLI Publications.

Parker, S. (2002). Quantifying the sonority hierarchy. Ph.D. dissertation, University of Massachusetts, Amherst, MA.

(2008). Sound level protrusions as phsycial correlates of sonority. *Journal of Phonetics*, 36, 55–90.

Pascual-Leone, A. & Hamilton, R. (2001). The metamodal organization of the brain. *Progress in Brain Research*, 134, 427–445.

Patel, A. D. (2008). *Music, Language, and the Brain*. Oxford; New York: Oxford University Press.

Patel, A. D., Iversen, J. R. & Rosenberg, J. C. (2006). Comparing the rhythm and melody of speech and music: the case of British English and French. *Journal of the Acoustical Society of America*, 119, 3034–3047.

Patel, A. D., Wong, M., Foxton, J., Lochy, A. & Peretz, I. (2008). Speech intonation perception deficits in musical tone deafness (congenital amusia). *Music Perception*, 25, 357–368.

Pater, J. (1997a). Metrical parameter missetting in second language acquisition. In S. J. Hannahs & M. Young-Scholten (eds.), *Focus on Phonological Acquisition* (pp. 235–261). Amsterdam: John Benjamins.

(1997b). Minimal violation and phonological development. *Language Acquisition*, 6, 201.

Pater, J. (2004). Bridging the gap between receptive and productive development with minimally violable constraints. In R. Kager, J. Pater & W. Zonneveld (eds.), *Constraints in Phonological Acquisition* (pp. 219–244). New York: Cambridge University Press.

Pater, J. & Barlow, J. A. (2003). Constraint conflict in cluster reduction. *Journal of Child Language*, 30, 487–526.

Pater, J., Stager, C. & Werker, J. (2004). The perceptual acquisition of phonological contrasts. *Language*, 80, 384–402.

Patterson, F. G. (1978). The gesture of a gorilla: language acquisition in another pongid. *Brain and Language*, 5, 72–97.

Paulesu, E., Demonet, J. F., Fazio, F., McCrory, E., Chanoine, V., Brunswick, N. & Frith, U. (2001). Dyslexia: cultural diversity and biological unity. *Science*, 291, 2165–2167.

Payne, R. S. & McVay, S. (1971). Songs of humpback whales. *Science*, 173, 585–597.

Peiffer, A. M., Friedman, J. T., Rosen, G. D. & Fitch, R. H. (2004). Impaired gap detection in juvenile microgyric rats. *Brain Research/Developmental Brain Research*, 152, 93–98.

Pennington, B. F. & Bishop, D. V. M. (2009). Relations among speech, language, and reading disorders. *Annual Review of Psychology*, 60, 283–306.
Peperkamp, S. (2007). Do we have innate knowledge about phonological markedness? Comments on Berent, Steriade, Lennertz, and Vaknin. *Cognition*, 104, 638–643.
Pepperberg, I. M. (2002). Cognitive and communicative abilities of grey parrots. *Current Directions in Psychological Science*, 11, 83.
Peretz, I. (2006). The nature of music from a biological perspective. *Cognition*, 100, 1–32.
Perfetti, C. A. (1985). *Reading Ability*. New York: Oxford University Press.
Perfetti, C. A. & Bell, L. (1991). Phonemic activation during the first 40 ms. of word identification: evidence from backward masking and priming. *Journal of Memory and Language*, 30, 473–485.
Perfetti, C. A. & Zhang, S. (1991). Phonological processes in reading Chinese words. *Journal of Experimental Psychology: Learning, Memory, and Cognition*, 17, 633–643.
  (1995). Very early phonological activation in Chinese reading. *Journal of Experimental Psychology: Learning Memory and Cognition*, 21, 24–33.
Perfetti, C. A., Bell, L. C. & Delaney, S. M. (1988). Automatic (pre-lexical) phonetic activation in silent reading: evidence from backward masking. *Journal of Memory and Language*, 32, 57–68.
Perfetti, C. A., Zhang, S. & Berent, I. (1992). Reading in English and Chinese: evidence for a "universal" phonological principle. In R. Frost & L. Katz (eds.), *Orthography, Phonology, Morphology, and Meaning* (pp. 227–248). Amsterdam: North-Holland.
Perlmutter, D. M. (1992). Sonority and syllable structure in American Sign Language. *Linguistic Inquiry*, 407–442.
Pertz, D. L. & Bever, T. G. (1975). Sensitivity to phonological universals in children and adolescents. *Language*, 51, 149–162.
Petitto, L. A., Zatorre, R. J., Gauna, K., Nikelski, E. J., Dostie, D. & Evans, A. C. (2000). Speech-like cerebral activity in profoundly deaf people processing signed languages: implications for the neural basis of human language. *Proceedings of the National Academy of Sciences of the United States of America*, 97, 13961–13966.
Phillips, C., Pellathy, T., Marantz, A., Yellin, E., Wexler, K., Poeppel, D., et al. (2000). Auditory cortex accesses phonological categories: an MEG mismatch study. *Journal of Cognitive Neuroscience*, 12, 1038–1055.
Pierrehumbert, J. B. (1975). The phonology and phonetics of English intonation. Ph.D. dissertation, MIT.
  (1990). Phonological and phonetic representation. *Journal of Phonetics*, 18, 375–394.
  (2001). Stochastic phonology. *GLOT*, 5(6), 1–13.
Pinker, S. (1994). *The Language Instinct*. New York: Morrow.
  (1997). *How the Mind Works*. New York: Norton.
  (1999). *Words and Rules: The Ingredients of Language*. New York: Basic Books.
  (2002). *The Blank Slate: The Modern Denial of Human Nature*. New York: Viking.
Pinker, S. & Bloom, P. (1994). Natural language and natural selection. *Behavioral and Brain Sciences*, 13, 707–784.
Pinker, S. & Jackendoff, R. (2005). The faculty of language: what's special about it? *Cognition*, 95, 201–236.
Pinker, S. & Prince, A. (1988). On language and connectionism: analysis of a parallel distributed processing model of language acquisition. *Cognition*, 28, 73–193.

Pitt, M. A. (1998). Phonological processes and the perception of phonotactically illegal consonant clusters. *Perception and Psychophysics*, 60, 941–951.

Poeppel, D. (2003). The analysis of speech in different temporal integration windows: cerebral lateralization as "asymmetric sampling in time." *Speech Communication*, 41, 245.

(2011). The biology of language. Talk delivered at the *Annual Meeting of the Cognitive Science Society*. Boston, MA.

Poeppel, D., Idsardi, W. J. & van Wassenhove, V. (2008). Speech perception at the interface of neurobiology and linguistics. *Philosophical Transactions of the Royal Society of London. Series B, Biological Sciences*, 363, 1071–1086.

Pons, F. & Toro, J. M. (2010). Structural generalizations over consonants and vowels in 11-month-old infants. *Cognition*, 116, 361–367.

Poole, J. H., Tyack, P. L., Stoeger-Horwath, A. S. & Watwood, S. (2005). Animal behaviour: elephants are capable of vocal learning. *Nature*, 434, 455–456.

Prasada, S. & Pinker, S. (1993). Generalization of regular and irregular morphological patterns. *Language and Cognitive Processes*, 8, 1–55.

Prieto, P. (2006). The relevance of metrical information in early prosodic word acquisition: a comparison of Catalan and Spanish. *Language and Speech*, 49, 231–259.

Prince, A. & Smolensky, P. (1993/2004). *Optimality Theory: Constraint Interaction in Generative Grammar*. Malden, MA: Blackwell.

Pycha, A., Nowak, P., Shin, E. & Shosted, R. (2003). Phonological rule-learning and its implications for a theory of vowel harmony. In G. Garding & M. Tsujimura (eds.), *Proceedings of the West Coast Conference on Formal Linguistics 22* (pp. 423–435). Somerville, MA: Cascadilla Press.

Pylyshyn, Z. (1984). *Computation and Cognition: Towards a Foundation for Cognitive Science*. Cambridge: MIT Press.

Rack, J. P., Snowling, M. J. & Olson, R. K. (1992). The nonword reading deficit in developmental dyslexia: a review. *Reading Research Quarterly*, 27(1), 28–53.

Ramus, F. (2001). Outstanding questions about phonological processing in dyslexia. *Dyslexia*, 7, 197–216.

Ramus, F. & Mehler, J. (1999). Language identification with suprasegmental cues: a study based on speech resynthesis. *Journal of the Acoustical Society of America*, 105, 512–521.

Ramus, F. & Szenkovits, G. (2006). What phonological deficit? *Quarterly Journal of Experimental Psychology*, 61, 129–141.

Ramus, F., Hauser, M. D., Miller, C., Morris, D. & Mehler, J. (2000). Language discrimination by human newborns and by cotton-top tamarin monkeys. *Science*, 288, 349–351.

Ramus, F., Rosen, S., Dakin, S. C., Day, B. L., Castellote, J. M., White, S., et al. (2003). Theories of developmental dyslexia: insights from a multiple case study of dyslexic adults. *Brain*, 126, 841–865.

Rapp, B. C. (1992). The nature of sublexical orthographic organization: The bigram trough hypothesis examined. *Journal of Memory and Language*, 31, 33–53.

Rapp, B. C. & Goldrick, M. (2006). Speaking words: contributions of cognitive neuropsychological research. *Cognitive Neuropsychology*, 23, 39–73.

Rapp, B. C., McCloskey, M., Rothlein, D., Lipka, K. & Vindiola, M. (2009). Vowel-specific synesthesia: evidence for orthographic consonants and vowels. Paper presented at the annual meeting of the Psychonomic Society. Boston, MA.

Read, C. (1971). Pre-school children's knowledge of English phonology. *Harvard Educational Review*, 41, 1–34.

Reali, F. & Christiansen, M. H. (2005). Uncovering the richness of the stimulus: structure dependence and indirect statistical evidence. *Cognitive Science: A Multidisciplinary Journal*, 29, 1007–1028.

Redford, M. A. (2008). Production constraints on learning novel onset phonotactics. *Cognition*, 107, 785–816.

Redford, M. A., Chen, C. C. & Miikkulainen, R. (2001). Constrained emergence of universals and variation in syllable systems. *Language and Speech*, 44, 27–56.

Ren, J., Gao, L. & Morgan, J. L. (2010). Mandarin speakers' knowledge of the sonority sequencing principle. Paper presented at the *20th Colloquium of Generative Grammar*, University of Pompeu Fabra, Barcelona.

Rice, K. (2007). Markedness in phonology. In P. de Lacy (ed.), *The Cambridge Handbook of Phonology* (pp. 79–97). Cambridge University Press.

Ridley, M. (2008). *Evolution* (3rd edn.). Oxford; New York: Oxford University Press.

Riede, T. & Zuberbühler, K. (2003). The relationship between acoustic structure and semantic information in Diana monkey alarm vocalization. *Journal of the Acoustical Society of America*, 114, 1132–1142.

Robinson, J. G. (1984). Syntactic structures in the vocalizations of wedge-capped capuchin monkeys, Cebus olivaceu. *Behaviour*, 90, 46–79.

Roelofs, A. & Meyer, A. S. (1998). Metrical structure in planning the production of spoken words. *Journal of Experimental Psychology: Learning, Memory, and Cognition*, 24, 922–939.

Rogers, H. (2005). *Writing Systems: A Linguistic Approach*. Malden, MA: Blackwell.

Romani, C. & Calabrese, A. (1998a). Syllabic constraints in the phonological errors of an aphasic patient. *Brain and Language*, 64, 83–121.

(1998b). Syllabic constraints on the phonological errors of an aphasic patient. *Brain and Language*, 64, 83–121.

Rumelhart, D. E. & McClelland, J. L. (1986). On learning the past tense of English verbs: implicit rules or parallel distributed processing? In D. Rumelhart, E. J. McClelland & the PDP Research Group (eds.), *Parallel Distributed Processing: Explorations in the Microstructure of Cognition* (Vol. II, pp. 216–271). Cambridge, MA: MIT Press.

Rutter, M., Caspi, A., Fergusson, D., Horwood, L. J., Goodman, R., Maughan, B., et al. (2004). Sex differences in developmental reading disability: new findings from 4 epidemiological studies. *JAMA: The Journal of the American Medical Association*, 291, 2007–2012.

Saffran, J. R. (2003a). Statistical language learning: mechanisms and constraints. *Current Directions in Psychological Sciences*, 12, 110–114.

(2003b). Musical learning and language development. *Annals of the New York Academy of Sciences*, 999, 397–401.

Saffran, J. R., Aslin, R. N. & Newport, E. L. (1996). Statistical learning by 8-month-old infants. *Science*, 274, 1926–1298.

Sahin, N. T., Pinker, S., Cash, S. S., Schomer, D. & Halgren, E. (2009). Sequential processing of lexical, grammatical, and phonological information within Broca's Area. *Science*, 326, 445–449.
Samuels, R. (2004). Innateness in cognitive science. *Trends in Cognitive Sciences*, 8, 136–141.
(2007). Is innateness a confused concept? In P. Carruthers, S. Laurence & S. Stich (eds.), *The Innate Mind: Foundations and the Future* (pp. 17–34). Oxford University Press.
Sandler, W. (1989). *Phonological Representation of the Sign: Linearity and Nonlinearity in American Sign Language*. Dordrecht: Foris.
(1993). A sonority cycle in American Sign Language. *Phonology*, 10, 242–279.
(2008). The syllable in sign language: considering the other natural language modality. In B. L. Davis & K. Zajdó (eds.), *The Syllable in Speech Production* (pp. 379–408). New York: Lawrence Erlbaum.
Sandler, W. (2011). The phonology of movement in sign language. In M. Oostendorp, C. Ewen, B. Hume & K. Rice (eds.), *The Blackwell Companion to Phonology*. Oxford: Wiley-Blackwell.
Sandler, W. & Lillo-Martin, D. C. (2006). *Sign Language and Linguistic Universals*. Cambridge University Press.
Sandler, W., Meir, I., Padden, C. & Aronoff, M. (2005). The emergence of grammar: systematic structure in a new language. *Proceedings of the National Academy of Sciences of the United States of America*, 102, 2661–2665.
Sandler, W., Aronoff, M., Meir, I. & Padden, C. (2011). The gradual emergence of phonological form in a new language. *Natural Language and Linguistic Theory*, 29, 505–543.
Saussure, F. de (1915/1959). *Course in General Linguistics*. New York: Philosophical Library.
Scerri, T. S. & Schulte-Körne, G. (2010). Genetics of developmental dyslexia. *European Child and Adolescent Psychiatry*, 19, 179–197.
Schachner, A., Brady, T. F., Pepperberg, I. M. & Hauser, M. D. (2009). Spontaneous motor entrainment to music in multiple vocal mimicking species. *Current Biology: CB*, 19, 831–836.
Schane, S. A., Tranel, B. & Lane, H. (1974). On the psychological reality of a natural rule of syllable structure. *Cognition*, 3, 351–358.
Scharff, C. & Haesler, S. (2005). An evolutionary perspective on FoxP2: strictly for the birds? *Current Opinion in Neurobiology*, 15(6), 694–703. doi: S0959-4388(05)00154-6 [pii].
Schiller, N. O. & Caramazza, A. (2002). The selection of grammatical features in word production: the case of plural nouns in German. *Brain and Language*, 81, 342–357.
Schlaug, G., Jäncke, L., Huang, Y. & Steinmetz, H. (1995). In vivo evidence of structural brain asymmetry in musicians. *Science*, 267, 699–701.
Schusterman, R. J., Reichmuth Kastak, C. & Kastak, D. (2003). Equivalence classification as an approach to social knowledge: from sea lions to simians. In F. B. M. DeWaal & P. L. Tyack (eds.), *Animal Social Complexity: Intelligence, Culture, and Individualized Societies* (pp. 179–206). Cambridge, MA: Harvard University Press.
Seidenberg, M. (1985). The time course of phonological code activation in two writing systems. *Cognition*, 10, 645–657.

(1987). Sublexical structures in visual word recognition: Access units of orthographic redundancy? In M. Coltheart (ed.), *Attention and Performance* (Vol. XII: *The Psychology of Reading*, pp. 245–263). Hillsdale, NJ: Lawrence Erlbaum.

Seidenberg, M. S., Waters, G. S., Barnes, M. A. & Tanenhaus, M. K. (1984). When does irregular spelling or pronunciation influence word recognition? *Journal of Verbal Learning and Verbal Behavior*, 23, 383–404.

Seidl, A. & Cristia, A. (2008). Developmental changes in the weighting of prosodic cues. *Developmental Science*, 11, 596–606.

Seidl, A. & Johnson, E. K. (2008). Boundary alignment enables 11-month-olds to segment vowel initial words from speech. *Journal of Child Language*, 35, 1–24.

Seidl, A., Cristià, A., Bernard, A. & Onishi, K. H. (2009). Allophonic and phonemic contrasts in infants' learning of sound patterns. *Language Learning & Development*, 5, 191–202.

Selkirk, E., O. (1984). On the major class features and syllable theory. In M. Aronoff & R. T. Oerhle (eds.), *Language Sound Structure: Studies in Phonology Presented to Morris Halle by His Teacher and Students* (pp. 107–136). Cambridge, MA: MIT Press.

Senghas, A. & Coppola, M. (2001). Children creating language: how Nicaraguan sign language acquired a spatial grammar. *Psychological Science*, 12, 323–328.

Senghas, A., Kita, S. & Ozyurek, A. (2004). Children creating core properties of language: evidence from an emerging sign language in Nicaragua. *Science*, 305, 1779–1782.

Serniclaes, W., Sprenger-Charolles, L., Carre, R. & Demonet, J.-F. (2001). Perceptual discrimination of speech sounds in developmental dyslexia. *Journal of Speech, Language, and Hearing Research*, 44, 384–399.

Sevald, C., Dell, G. & Cole, J. (1995). Syllable structure in speech production: are syllables chunks or schemas? *Journal of Memory and Language*, 34, 807–820.

Seyfarth, R. M., Cheney, D. L. & Marler, P. (1980). Monkey responses to three different alarm calls: evidence of predator classification and semantic communication. *Science (New York, N.Y.)*, 210, 801–803.

Shannon, C. E. S. (1948). A mathematical theory of communication. *The Bell System Technical Journal*, 27, 379–423.

Share, D. L. (2008). On the Anglocentricities of current reading research and practice: the perils of overreliance on an "outlier" orthography. *Psychological Bulletin*, 134, 584–615.

Sharma, A., Kraus, N., McGee, T., Carrell, T. & Nicol, T. (1993). Acoustic versus phonetic representation of speech as reflected by the mismatch negativity event-related potential. *Electroencephalography and Clinical Neurophysiology*, 88, 64–71.

Shastry, B. S. (2007). Developmental dyslexia: an update. *Journal of Human Genetics*, 52, 104–109.

Shattuck-Hufnagel, S. (1992). The role of word structure in segmental serial ordering. *Cognition*, 42, 213–259.

(2011). The role of the syllable in speech production in American English: a fresh consideration of the evidence. In C. E. Cairns & E. Raimy (eds.), *Handbook of the Syllable* (pp. 197–224). Leiden: E. J. Brill.

Shaywitz, S. (1998). Dyslexia. *The New England Journal of Medicine*, 338, 307–312.

Shriberg, L. D., Ballard, K. J., Tomblin, J. B., Duffy, J. R., Odell, K. H. & Williams, C. A. (2006). Speech, prosody, and voice characteristics of a mother and daughter with a 7;13 translocation affecting FOXP2. *Journal of Speech, Language & Hearing Research*, 49, 500–525.

Shu, W., Cho, J. Y., Jiang, Y., Zhang, M., Weisz, D., Elder, G. A., et al. (2005). Altered ultrasonic vocalization in mice with a disruption in the Foxp2 gene. *Proceedings of the National Academy of Sciences of the United States of America*, 102, 9643–9648.

Simos, P. G., Breier, J. I., Fletcher, J. M., Foorman, B. R., Castillo, E. M. & Papanicolaou, A. C. (2002). Brain mechanisms for reading words and pseudowords: an integrated approach. *Cerebral Cortex*, 12, 297–305.

Simos, P. G., Pugh, K., Mencl, E., Frost, S., Fletcher, J. M., Sarkari, S., et al. (2009). Temporal course of word recognition in skilled readers: a magnetoencephalography study. *Behavioural Brain Research*, 197, 45–54.

Simpson, G. B. & Kang, H. (2004). Syllable processing in alphabetic Korean. *Reading and Writing*, 17, 137–151.

Siok, W. T., Niu, Z., Jin, Z., Perfetti, C. A. & Tan, L. H. (2008). A structural-functional basis for dyslexia in the cortex of Chinese readers. *Proceedings of the National Academy of Sciences of the United States of America*, 105, 5561–5566.

Skoruppa, K., Pons, F., Christophe, A., Bosch, L., Dupoux, E., Sebastián-Gallés, N., Limissuri, R. A. & Peperkamp, S. (2009). Language-specific stress perception by 9-month-old French and Spanish infants. *Developmental Science*, 12(6), 914–919.

Smith, J. L. (2005). *Phonological Augmentation in Prominent Positions*. New York: Routledge.

Smith, N. (2009). *Acquiring Phonology: A Cross-Generational Case-Study*. Leiden: Cambridge University Press.

Smolensky, P. (1996). On the comprehension production dilemma in child language. *Linguistic Inquiry*, 27, 720–731.

(2006). Optimality in phonology II: harmonic completeness, local constraint conjunction, and feature domain markedness. In P. Smolensky & G. Legendre (eds.), *The Harmonic Mind: From Neural Computation to Optimality-Theoretic Grammar* (Vol. II: *Linguistic and Philosophical Implications*, pp. 27–160). Cambridge, MA: MIT Press.

Smolensky, P. & Legendre, G. (2006). Principles of integrated connectionist/symbolic cognitive architecture. In P. Smolensky & G. Legendre (eds.), *The Harmonic Mind: From Neural Computation to Optimality-Theoretic Grammar* (Vol. I: *Cognitive Architecture*, pp. 63–97). Cambridge, MA: MIT Press.

Soha, J. A. & Marler, P. (2001). Vocal syntax development in the white-crowned sparrow (Zonotrichia leucophrys). *Journal of Comparative Psychology*, 115, 172–180.

Spaepen, E., Coppola, M., Spelke, E. S., Carey, S. E. & Goldin-Meadow, S. (2011). Number without a language model. *Proceedings of the National Academy of Sciences of the United States of America*, 108, 3163–3168.

Spelke, E. S. (1994). Initial knowledge: six suggestions. *Cognition*, 50, 431–445.

(2000). Core knowledge. *American Psychologist*, 55, 1233–1243.

Spelke, E. S. & Kinzler, K. D. (2007). Core knowledge. *Developmental Science*, 10, 89–96.

Spinks, J. A., Liu, Y., Perfetti, C. A. & Tan, L. H. (2000). Reading Chinese characters for meaning: the role of phonological information. *Cognition*, 76, B1–B11.

Stampe, D. (1973). A dissertation on natural phonology. Ph.D. dissertation, University of Chicago.
Stein, J. & Walsh, V. (1997). To see but not to read: the magnocellular theory of dyslexia. *Trends in Neurosciences*, 20, 147–152.
Stemberger, J. P. (1984). Length as a suprasegmental: evidence from speech errors. *Language*, 60, 895–913.
Stemberger, J. P. & Treiman, R. (1986). The internal structure of word-initial consonant clusters. *Journal of Memory and Language*, 25, 163–180.
Stenneken, P., Bastiaanse, R., Huber, W. & Jacobs, A. M. (2005). Syllable structure and sonority in language inventory and aphasic neologisms. *Brain and Language*, 95, 280–292.
Steriade, D. (1982). Greek prosodies and the nature of syllabification. Ph.D. dissertation, MIT, Cambridge, MA (available from MITWPL, Department of Linguistics and Philosophy, MIT).
  (1997). Phonetics in phonology: the case of laryngeal neutralization. Unpublished manuscript.
  (1999). Alternatives to the syllabic interpretation of consonantal phonotactics. In O. Fujimura, B. Joseph & B. Palek (eds.), *Proceedings of the 1998 Linguistics and Phonetics Conference* (pp. 205–242). Prague: The Karolinum Press.
  (2001). The phonology of perceptibility effects: the P-map and its consequences for constraint organization. Unpublished manuscript.
  (2007). Contrast. In P. de Lacy (ed.), *The Cambridge Handbook of Phonology* (pp. 139–157). Cambridge University Press.
Stokoe, W. C., Jr. (1960). Sign language structure: an outline of the visual communication systems of the American Deaf. *Journal of Deaf Studies and Deaf Education*, 10, 3–37.
Stroop, J. R. (1935). Studies of interference in serial verbal reactions. *Journal of Experimental Psychology*, 18, 643–662.
Suarez, R., Golby, A., Whalen, S., Sato, S., Theodore, W. H., Kufta, C. V., Devinsky, O., Balish, M., Bromfield, E. B. (2010). Contributions to singing ability by the posterior portion of the superior temporal gyrus of the non-language-dominant hemisphere: first evidence from subdural cortical stimulation, Wada testing, and fMRI. *Cortex*, 46, 343–353.
Suge, R. & Okanoya, K. (2010). Perceptual chunking in the self-produced songs of Bengalese finches (Lonchura striata var. domestica). *Animal Cognition*, 13, 515–523.
Suthers, R. A. & Zollinger, S. A. (2004). Producing song: the vocal apparatus. *Annals of the New York Academy of Sciences*, 1016, 109–129.
Suzuki, K. (1998). A typological investigation of dissimilation. Ph.D. dissertation, University of Arizona, Tucson, AZ.
Suzuki, R., Buck, J. R. & Tyack, P. L. (2006). Information entropy of humpback whale songs. *Journal of the Acoustical Society of America*, 119, 1849–1866.
Szenkovits, G., Darma, Q., Darcy, I. & F., R. (2011). Exploring dyslexics' phonological deficit II: phonological grammar. Unpublished manuscript.
Tallal, P. (2004). Improving language and literacy is a matter of time. *Nature Reviews Neuroscience*, 5, 721–728.
Tallal, P. & Piercy, M. (1973). Defects of non-verbal auditory perception in children with developmental aphasia. *Nature*, 241, 468–469.

Tan, L. H. & Perfetti, C. A. (1997). Visual Chinese character recognition: does phonological information mediate access to meaning? *Journal of Memory and Language*, 37, 41–57.

Temple, E., Poldrack, R., Protopapas, A., Nagarajan, S., Salz, T., Tallal, P., et al. (2000). Disruption of the neural response to rapid acoustic stimuli in dyslexia: evidence from functional MRI. *Proceedings of the National Academy of Sciences of the United States of America*, 97, 13907–13912.

Temple, E., Deutsch, G. K., Poldrack, R. A., Miller, S. L., Tallal, P., Merzenich, M. M., et al. (2003). Neural deficits in children with dyslexia ameliorated by behavioral remediation: evidence from functional MRI. *Proceedings of the National Academy of Science*, 100, 2860–2865.

Tervaniemi, M., Kujala, A., Alho, K., Virtanen, J., Ilmoniemi, R. J. & Naatanen, R. (1999). Functional specialization of the human auditory cortex in processing phonetic and musical sounds: a magnetoencephalographic (MEG) study. *Neuroimage*, 9, 330–336.

Theodore, R. M. & Miller, J. L. (2010). Characteristics of listener sensitivity to talker-specific phonetic detail. *Journal of the Acoustical Society of America*, 128, 2090–2099.

Theodore, R. M. & Schmidt, A. M. (2003). Perceptual prothesis in native Spanish speakers. *Journal of the Acoustical Society of America*, 113, 256.

Theodore, R. M., Miller, J. L. & DeSteno, D. (2009). Individual talker differences in voice-onset-time: contextual influences. *Journal of the Acoustical Society of America*, 125, 3974–3982.

Tillmann, B., Koelsch, S., Escoffier, N., Bigand, E., Lalitte, P., Friederici, A. D., et al. (2006). Cognitive priming in sung and instrumental music: activation of inferior frontal cortex. *Neuroimage*, 31, 1771–1782.

Tincoff, R., Hauser, M., Tsao, F., Spaepen, G., Ramus, F. & Mehler, J. (2005). The role of speech rhythm in language discrimination: further tests with a non-human primate. *Developmental Science*, 8, 26–35.

Toro, J. M. & Trobalón, J. B. (2005). Statistical computations over a speech stream in a rodent. *Perception and Psychophysics*, 67, 867–875.

Toro, J. M., Nespor, M., Mehler, J. & Bonatti, L. L. (2008). Finding words and rules in a speech stream: functional differences between vowels and consonants. *Psychological Science*, 19, 137–144.

Trainor, L. J., McDonald, K. L. & Alain, C. (2002). Automatic and controlled processing of melodic contour and interval information measured by electrical brain activity. *Journal of Cognitive Neuroscience*, 14, 430–442.

Treiman, R. (1984). On the status of final consonant clusters in English syllables. *Journal of Verbal Learning and Verbal Behavior*, 23, 343–356.

(1986). The division between onsets and rimes in English Syllables. *Journal of Memory and Language*, 25, 476–491.

(2004). Phonology and spelling. In P. Bryant & T. Nunes (eds.), *Handbook of Children's Literacy* (pp. 31–42). Dordrecht: Kluwer.

Treiman, R. & Cassar, M. (1997). Spelling acquisition in English. In C. Perfetti, A. L. Rieben & M. Fayol (eds.), *Learning to Spell: Research, Theory and Practice across Languages*. Mahwah, NJ: Lawrence Erlbaum.

Treiman, R. & Danis, C. (1988). Syllabification of intervocalic consonants. *Journal of Memory and Language*, 27, 87–104.

Treiman, R. & Kessler, B. (1995). In defense of an onset-rime syllable structure for English. *Language and Speech*, 38 (Pt 2), 127–142.

Treiman, R., Bowey, J. & Bourassa, D. (2002). Segmentation of spoken words into syllables by English-speaking children as compared to adults. *Journal of Experimental Child Psychology*, 83, 213–238.

Trout, J. (2003). Biological specialization for speech: what can the animals tell us? *Current Directions in Psychological Sciences*, 12, 155–159.

Trubetzkoy, N. S. (1969). *Principles of Phonology* (trans. C. A. M. Baltaxe). Berkeley: University of California Press.

Turing, A. M. (1936). On computable numbers, with an application to the Entscheidungsproblem. *Proceedings of the London Mathematical Society*, 24, 230–265.

Tzelgov, J., Henik, A., Sneg, R. & Baruch, O. (1996). Unintentional word reading via the phonological route: the Stroop effect with cross-script homophones. *Journal of Experimental Psychology: Learning, Memory, and Cognition*, 22, 336–349.

Ullman, M. T. & Pierpont, E. I. (2005). Specific language impairment is not specific to language: the procedural deficit hypothesis. *Cortex: A Journal Devoted to the Study of the Nervous System and Behavior*, 41, 399–433.

van der Hulst, H. (2000). Modularity and modality in phonology. In N. Burton-Roberts, P. Carr & G. Docherty (eds.), *Phonological Knowledge: Conceptual and Empirical Issues* (pp. 207–243). Oxford University Press.

(2009). Two phonologies. In J. Grijzenhout & B. Kabak (eds.), *Phonological Domains: Universals and Deviations* (pp. 315–352). New York and Berlin: Mouton de Gruyter.

van der Lely, H. K. (2005). Domain-specific cognitive systems: insight from Grammatical-SLI. *Trends in Cognitve Sciences*, 9, 53–59.

van der Lely, H. K., Rosen, S. & Adlard, A. (2004). Grammatical language impairment and the specificity of cognitive domains: relations between auditory and language abilities. *Cognition*, 94, 167–183.

van Heijningen, C. A., de Visser, J., Zuidema, W. & ten Cate, C. (2009). Simple rules can explain discrimination of putative recursive syntactic structures by a songbird species. *Proceedings of the National Academy of Sciences of the United States of America*, 106, 20538–20543.

van Leeuwen, T., Been, P., van Herten, M., Zwarts, F., Maassen, B. & van der Leij, A. (2007). Cortical categorization failure in 2-month-old infants at risk for dyslexia. *Neuroreport*, 18, 857–861.

Van Orden, G. C. (1987). A ROWS is a ROSE: spelling, sound and reading. *Memory and Cognition*, 15, 181–190.

Van Orden, G. C., Johnston, J. C. & Hale, B. L. (1988). Word identification in reading proceeds from spelling to sound to meaning. *Journal of Experimental Psychology: Learning, Memory, and Cognition*, 14, 371–386.

Van Orden, G. C., Pennington, B. F. & Stone, G. O. (1990). Word identification in reading and the promise of subsymbolic psycholinguistics. *Psychological Review*, 97, 488–522.

Vargha-Khadem, F., Watkins, K., Alcock, K., Fletcher, P. & Passingham, R. (1995). Praxic and nonverbal cognitive deficits in a large family with a genetically transmitted speech and language disorder. *Proceedings of the National Academy of Sciences of the United States of America*, 92, 930–933.

Vargha-Khadem, F., Watkins, K. E., Price, C. J., Ashburner, J., Alcock, K. J., Connelly, A., et al. (1998). Neural basis of an inherited speech and language disorder. *Proceedings of the National Academy of Sciences of the United States of America*, 95, 12695–12700.

Vargha-Khadem, F., Gadian, D. G., Copp, A. & Mishkin, M. (2005). FOXP2 and the neuroanatomy of speech and language. *Nature Reviews Neuroscience*, 6, 131–138.

Vaux, B. (2002). Consonantal epenthesis and the problem of unnatural phonology. Unpublished manuscript.

Vennemann, T. (1972). On the theory of syllable phonology. *Linguistische Berichte*, 18, 1–18.

Vouloumanos, A. & Werker, J. F. (2007). Listening to language at birth: evidence for a bias for speech in neonates. *Developmental Science*, 10, 159–164.

Vouloumanos, A., Kiehl, K. A., Werker, J. F. & Liddle, P. F. (2001). Detection of sounds in the auditory stream: event-related fMRI evidence for differential activation to speech and nonspeech. *Journal of Cognitive Neuroscience*, 13, 994–1005.

Vouloumanos, A., Hauser, M. D., Werker, J. F. & Martin, A. (2010). The tuning of human neonates' preference for speech. *Child Development*, 81, 517–527.

Walker, R. L. (1998). Nasalization, neutral segments, and opacity effects. Ph.D. dissertation, University of California, Santa Cruz.

Wang, K., Mecklinger, A., Hofmann, J. & Weng, X. (2010). From orthography to meaning: an electrophysiological investigation of the role of phonology in accessing meaning of Chinese single-character words. *Neuroscience*, 165, 101–106.

Warren, W. C., Clayton, D. F., Ellegren, H., Arnold, A. P., Hillier, L. W., Künstner, A., et al. (2010). The genome of a songbird. *Nature*, 464, 757–762.

Werker, J. F. & Tees, R. C. (1984). Cross-language speech perception: evidence for perceptual reorganization during the first year of life. *Infant Behavior and Development*, 7, 49–63.

Wilkinson, K. (1988). Prosodic structure and Lardil phonology. *Linguistic Inquiry*, 19, 325–334.

Wilson, C. (2003). Experimental investigation of phonological naturalness. In G. Garding & M. Tsujimura (eds.), *Proceedings of the 22nd West Coast Conference on Formal Linguistics* (pp. 533–546). Somerville, MA: Cascadilla.

(2006). Learning phonology with substantive bias: an experimental and computational study of velar palatalization. *Cognitive Science*, 30, 945–982.

Wilson, S. M. & Iacoboni, M. (2006). Neural responses to non-native phonemes varying in producibility: evidence for the sensorimotor nature of speech perception. *Neuroimage*, 33, 316–325.

Wilson, S. J., Lusher, D., Wan, C. Y., Dudgeon, P., & Reutens, D. C. (2009). The neurocognitive components of pitch processing: insights from absolute pitch. *Cerebral Cortex* (New York, N.Y.: 1991), 19, 724–732.

Wolf, M., Bally, H. & Morris, R. (1986). Automaticity, retrieval processes, and reading: a longitudinal study in average and impaired readers. *Child Development*, 57, 988–1000.

Wolmetz, M., Poeppel, D. & Rapp, B. (2011). What does the right hemisphere know about phoneme categories? *Journal of Cognitive Neuroscience*, 23, 552–569.

Wong, P. C., Skoe, E., Russo, N. M., Dees, T. & Kraus, N. (2007). Musical experience shapes human brainstem encoding of linguistic pitch patterns. *Nature Neuroscience*, 10, 420–422.
Wright, R. (2004). A review of perceptual cues and robustness. In D. Steriade, R. Kirchner & B. Hayes (eds.), *Phonetically-Based Phonology* (pp. 34–57). Cambridge University Press.
Wydell, T. N., Patterson, K. E. & Humphreys, G. W. (1993). Phonologically mediated access to meaning for Kanji: is a rows still a rose in Japanese Kanji? *Journal of Experimental Psychology: Learning, Memory, and Cognition*, 19, 491–514.
Wyllie-Smith, L., McLeod, S. et al. (2006). Typically developing and speech-impaired children's adherence to the sonority hypothesis. *Clinical Linguistics & Phonetics* 20(4), 271–291.
Wyttenbach, R. A., May, M. L. & Hoy, R. R. (1996). Categorical perception of sound frequency by crickets. *Science*, 273, 1542–1544.
Yang, C. D. (2004). Universal Grammar, statistics or both? *Trends in Cognitive Sciences*, 8, 451–456.
Yavas, M., Ben-David, A., Gerrits, E., Kristoffersen, K. E. & Simonsen, H. G. (2008). Sonority and cross-linguistic acquisition of initial s-clusters. *Clinical Linguistics & Phonetics*, 22, 421–441.
Yip, M. (1989). Feature geometry and cooccurrence restrictions. *Phonology*, 6, 349–374.
  (2006). The search for phonology in other species. *Trends in Cognitive Sciences*, 10, 442–446.
Yun, Y. (2004). Glides and high vowels in Korean syllables. Ph.D. dissertation, University of Washington.
Zec, D. (2007). The syllable. In P. de Lacy (ed.), *The Cambridge Handbook of Phonology* (pp. 161–194). Cambridge University Press.
Zhang, J. (2004). The role of contrast-specific and language-specific phonetics in contour tone distribution. In B. Hayes, R. Kirchner & D. Steriade (eds.), *Phonetically-Based Phonology* (pp. 157–190). Cambridge University Press.
Zhang, S. & Perfetti, C. A. (1993). The tongue-twister effect in reading Chinese. *Journal of Experimental Psychology: Learning, Memory, and Cognition*, 19, 1082–1093.
Zhao, X. & Berent, I. (2011). Are markedness constraints universal? Evidence from Mandarin Chinese speakers. Paper presented at the Boston University Conference on Language Development. Boston, MA.
Ziegler, J. C., Benraïss, A. & Besson, M. (1999). From print to meaning: an electrophysiological investigation of the role of phonology in accessing word meaning. *Psychophysiology*, 36, 775–785.
Ziegler, J. C., Pech-Georgel, C., George, F., Alario, F. X. & Lorenzi, C. (2005). Deficits in speech perception predict language learning impairment. *Proceedings of the National Academy of Sciences of the United States of America*, 102, 14110–14115.
Ziegler, J. C., Pech-Georgel, C., George, F. & Lorenzi, C. (2009). Speech-perception-in-noise deficits in dyslexia. *Developmental Science*, 12, 732–745.
Zsiga, E. C. (2000). Phonetic alignment constraints: consonant overlap and palatalization in English and Russian. *Journal of Phonetics*, 28, 69–102.
Zuraw, K. (2007). The role of phonetic knowledge in phonological patterning: corpus and survey evidence from Tagalog infixation. *Language*, 83, 277–316.

Index

absolute pitch, 54, 271
across-the-board generalizations, 104, 105, 106, 107, 109, 110, 117
Adam, Galit, 212
addressed phonology, 291
affricates, 24
agreement
  voicing
    generalization to nonnative phonemes, 102
algebraic
  computation, 36
  machinery, 110, 111, 307
  operations, 111
  rules
    by animals, 234
    in birdsong, 236
    in infancy, 202
    structure, 113
algebraic optimization, 27, 231, 245
  in animal communication, 29, 246
  uniqueness of, 309
alignment of feature domains, 172
Allocco, Theresa, 214
allophonic, 158
  distinctions, 208
  by infants, 208
Al-Sayyid Bedouin Sign Language, 13, 118, 311
  assimilation, 15
  movement, 14
  regenesis, 14
American Sign Language (ASL), 143, 144, 267
amusia, 271
Ancient Greek, 167, 168
animal communication
  algebraic optimization, 29, 246
  algebraic restrictions on, 246
  equivalence classes
    in birdsong, 233
    in natural communication, 232
    in Swamp Sparrow song, 237
  hierarchical structure, 228
  motor restrictions on, 245

regenesis
  in birdsong, 244
  restrictions on
    learned vs. innate, 228, 242
  rule-learning, 235
    birdsong, 236
    primates, 237
    Swamp Sparrows, 236
  substantive constraints, 228
animals
  categorical perception, 228
  combinatorial structure, 29
  discrimination of human languages, 229
  equivalence classes
    learned in lab, 232
  perception of human speech, 228
  rule-learning, 234
  identity rules, 239
  primates, 238
  recursion, 240
  rule-learning in lab
    rhesus monkeys, 238
  statistical-learning, 229
aphasia, 81–82
  and consonant–vowel distinction, 261
  and syllable markedness, 263–264
Arabic
  Egyptian, 37
  identity restrictions, 86, 96
aspiration, 65
assimilation, 74
  in infancy, 215
  in sign language, 15, 146
associationism, 50, 106, 109, 113, 114. See connectionism
associative network, 108
autonomy
  of phonology, 26, 140
  sign language, 142
  of processing, 45
Axininca Campa, 127, 129
  onset constraint, 127

352

Index

backness, 152
Badecker, William, 156
Balaban, Evan, 136, 190
Bat-El, Outi, 89, 90, 212
Bemis, Doug, 110
Bertoncini, Josiane, 218
birdsong
  motor restrictions on, 245, 248
  patterning of, 29
  restrictions on
    learned vs. innate, 242
  Zebra Finches, 45
Bonatti, Luca, 80
brain
  computation, 276
  specialization
    weak vs. strong views, 252
brain network
  for consonant–vowel distinction, 260
  for music, 271
  for phonological contrast, 256, 257, 259
  for phonology, 308, 314
  for reading, 295
  for speech, 254
  for syllable markedness, 262–263
  in sign language, 267
  MEG, 257
Brentari, Diane, 16, 147
British Sign Language, 267

Calabrese, Andrea, 264
Caramazza, Alfonso, 81, 260
Carey, Susan, 46
categorical perception
  and statistical learning, 53
  and the consonant–vowel distinction, 75
  by nonhuman animals, 53, 228
  in infancy
    phonological vs. phonetic sources, 89, 192, 206
  of speech, 53
Chamicuro, 129
  repair, 129
child phonology
  and speech segmentation, 211
  and word simplification, 219
  computational machinery, 202
  consonant–vowel distinction, 209
  features, 206
  language discrimination, 209
  markedness restrictions, 215, 224
  phonetic contrast, 208
  phonological primitives, 206, 224
  prosodic structure, 212
  sonority restrictions, 220, 221, 224

syllable
  as an equivalence class, 212
  syllable markedness, 217
  syllable structure, 211
  universal grammatical restrictions, 204
  vowel features, 206
Chinese characters
  and syllables, 283
Chomsky, Noam, 48, 49, 52, 55
classifiers
  handling, 16
  morphological, 16
  object, 16
cluster reduction, 145, 220
COD $\equiv$ ALIGN-L (C, σ), 172
coda, 27
Cole, Jennifer, 76
combinatorial principles
  concatenation, 85
  relations, 85
computation
  algebraic, 36
  computational machinery, 230
  in infancy, 202
Computational Theory of Mind, 38
connectionism, 50, 109. *See* associationism
connectionist networks, 109
  feed-forward networks, 109
  simple recurrent networks, 109
Conrad, Markus, 72–73
consonant epenthesis, 141
consonant–vowel distinction, 209
  and language discrimination by adults, 209
  and rhythm, 209
  and statistical-learning, 80
  as equivalence classes, 75
  brain network for, 260
  categorical perception, 75
  computational properties of, 79
  connectionist networks, 81
  in aphasia, 81, 261
  in conveying grammatical information, 74
  in infancy,
  in word identification, 73
  information content, 74
  marking syntactic head/complement, 74
  preservation of, 74
  prototype effect, 75
  rule-learning, 80
  statistical distinctions, 75
constraint
  COD $\equiv$ ALIGN-L (C, σ), 172
  conflation, 134
  faithfulness, 124
  FILL, 124

constraint (cont.)
  in reading, 296
  in silent reading, 298
  markedness, 123
  NoCODA, 123, 124
  ONSET, 123
  PARSE, 124
  ranking, 125
    Axininca Campa vs. English, 127
    English vs. Lardil, 125
  Surface Correspondence by Identity
    (SCorrI), 89
  Surface Correspondence by Position
    (SCorrP), 89
  universality of, 123
constraint ranking, 214
onset, 127
contrast, 256
  phonemic, 21
  phonemic vs. phonetic, 257
  phonetic, 22
  phonological
    brain network, 256, 257, 259
    transcranial magnetic stimulation, 259
core knowledge, 36, 46, 118, 175, 311
  agents, 4
  and reading, 281
  as a scaffold, 281
  in adulthood, 281
  in infancy, 280
  in the initial states
  tests of, 204
  language, 4
  living things, 4
  object, 4
  properties of, 46
critical period
  across modalities, 268
cross-modal phonological transfer, 268–269
cultural evolution, 309
CV skeleton, 76, 88
  and equivalence classes, 79
  and speech production, 76
  speech errors, 76

Davidson, Lisa, 177
de Lacy, Paul, 127, 129, 134, 141, 175
Dell, Gary, 76
descriptive adequacy, 107
design, 18, 27, 308
  adaptive advantage, 49, 248
  innateness, 47
  uniqueness, 28
  universal, 47
discrete infinity, 85, 111

dissociations
  double, 79
  neurological disorders, 81
  phonology vs. phonetics, 142
domain specificity. See specialization
  and brain structure, 265
  cartoon version of, 45
  functional vs. hardware accounts,
    252, 276
domain-general mechanisms, 52
dorsals, 24
duality of patterning, 4
  in sign language, 13
dual-route architecture, 114
dual-route model
  of reading, 290
dyslexia, 33, 274
  and neural migration, 304
  and phonological grammar, 302
  and spoken language processing, 301
  etiology of, 303
  genetic basis of, 301
  phonological decoding
    in reading, 300
  precursors in infancy, 301

Egyptian Arabic, 25, 37, 140
E-language, 143
emergence, 106, 107, 109, 139
  computational challenges to, 140
  of universal grammar, 310
  challenges for, 250
English, 125, 167, 214
epenthesis. See repair
E-phonology, 133
equivalence classes, 41, 63, 73, 75, 117
  relations among classes, 106
  rules, 106
evolution, 314
  cultural, 52
  of the phonological mind, 227
  universal grammar, 249
explanatory adequacy, 107
expletive affixation, 66

faculty of language
  broad, 55
  narrow, 55
faithfulness constraints, 124
feature
  affricate, 24
  backness, 152
  in infancy, 206
  in silent reading, 297
  in writing systems, 285

Index

agreement, 151
   vowel height, 151
approximants, 24
   as a phonological primitive, 119
   aspiration, 65
   coronals, 24
   directional interactions, 156
   dorsals, 24
   fricative, 24
   in child phonology, 206
   labials, 23, 24
   manner, 24
   nasals, 24
   obstruents, 24
   place of articulation, 24
   sonorants, 24
   stops, 24
   tongue tip constriction area, 104
   velars, 23
   voicing, 24
feature agreement
   sign language, 146
feature harmony
   phonetically grounded, 160
feed-forward networks, 109
FILL, 124
Finley, Sara, 156
Fitch, Tecumseh, 55
fMRI study, 260
Fodor, Jerry, 38
FOXP2, 272
frame. *See* CV skeleton
French, 10
fricative, 24
functional specialization. *See* specialization

Galaburda, Albert, 304
Gallistel, Randy, 277
geminates, 26, 37
   representation of, 37
   typology, 51
generalizations, 110
   across-the-board, 98, 101
      to nonnative phonemes, 102
   beyond the training space, 100, 104, 110
   open-ended, 86
   outside the training space, 107
      Maxent model, 110
   scope of, 98, 101, 104
   to nonnative phonemes, 102
   within the training space, 110
      Maxent model, 110
Gervain, Judit, 203
Gnanadesikan, Amalia, 212
Goldrick, Matt, 263

grammar, 114
   final state of, 201
   initial state of, 201, 314
grammatical universals, 18, 27, 51, 123, 308
grounded phonology, 139

Halle, Morris, 102
Hangul, 285
hardware independence, 45
hardware segregation, 275
Hauser, Marc, 55
Hayes, Bruce, 23, 109
Hebrew, 87, 103, 112
height, 152
Hickok, Gregory, 254
hierarchies
   containment, 30
   in birdsong, 233
   stability, 31
home signs, 15
homophony effects
   in silent reading, 292
Huichol, 169

identity, 42
identity avoidance
   in sign language, 147
   in spoken language, 146
identity function, 85, 108, 109, 110
identity relations, 107, 110
identity restrictions, 112
   and feature co-occurrence, 95
   as algebraic rules, 95
   generalization to novel roots, 92
   homorganic consonants, 95
   in animals, 239
   in infancy, 202, 203
      on musical notes, 203
   in silent reading, 298
   morphological domain of, 93
   statistical learning of, 94
I-language, 133, 143
illusory conjunctions, 67, 70, 71
Imdlawn Tashlhiyt Berber, 121
implicational asymmetries, 121
   peak/margin, 121
   sonority, 122
   sonority restrictions, 170
   syllable structure, 121
infancy
   and allophonic distinctions, 208
   and dyslexia, 301
   and place assimilation, 215
   and rule-learning, 238
   and syllable markedness, 218

information, 19
information processing, 19
innate knowledge
  agents, 4
  language, 4
  living things, 4
  object, 4
innateness, 110
  as experience independence, 45
  of algebraic machinery, 107
I-phonology, 133
irregular verbs, 111
Israeli Sign Language, 144

Jacquemot, Charlotte, 262
Jakobson, Roman, 263
Japanese, 10
Japanese Kana, 284
  and moras, 284
Jusczyk, Peter, 214, 215

Kingston, John, 141
Korean, 194–195
Krakauer, David, 49, 101

labials, 24
language discrimination
  by cotton-top tamarin monkeys, 55
  by infants, 55
language diversity
  grammatical reasons for, 133
    constraint conflation, 134
    constraint ranking, 133
  non-grammatical reasons for, 132
  phonetic reasons for, 135
language universals
  and cultural evolution, 131
  and sensorimotor constraints, 131
  counterexamples to, 131
Lardil, 125
Lenneberg, Eric, 268
Lennertz, Tracy, 180, 190
lexical analogy, 112, 113
lexical decision, 92
lexicon, 71
linguistic productivity, 111
linguistic tone
  genetic basis of, 271

MacNeilage, Peter, 52
magnetoencephalography, 257
majority rule, 156
Mandarin, 195
Marcus, Gary, 98, 107, 110, 202, 252
markedness. *See* universal grammar;
  grammatical universals
  and repair, 178
  in infancy
    experimental evidence, 215
  in reading, 299
  in sign language, 143
  sonority distance, 171
markedness constraints, 123
markedness restrictions
  in child language, 224
  in infancy, 215
  their role in the initial state, 213
Marom, Michal, 77
Maxent model, 109, 110
  generalization outside the training
    space, 110
  generalization within the training
    space, 110
Mayberry, Rachel, 268
McCarthy, John, 87, 88
McClelland, James, 50, 106
MEG, 257
Mehler, Jacques, 209, 218
metrical feet, 66
misidentification
  and stimuli artifacts, 187
  grammatical vs. phonetic reasons, 187
  phonological vs. phonetic reasons
    experimental tests of, 190; printed
    forms, 89
Monaghan, Padraic, 81, 106
moras
  in writing systems, 284
Moreton, Elliott, 151
morphological evidence, 114
music, 30, 53, 270
  brain network for, 271
  containment hierarchies, 30
  rhythm, 54
  stability hierarchies, 31
musical notes, 203

nasal harmony
  experimental evidence, 155
  in infancy, 215
nasal place assimilation, 215
nasals, 24
nativism, radical, 45, 310
natural phonology, 51
Nazzi, Thierry, 209
near infra-red spectroscopy, 203
Nespor, Marina, 73, 79
Nicaraguan Sign Language, 243
NoCODA, 123, 124
nonlinguistic tonal material, 271
Nowak, Martin, 49, 101
nucleus, 64, 142

Index

Obligatory Contour Principle, 88
  and variables, 90
obstruents, 24
Ohala, Diane, 220
onset, 27, 123
  in English, 127
  repair of, 127
onset consonant, 129
onset restrictions
  and constraint ranking, 214
  learnability of, 214
ontogeny, 139
operations over variables, 107, 109, 110, 111
Optimality Theory, 58, 123
orthography, 32, 71
  opaque, 71
  transparent, 71

palatalization
  phonetic vs. phonological, 159, 160
PARSE, 124
Patel, Aniruddh, 54
Patterson, Karalyn, 106
Perfetti, Charles, 293
Petitto, Laura Ann, 267
Phillip, Adam, 277
Phillips, Colin, 257
phonemes, 21
  as the basis for writing systems, 284
  brain networks, 256
phonetic contrast
  adults, 208
  infants, 208
phonetic optimization, 49
phonetic triggering hypothesis, 135, 139, 311
phonology–phonetics link, 26, 50, 135, 139, 313
  in ontogeny, 139
  in phylogeny, 139
phonological biases
  role of complexity, 162
  role of language experience, 161
  role of phonetic factors, 163
phonological borrowing, 101, 102
phonological competence
  genetic basis of, 271, 314
phonological decoding
  in reading, 282, 288
  of single isolated words, 290
  of text, 289
phonological features, 207
phonological generalizations
  to nonnative phonemes, 102
phonological grammar, 36
  as an algebraic system, 110
  computational properties of, 85
  in dyslexia, 302

phonological interference effects
  in silent reading, 294
phonological knowledge
  tacit, 9
phonological optimization, 308
phonological patterns
  properties of, 8
phonological predictability effects
  in silent reading, 291
phonological priming, 293
phonological primitives
  consonants and vowels, 120
  features, 119
  in silent reading, 297
  in invented spellings, 285
  in typology, 119
  in writing systems, 283
  phonemes, 119
  sign language
    features, 142
    nucleus, 142
    syllables, 142
  syllables, 120
    in silent reading, 297
phonological recoding, 32
phonological universals, 27, 119
  and specialization, 118
  combinatorial principles, 120
  primitives
    consonants and vowels, 120
  syllable structure, 120
phonology, 5
phonology-phonetic dissociation
  sign language, 142
phonotactic knowledge
  increased interest, 109
phonotactics, 58, 109
Pinker, Steven, 48, 114
Pitt, Mark, 178
place of articulation, 121
place of articulation hierarchy, 129
  and onset repair, 129
  in typology, 122
plasticity, 275
Poeppel, David, 254
Polish, 167
poverty of the stimulus, 52
  as experimental logic, 150
primitives
  phonological. *See* phonological primitives
productivity, 11, 202
  by associations, 114
  by rules, 114
Pycha, Anne, 154

Quebec Sign Language, 267

Rabagliati, Hugh, 252
Rapp, Brenda, 263
reading
   and core knowledge, 281, 308, 314
   and grammatical constraints, 296, 298
   and scaffolding, 281
   and syllable markedness, 299
   and syllable structure, 297
   assembled vs. addressed phonology, 291
   brain networks for, 295
   dual route model, 290
   grammatical constraints
      syllable markedness, 299
   phonological decoding, 32, 282, 288
      and semantic categorization, 294
      homophony effect, 292
      in dyslexia, 300
      in the Stroop task, 294
      of phonological primitives, 297
      of single words, 290
      of text, 289
      predictability effects, 291
      priming effects, 293
recursion, 110
reduplication, 65, 86
   and sonority restrictions, 168
regenesis, 12, 16, 311
   in birdsong, 244
   in sign language, 243
   in Zebra Finches, 47
   Nicaraguan Sign Language, 243
relations, 85, 117
Ren, Jie, 195
repair, 10, 127
   and the place of articulation hierarchy, 122
   as a markedness reflex, 178
   as a stochastic markedness effect, 179
   Chamicuro, 129
   choice of, 129
   epenthesis vs. prothesis, 184
   grammatical vs. phonetic reasons, 187
   in French, 10
   in Japanese, 10
   in Spanish, 11
   of coda, 125
   phonological vs. phonetic reasons
      experimental tests of, 190
representation
   as symbols, 39
   complexity, 40
   discrete, 21
   structure of, 39
rhythm
   and consonant–vowel distinction, 209
   music, 54

richness of the base, 214
Romani, Cristina, 264
root, 87, 88, 110
rounding, 152
rule learning
   by animals, 234
   by rhesus monkeys, 238
   consonant–vowel distinction, 80
   in infancy, 202, 203, 204, 206, 212, 238
   on musical notes, 203
Rumelhart, David, 50
Russian, 167

Sandler, Wendy, 13, 145
Seidl, Amanda, 206
self-organization
   and univeral grammar, 249
semantic categorization, 294
Semitic languages, 87, 89
Sevald, Christine, 76
Shillcock, Richard, 81, 106
Shimron, Joseph, 95
sign language, 12
   assimilation, 243
   brain network for, 267
   combinatorial structure, 30
   features, 30, 142
   identity avoidance, 147
   identity restrictions, 30
   markedness, 143
   sonority, 145
   syllables, 30
   syllables vs. morphemes, 142
   the monosyllabic bias, 147
Sign Language of the Netherlands, 145
signal
   analog, 19
   digital, 20
   discrete, 19
silent reading
   and identity restrictions, 298
   syllable, 69
simple recurrent networks, 109
sisem–simem asymmetry, 87, 90, 112
   algebraic explanation for, 90
   counterexamples for, 112
   statistical explanation for, 90
SLI, 273
Smolensky, Paul, 171, 172, 214
Somali, 175
sonorants, 24
sonority, 81, 122, 166, 220
   and cluster reduction, 145
   and coda voicing, 175
   and phonetic pressures, 175

aphasia, 81
distance, 171
  and repair, 178
  markedness of, 171
  minimal allowed, 167
  in sign language, 145, 146
  in word simplification
    by children, 219
  restrictions on onsets, 167
  scale, 166
sonority restrictions, 167, 175. *See* sonority
  and cluster reduction by children, 220
  and feature alignment, 175
  and implicational asymmetries, 170
  and reduplication, 168
  and syllabification, 168
  and syllable structure, 167
  as core knowledge, 175
  as feature alignment restrictions, 171
  by children, 220, 221
  choice of repair, 184
  counterexamples to, 169
  experimental evidence, 176
  functional explanations, 175
  in nasal-initial cluster, 186
  in syllable count, 181
  in typology, 169
  on onsets, 167
  on unattested onsets
    experimental evidence, 180
  the role of lexical analogy, 194
sound patterns
  knowledge of, 9
Spanish, 11, 119, 194
specialization, 44, 265, 308. *See also* domain-
    specificity
  and brain structure, 265
  evidence from design, 118
  functional vs. hardware, 252, 276
    weak vs. strong views, 252
  of phonological brain networks, 265
species-specificity
  algebraic optimization, 231
  computational machinery, 230
  substantive constraints, 230
Specific Language Impairment
  genetic basis of, 272
speech
  brain network of, 254
  perception, 254
  production
    and syllable, 69
  segmentation
    by infants, 211
Spelke, Elizabeth, 46

spelling
  invented, 285
  and phonological over-specification, 287
  and phonological under-specification, 286
stability, 31
statistical learning, 53
  and categorical perception, 53
  by nonhuman animals, 229, 233
  in infants, 53
  irrelevance of, 70
  of roots, 112
  role in sonority restrictions, 194
stem, 89
Steriade, Donca, 68, 168, 180
Stokoe, William, 12
stop consonants, 23
stops
  production difficulty, 25
  voiceless, 25
Stroop, 77, 294
  and CV skeleton, 77
  and identity restrictions, 92
  phonological decoding, 294
substantive constraints
  species-specificity, 230
Surface Correspondence by Identity
    (SCorrI), 89
Surface Correspondence by Position (SCorrP)
    constraint, 89
syllabification
  and sonority restrictions, 168
syllable markedness
  brain network for, 262–263
  in aphasia, 263–264
syllable structure
  coda, 27
  implicational asymmetries, 121
  in infancy, 211
  in reading, 297
  onset, 27
  sonority restrictions, 167, 183
syllables
  and phonotactic restrictions, 65
  and the trochaic bias, 212
  as a chunk, 64, 68
  as an equivalence class, 64, 68, 73
    in childhood, 212
  as bigram trough, 69
  illusory conjunctions, 67, 70, 71
  in child phonology, 212
  in Chinese characters, 283
  in sign language, 142
  in silent reading, 69
  in speech production, 69
  in writing-systems, 283

syllables (cont.)
  initial-syllable effect, 72
  lexical access, 71
  markedness
    in infancy, 218
    in reading, 299
    order of acquisition, 217
    phonetic realization, 65
symbols, 39
syntactic relations, 110
syntax, 111
systematicity, 20, 37
  of phonological borrowing, 101
systematicity-transmissibility dilemma, 20
systems
  blending, 19
  combinatorial, 19, 23

tableau, Optimality Theoretic, 89, 123
tone, 137
  and vowel length, 137
  linguistic, 53
  linguistic and absolute pitch, 54
  linguistic and musical training, 54
  phonetic restrictions on, 137
tongue-twisters, 289
Toro, Juan, 80
training independence, 108
training space, 100, 107, 110
  features, 104
  of phonemes, 103
transcranial magnetic stimulation, 259
trochaic bias
  in infancy, 212
Turing machine, 38
Turing, Alan, 38
typology
  implicational asymmetry, 121
  universals, 51, 119
    primitives, 119
    vs. individual speakers' preferences, 150

uniqueness
  design, 28
  of human phonology, 228
  of phonological design, 308
universal
  combinatorial principles, 27
  primitives, 27

universal grammar, 49, 52, 249. *See* markedness; grammatical universals
  and phonetic triggering, 139
  as a genotype, 136, 312
  emergence by self-organization, 249
  evolution, 249
  in infancy, 204
    hallmarks of, 204
universals, 51
  in typology,

Vaknin-Nusbaum, Vered, 95, 180
variables, 41–42, 85, 105, 109, 110
  and identity restrictions, 86
velar palatalization
  experimental evidence, 158
Vietnamese, 214
voice onset time (VOT), 22
voicing, 22
vowel agreement
  backness
    experimental evidence for, 154
vowel harmony, 154
  directional, 156
  experimental evidence, 156
vowel height
  agreement, 151
  and phonetic interactions, 152
  experimental evidence, 152
vowel reduction, 74
vowels, 152
  information, 74
  orals vs. nasals, 206
  prototype, 75

Wilson, Colin, 109, 110, 154, 158
word pattern, 87
writing features
  and phonological features, 285
writing systems, 32
  and moras, 284
  and phonemes, 284
  invention of, 283
  phonological primitives, 283
  syllable-based, 283

Zebra Finches, 45
Zhang, Jie, 137
Zhao, Xu, 196